Lecture Notes in Mathematics 1697

Editors:
A. Dold, Heidelberg
F. Takens, Groningen
B. Teissier, Paris

Subseries: Fondazione C. I. M. E., Firenze
Advisor: Roberto Conti

T0190308

Springer
Berlin
Heidelberg
New York
Barcelona
Hong Kong
London
Milan
Paris
Singapore
Tokyo

B. Cockburn   C. Johnson
C.-W. Shu   E. Tadmor

Advanced Numerical
Approximation
of Nonlinear Hyperbolic
Equations

Lectures given at the 2nd Session of the
Centro Internazionale Matematico Estivo
(C.I.M.E.) held in Cetraro, Italy,
June 23–28, 1997

Editor: Alfio Quarteroni

Fondazione
C.I.M.E.

 Springer

Authors

Bernardo Cockburn
School of Mathematics
University of Minnesota
Minneapolis, MN 55455, USA
e-mail: cockburn@math.umn.edu

Chi-Wang Shu
Division of Applied Mathematics
Brown University
Providence, RI 02912, USA
e-mail: shu@cfm.brown.edu

Claes Johnson
Mathematics Department
Chalmers University of Technology
S-41296 Göteborg, Sweden
e-mail: claes@math.chalmers.se

Eitan Tadmor
Department of Mathematics UCLA
Los Angeles, CA 90095, USA
and
School of Mathematical Sciences
Tel-Aviv University
Tel-Aviv, Israel
e-mail: tadmor@math.ucal.edu

Editor

Alfio Quarteroni
Dipartimento di Matematica
Politecnico di Milano
Via Bonardi 9
I-20133 Milano, Italy
e-mail: aq@mate.polimi.it

Library of Congress Cataloging-in-Publication Data

Advanced numerical approximation of nonlinear hyperbolic equations /
B. Cockburn ... [et al.].
    p.   cm. -- (Lecture notes in mathematics ; 1697)
    "Lectures given at the 2nd session of the Centro internazionale
matematico estivo (C.I.M.E.) held in Cetraro, Italy, June 23-28,
1997."
    Includes bibliographical references and index.
    ISBN 3-540-64977-8 (softcover : alk. paper)
    1. Differential equations, Hyperbolic--Numerical solutions-
-Congresses. 2. Differential equations, Nonlinear--Numerical
solutions--Congresses.   I. Cockburn, B. (Bernardo) II. Centro
internazionale matematico estivo. III. Series: Lecture notes in
mathematics (Springer-Verlag) ; 1697.
QA3.L28 no. 1697
[QA377]
510 s--dc21
[515'.353]
                                                    98-39224
                                                    CIP

Mathematics Subject Classification (1991): 76N10, 76N15, 65M06, 65M60, 65M70

ISSN 0075-8434
ISBN 3-540-64977-8 Springer-Verlag Berlin Heidelberg New York

Typesetting: Camera-ready TeX output by the authors
SPIN: 10650116   46/3143-543210 - Printed on acid-free paper

Preface

The C.I.M.E. School on *Advanced Numerical Approximation on Nonlinear Hyperbolic Equations*, held in Cetraro (Italy) from June 23rd to June 28th, 1997, aimed to provide a comprehensive and up-to-date presentation of numerical methods which are used nowadays to solve nonlinear partial differential equations of hyperbolic type, developing shock discontinuities. The lectures were given by four outstanding scientists in the field and reflect the state of the art of a broad spectrum of topics. The most modern and effective methodologies in the framework of finite elements, finite differences, finite volumes, spectral methods and kinetic methods are addressed. In particular, the following approaches are considered: high-order shock capturing techniques, discontinuous space-time finite elements, discontinuous Galerkin methods, adaptive techniques based upon a-posteriori error analysis. The theoretical properties of each method and its algorithmic complexity are addressed. A wide variety of applications to the solution of systems of conservation laws arising from fluid dynamics and other fields is considered.

This volume collects the texts of the four series of lectures presented at the Summer School. While they are ordered alphabetically by author, the book opens with the lecture of Prof. Eitan Tadmor, as it contains an introductory overview to the subject which can serve as an introduction for the other lectures in this volume.

It is my pleasure as editor of these Lecture Notes to thank the Director and the Members of the C.I.M.E. Scientific Committee, in particular Prof. Arrigo Cellina for the invitation to organize the School and their support during the organization and to the C.I.M.E. staff, lead by Prof. Pietro Zecca. My very sincere thanks go to the lecturers for their excellent job of preparing and teaching the Course and a preliminary version of the lectures to be distributed among the participants. Particular thanks go to all the participants for having created an extraordinarily friendly and stimulating atmosphere, and to those who have contributed with short communications: Tim Barth, Angelo Iollo, Stefano Micheletti, Gabriella Puppo, Giovanni Russo, Riccardo Sacco, Fausto Saleri and Alessandro Veneziani. Finally, I would like to thank the Director and staff of the "Grand Hotel San Michele" in Cetraro (Cosenza) for the kind hospitality and efficiency and the following collaborators for their invaluable help: Simona Lilliu from CRS4 (our scientific secretary), Francesco Bosisio, Simona Perotto and Alessandro Veneziani from the Politecnico di Milano, for their careful editing of the manuscripts.

Milan, March 1998   Alfio Quarteroni

Table of Contents

Chapter 1

Approximate Solutions of Nonlinear Conservation Laws

Eitan Tadmor

Department of Mathematics
UCLA, Los-Angeles CA 90095
and
School of Mathematical Sciences
Tel-Aviv University, Tel-Aviv 69978
E-mail: tadmor@math.ucla.edu

ABSTRACT

This is a summary of five lectures delivered at the CIME course on "Advanced Numerical Approximation of Nonlinear Hyperbolic Equations" held in Cetraro, Italy, on June 1997.

Following the introductory lecture I — which provides a general overview of approximate solution to nonlinear conservation laws, the remaining lectures deal with the specifics of four complementing topics:

- *Lecture II. Finite-difference methods – non-oscillatory central schemes;*
- *Lecture III. Spectral approximations – the Spectral Viscosity method;*
- *Lecture IV. Convergence rate estimates – a Lip' convergence theory;*
- *Lecture V. Kinetic approximations – regularity of kinetic formulations.*

Acknowledgements. I thank Alfio Quarteroni for the invitation, B. Cockburn, C. Johnson & C.-W Shu for the team discussions, the Italian participants for their attention, and the hosting Grand Hotel at San Michele for its remarkably unique atmosphere. Research was supported in part by ONR grant #N00014-91-J-1076 and NSF grant #97-06827.
MSC Subject Classification (1991): Primary 35L65, 35L60. Secondary 65M06, 65M12, 65M15, 65M60, 65M70

4

1   A General Overview

Abstract. In this introductory lecture, we overview the development of modern, high-resolution approximations to hyperbolic conservation laws and related non-linear equations. Since this overview also serves as an introduction for the other lectures in this volume, it is less of a comprehensive overview, and more of a bird's eye view of topics which play a pivotal role in the lectures ahead. It consists of a dual discussion on the various mathematical concepts and the related discrete algorithms which are the required ingredients for these lectures.

I start with a brief overview on the mathematical theory for nonlinear hyperbolic conservation laws. The theory of the continuum (– and in this case, the discontinuum), is intimately related to the construction, analysis and implementation of the corresponding discrete approximations. Here, the basic the notions of viscosity regularization, entropy, monotonicity, total variation bounds and Riemann's problem are introduced. Then follow the the basic ingredients of the discrete theory: the Lax-Wendroff theorem, and the pivotal finite-difference schemes of Godunov, Lax-Friedrichs, and Glimm.

To proceed, our dual presentation of high-resolution approximations is classified according to the analytical tools which are used in the development of their convergence theories. These include classical compactness arguments based on Total Variation (TV) bounds, e.g., TVD finite-difference approximations. The use of compensated compactness arguments based on H^{-1}-compact entropy production is demonstrated in the context of streamline diffusion finite-element method and spectral viscosity approximations. Measure valued solutions – measured by their negative entropy production, are discussed in the context of multidimensional finite-volume schemes. Finally, we discuss the recent use of averaging lemmas which yield new compactness and regularity results for approximate solutions of nonlinear conservation laws (as well as some related equations), which admit an underlying kinetic formulation, e.g., finite-volume and relaxation schemes.

1.1   Introduction

The lectures in this volume deal with modern algorithms for the accurate computation of shock discontinuities, slip lines, and other similar phenomena which could be characterized by spontaneous evolution of change in scales. Such phenomena pose a considerable computational challenge, which is answered, at least partially, by these newly constructed algorithms. New modern algorithms were devised, that achieve one or more of the desirable properties of high-resolution, efficiency, stability — in particular, lack of spurious oscillations, etc. The impact of these new algorithms ranges from the original impetus in the field of Computational Fluid Dynamics (CFD), to the fields oil recovery, moving fronts, image processing,... [75], [138], [132], [1].

In this introduction we survey a variety of these algorithms for the approximate solution of nonlinear conservation laws. The presentation is neither comprehensive nor complete — the scope is too wide for the present framework Instead, we discuss the analytical tools which are used to study the stability and convergence of these modern algorithms. We use these analytical issues as our 'touring guide' to provide a readers' digest on the relevant approximate methods, while studying there convergence properties. They include

- *Finite-difference methods.* These are the most widely used methods for solving nonlinear conservation laws. Godunov-type difference schemes play a pivotal role. Two canonical examples include the upwind ENO schemes (discussed in C.-W. Shu's lectures) and a family of high-resolution non-oscillatory central schemes (discussed in Lecture II);
- *Finite element schemes.* Here, the streamline diffusion method and its extensions are canonical example, discussed in C. Johnson's lectures;
- *Spectral approximations.* The Spectral Viscosity (SV) methods is discussed in Lecture III.
- *Finite-volume schemes.* Finite-Volume (FV) schemes offer a particular advantage for integration over multidimensional general triangulation, beyond the Cartesian grids. More can be found in B. Cockburn's lectures.
- *Kinetic formulations.* Compactness and regularizing effects of approximate solutions is quantified in terms of their underlying kinetic formulations, Lecture V.

Some general references are in order. The theory of hyperbolic conservation laws is covered in [94], [178],[157], [149]. For the theory of their numerical approximation consult [102],[58],[59],[159]. We are concerned with analytical tools which are used in the convergence theories of such numerical approximations. The monograph [50] could be consulted on recent development regarding weak convergence. The reviews of [171], [123,124] are recommended references for the theory of compensated compactness, and [40,41],[17] deal with applications to conservation laws and their numerical approximations. Measure-valued solutions in the context of nonlinear conservation laws were introduced in [42]. The articles [62], [53], [45] prove the averaging lemma, and [111],[112],[77] contain applications in the context of kinetic formulation for nonlinear conservation laws and related equations.

A final word about notations. Different authors use different notations. In this introduction, the conservative variable are denoted by the "density" ρ, the spatial flux is $A(\cdot)$, (η, F) are entropy pairs, etc. In later lectures, these are replaced by the more generic notations: conservative variables are $u, v, \ldots$, fluxes are denoted by $f, g, \ldots$, the entropy function is denoted U, etc.

1.2 Hyperbolic Conservation Laws

A very brief overview — m equations in d spatial dimensions
The general set-up consists of m equations in d spatial dimensions

$$\partial_t \rho + \nabla_x \cdot A(\rho) = 0, \quad (t, x) \in \mathbb{R}_t^+ \times \mathbb{R}_x^d. \tag{1.1}$$

Here, $A(\rho) := (A_1(\rho), \ldots, A_d(\rho))$ is the d-dimensional flux, and $\rho := (\rho_1(t, x), \ldots, \rho_m(t, x))$ is the unknown m-vector subject to initial conditions $\rho(0, x) = \rho_0(x)$.

The basic facts concerning such nonlinear hyperbolic systems are, consult [94],[113], [35],[157],[58],[149],

- The evolution of *spontaneous* shock discontinuities which requires weak (distributional) solutions of (1.1);
- The existence of possibly infinitely *many* weak solutions of (1.1);
- To *single* out a unique 'physically relevant' weak solution of (1.1), we seek a solution, $\rho = \rho(t, x)$, which can be realized as a viscosity limit solution, $\rho = \lim \rho^\varepsilon$,

$$\partial_t \rho^\varepsilon + \nabla_x \cdot A(\rho^\varepsilon) = \varepsilon \nabla_x \cdot (Q \nabla_x \rho^\varepsilon), \quad \varepsilon Q > 0; \tag{1.2}$$

- *The entropy condition.* The notion of a viscosity limit solution is intimately related to the notion of an *entropy solution*, ρ, which requires that for all convex entropy functions, $\eta(\rho)$, there holds, [93], [88, §5]

$$\partial_t \eta(\rho) + \nabla_x \cdot F(\rho) \leq 0. \tag{1.3}$$

A scalar function, $\eta(\rho)$, is *an entropy function* associated with (1.1), if its Hessian, $\eta''(\rho)$, symmetrizes the spatial Jacobians, $A'_j(\rho)$,

$$\eta''(\rho) A'_j(\rho) = A'_j(\rho)^\top \eta''(\rho), \qquad j = 1, \dots, d.$$

It follows that in this case there exists an entropy flux, $F(\rho) := (F_1(\rho), \dots, F_d(\rho))$, which is determined by the compatibility relations,

$$\eta'(\rho)^\top A'_j(\rho) = F'_j(\rho)^\top, \ j = 1, \dots, d. \tag{1.4}$$

What is the relation between the entropy condition (1.3) and the viscosity limit solution in (1.2)? multiply the latter, on the left, by $\eta'(\rho^\varepsilon)$; the compatibility relation (1.4) implies that the resulting two terms on the left of (1.2) amount to the sum of *perfect derivatives*, $\partial_t \eta(\rho)^\varepsilon + \nabla_x \cdot F(\rho^\varepsilon)$. Consider now the right hand side of (1.2) (for simplicity, we assume the viscosity matrix on the right to be the identity matrix, $Q = I$). Here we invoke the identity

$$\varepsilon \eta'(\rho^\varepsilon) \Delta_x \rho^\varepsilon \equiv \varepsilon \Delta_x \eta(\rho^\varepsilon) - \varepsilon (\nabla_x \rho^\varepsilon)^\top \eta''(\rho^\varepsilon) \nabla_x \rho^\varepsilon.$$

The first term tends to zero (in distribution sense); the second term is nonpositive thanks to the convexity of η, and hence tend to a nonpositive measure. Thus, a viscosity limit solution must satisfy the entropy inequality (1.3). The inverse implication: (1.3) $\implies \rho = \lim \rho^\varepsilon$ of viscosity solutions ρ^ε satisfying (1.2), holds in the scalar case; the question requires a more intricate analysis for systems, consult [93],[157] and the references therein.

Indeed, the basic questions regarding the existence, uniqueness and stability of entropy solutions for general systems are open. Instead, the present trend seems to concentrate on special systems with additional properties which enable to answer the questions of existence, stability, large time behavior, etc. One-dimensional 2×2 systems is a notable example for such systems: their properties can be analyzed in view of the existence of Riemann invariants and a family of entropy functions, [56], [94, §6], [157], [40,41]. The system of $m \geq 2$ chromatographic equations, [77], is another example for such systems.

The difficulty of analyzing general systems of conservation laws is demonstrated by the following negative result due to Temple, [174], which states that already for systems with $m \geq 2$ equations, there exists no metric, $\mathcal{D}(\cdot; \cdot)$, such that the problem (1.1), (1.3) is contractive, i.e.,

$$\mathcal{AD}: \quad \mathcal{D}(\rho^1(t, \cdot); \rho^2(t, \cdot)) \leq \mathcal{D}(\rho^1(0, \cdot); \rho^2(0, \cdot)), \quad 0 \leq t \leq T, \quad (m \geq 2). \tag{1.5}$$

In this context we quote from [168] the following

Theorem 1.1 *Assume the system (1.1) is endowed with a one-parameter family of entropy pairs, $(\eta(\rho; c), F(\rho; c))$, $c \in \mathbf{R}^m$, satisfying the symmetry property*

$$\eta(\rho; c) = \eta(c; \rho), \quad F(\rho; c) = F(c; \rho). \tag{1.6}$$

Let ρ^1, ρ^2 be two entropy solutions of (1.1). Then the following a priori estimate holds

$$\int_x \eta(\rho^1(t, x); \rho^2(t, x)) dx \leq \int_x \eta(\rho_0^1(x); \rho_0^2(x)) dx. \tag{1.7}$$

Theorem 1.1 is based on the observation that the symmetry property (1.6) is the key ingredient for Kružkov's penetrating ideas in [88], which extends his scalar arguments into the case of general systems. This extension seems to be part of the 'folklore' familiar to some, [36],[150]); a sketch of the proof can be found in [168].

Remark 1.1 Theorem 1.1 seems to circumvent the negative statement of (1.5). This is done by replacing the metric $\mathcal{D}(\cdot; \cdot)$, with the weaker topology induced by a *family* of convex entropies, $\eta(\cdot; \cdot)$. Many physically relevant systems are endowed with at least one convex entropy function (– which in turn, is linked to the hyperbolic character of these systems, [61],[52],[120]). Systems with "rich" families of entropies like those required in Theorem 1.1 are rare, however, consult [148]. The instructive (yet exceptional...) scalar case is dealt in §1.2. If we relax the contractivity requirement, then we find a uniqueness theory for one-dimensional systems which was recently developed by Bressan and his co-workers, [11]-[14]; Bressan's theory is based on the L^1-stability (rather than contractivity) of the entropy solution operator of one-dimensional systems.

Scalar conservation laws ($m = 1, d \geq 1$)

In the scalar case, the Jacobians $A'_j(\rho)$ are just scalars that can be always symmetrized, so that the compatibility relation (1.4) in fact *defines* the entropy fluxes, $F_j(\rho)$, for all convex η's. Consequently, the family of admissible entropies in the scalar case consists of *all* convex functions, and the envelope of this family leads to Kružkov's entropy pairs [88]

$$\eta(\rho; c) = |\rho - c|, \quad F(\rho; c) = sgn(\rho - c)(A(\rho) - A(c)), \qquad c \in R. \quad (1.8)$$

Theorem 1.1 applies in this case and (1.7) now reads

- L^1-*contraction.* If ρ^1, ρ^2 are two entropy solutions of the scalar conservation law (1.1), then

$$\|\rho^2(t, \cdot) - \rho^1(t, \cdot)\|_{L^1(x)} \le \|\rho_0^2(\cdot) - \rho_0^1(\cdot)\|_{L^1(x)}. \quad (1.9)$$

Thus, the entropy solution operator associated with scalar conservation laws is L^1-contractive (– or non-expansive to be exact), and hence, by the Crandall-Tartar lemma (discussed below), it is also monotone

$$\rho_0^2(\cdot) \ge \rho_0^1(\cdot) \implies \rho^2(t, \cdot) \ge \rho^1(t, \cdot). \quad (1.10)$$

The notions of conservation, L^1 contraction and monotonicity play an important role in the theory of nonlinear conservation laws, at least in the scalar case. We discuss the necessary details of these notions, by proving the *inverse* implication: the monotonicity property (1.10) implies the all important Kružkov's entropy pairs (1.8) satisfying (1.3).

Monotonicity and Kružkov's entropy pairs

An operator T is called monotone (or order preserving) if the following implication holds for all ρ's (in some unspecified measure subspace of L^1_{loc})

$$\rho_2 \ge \rho_1 \quad \text{a.e.} \implies T(\rho_2) \ge T(\rho_1) \quad \text{a.e.} \quad (1.11)$$

We use the terminology that if ρ_2 dominates (pointwise, a.e.) ρ_1, then $T(\rho_2)$ dominates $T(\rho_1)$.

The following lemma due to Crandall & Tartar, [32], provides a useful characterization for such monotone operators.

Lemma 1.1 (*Crandall-Tartar [32]*) *Consider an operator T, which is conservative in the sense that $\int T(\rho) = \int \rho$, $\forall \rho$'s. Then T is monotone iff it is an L^1-contraction,*

$$\int |T(\rho_2) - T(\rho_1)| \le \int |\rho_2 - \rho_1|. \quad (1.12)$$

Proof. The standard notations, $\rho_1 \vee \rho_2 := \max(\rho_1, \rho_2)$ and $\rho_1 \wedge \rho_2 := \min(\rho_1, \rho_2)$ will be used. Since $|\rho_1 - \rho_2| \equiv \rho_1 \vee \rho_2 - \rho_1 \wedge \rho_2$, we find by conservation that

$$\int |\rho_1 - \rho_2| = \int \rho_1 \vee \rho_2 - \int \rho_1 \wedge \rho_2 = \int T(\rho_1 \vee \rho_2) - \int T(\rho_1 \wedge \rho_2).$$
(1.13)

Now, $\rho_1 \vee \rho_2$ dominates (pointwise a.e.) both ρ_1 and ρ_2; hence, if T is order preserving, then $T(\rho_1 \vee \rho_2)$ dominates both $T(\rho_1)$ and $T(\rho_2)$, that is, $T(\rho_1 \vee \rho_2) \geq T(\rho_1) \vee T(\rho_2)$; similarly, $-T(\rho_1 \wedge \rho_2) \geq -T(\rho_1) \wedge T(\rho_2)$. We conclude that T is an L^1-contraction, for

$$\int |\rho_1 - \rho_2| \geq \int T(\rho_1) \vee T(\rho_2) - \int T(\rho_1) \wedge \int T(\rho_2) = \int |T(\rho_1) - T(\rho_2)|.$$
(1.14)

The inverse implication (attributed to Stampacchia, I believe) starts with the identity $2w_+ \equiv |w| + w$, where w_+ denotes, as usual, the 'positive part of', $w_+ := w \vee 0$. Setting $w = T(\rho_1) - T(\rho_2)$ the integrated version of this identity reads

$$2 \int (T(\rho_1) - T(\rho_2))_+ = \int |T(\rho_1) - T(\rho_2)| + \int T(\rho_1) - T(\rho_2).$$

Given that T is L^1-contractive, then together with conservation it yields that the two integrals on the right do not exceed

$$2 \int (T(\rho_1) - T(\rho_2)) \vee 0 \leq \int |\rho_1 - \rho_2| + \int \rho_1 - \rho_2.$$
(1.15)

Now, if ρ_2 dominates ρ_1, i.e., $\rho_1 \leq \rho_2$ a.e., then the sum of the two integrals on the RHS vanishes, and consequently, the non-negative integrand on the LHS vanishes as well, i.e., $T(\rho_1) - T(\rho_2) \leq 0$.

Remark 1.2 For *linear* operators T, monotonicity coincides with positivity, $T(\rho) \geq 0$, $\forall \rho \geq 0$. Positive operators play a classical role in various branches of analysis. They are encountered frequently, for example, in approximation theory, e.g., [37, §9.4]. A well known example is provided by Bernstein polynomials, $B_n(\rho)(x) := \sum_{k \leq n} \binom{n}{k} x^k (1 - x)^{n-k} \rho(\frac{k}{n})$. They produce a positive linear map(s), $\rho \to B_n(\rho)$ of $C[0, 1]$ into the space of n-degree polynomials. Linear monotone operators, like Bernstein projections, are at most first-order accurate.

We turn to discuss the relation between monotone operators and the entropy condition. Let $\{T_t, \ t \geq 0\}$ be a one-parameter family of operators which form a semi-group of constant-preserving, monotone operators. Thus we make

- **Semi-group** It is assumed that $\{T_t\}$ satisfies the basic semi-group 'closure' (causality) relations,

$$T_{t+s} = T_t T_s, \quad T_0 = \text{the identity mapping}, \tag{1.16}$$

and that it has an infinitesimal generator,

$$\nabla_x \cdot A(\rho) := \lim_{\Delta t \downarrow 0} (\Delta t)^{-1} (T_t(\rho) - \rho).$$

Remark 1.3 The existence of such a generator is outlined by the Hille-Yosida linear theory. Extensions within the context of *nonlinear* evolution equations are available: Kato's approach in semi-Hilbert spaces and Crandall-Liggett approach in Banach spaces, consult [179, §XIV.6&7]. A concise informative account of this theory, which was specifically 'tailored' within the L^1-setup for quasilinear evolution equations can be found in [28].

With this in mind we may identify, $T_t \rho_0 =: \rho(t)$, as the solution of the abstract Cauchy problem

$$\rho_t + \nabla_x \cdot A(\rho(t)) = 0, \tag{1.17}$$

subject to given initial conditions $\rho(0) = \rho_0$. We assume that the following two basic properties hold.

- **Constant-preserving** T_t preserves constants, namely

$$T_t[\rho \equiv Const.] = Const. \tag{1.18}$$

Finally, we bring in the key assumption of monotonicity.

- **Monotonicity** Our basic assumption is the monotonicity of the solution operators associated with (1.17),

$$\rho_0^2(\cdot) \geq \rho_0^1(\cdot) \Longrightarrow \rho^2(t, \cdot) \geq \rho^1(t, \cdot), \quad \forall t \geq 0. \tag{1.19}$$

The main result of this section, following the ingredients in [116] and in particular, [142], states that *monotone, constant-preserving* solution operators of the Cauchy problem (1.17), are uniquely identified by the following entropy condition.

Theorem 1.2 (Kružkov's Entropy Condition) *Assume $\{\rho(t), t \geq 0\}$ is a family of solutions for the Cauchy problem (1.17) which is constant-preserving, (1.18), and satisfies the monotonicity condition (1.19). Then the following entropy inequality holds:*

$$\partial_t |\rho(t) - c| + \nabla_x \cdot \{sgn(\rho - c)(A(\rho) - A(c))\} \leq 0, \quad \forall c's. \tag{1.20}$$

Thus, monotonicity (+constant preserving) recover Kružkov entropy pairs.

Proof. Starting with $\rho(t)$ at arbitrary $t \geq 0$, we compare $\rho(t+\Delta t) := T_{\Delta t}\rho(t)$ with $\zeta(t + \Delta t) := T_{\Delta t}\zeta(t)$, where $\zeta(t) := (\rho(t) \vee c)$ which is cut-off at an arbitrary constant level, c. Since $\rho(t) \vee c$ dominates both $\rho(t)$ and c, the monotonicity of $T_{\Delta t}$ implies that at later times, $\zeta(t + \Delta t)$ should dominate both, $T_{\Delta t}\rho(t)$ ($= \rho(t + \Delta t)$ by to our notations), and $T_{\Delta t}(c)$ ($= c$ since $T_{\Delta t}$ is assumed constant-preserving). Thus $\zeta(t + \Delta t) \geq \rho(t + \Delta t) \vee c$ and hence

$$\frac{\rho(t+\Delta t) \vee c - \rho(t) \vee c}{\Delta t} \leq \frac{\zeta(t + \Delta t) - \zeta(t)}{\Delta t}.$$

Let $\Delta t \downarrow 0$. By definition, the LHS gives $\partial_t(\rho(t) \vee c)$; the RHS is governed by its Cauchy problem, $\partial_t\zeta(t) = -\nabla_x \cdot A(\zeta(t)) = \nabla_x \cdot A(\rho(t) \vee c)$. We conclude that an arbitrary $t \geq 0$

$$\partial_t(\rho(t) \vee c) + \nabla_x \cdot A(\rho(t) \vee c) \leq 0. \tag{1.21}$$

Similar arguments yield

$$-\partial_t(\rho(t) \wedge c) + \nabla_x \cdot A(\rho(t) \wedge c) \leq 0. \tag{1.22}$$

Together, the last two inequalities add up to the entropy inequality (1.20). ∎

Early constructions of approximate solutions for scalar conservation laws, most notably — finite-difference approximations, utilized this monotonicity property to construct convergent schemes, [30], [143]. Monotone approximations are limited, however, to first-order accuracy [72]. At this stage we note that the limitation of first-order accuracy for monotone approximations, can be avoided if L^1-contractive solutions are replaced with (the weaker) requirement of bounded variation solutions.

– *TV bound.* The solution operator associated with (1.1) is translation invariant. Comparing the scalar entropy solution, $\rho(t, \cdot)$, with its translate, $\rho(t, \cdot + \Delta x)$, the L^1-contraction statement in (1.9) yields the TV bound, [177],

$$\|\rho(t,\cdot)\|_{BV} \leq \|\rho_0(\cdot)\|_{BV}, \quad \|\rho(t,\cdot)\|_{BV} := \sup_{\Delta x \neq 0} \frac{\|\rho(t,\cdot + \Delta x) - \rho(t,\cdot)\|_{L^1}}{\Delta x}. \tag{1.23}$$

Construction of scalar entropy solutions by TV-bounded approximations were used in the pioneering works of Oleĭnik [129], Vol'pert [177], Kružkov [88] and Crandall [28]. In the one-dimensional case, the TVD property (1.23) enables to construct convergent difference schemes with high-order (> 1) resolution; Harten initiated the construction of high-resolution TVD schemes in [70], following the earlier works [6], [98]. A whole generation of TVD schemes was then developed during the beginning of the '80s; some aspects of these developments can be found in §1.3.

One dimensional systems $(m \geq 1, d = 1)$

We focus our attention on one-dimensional hyperbolic systems governed by

$$\partial_t \rho + \partial_x A(\rho) = 0, \quad (t, x) \in \mathbb{R}_t^+ \times \mathbb{R}_x, \tag{1.24}$$

and subject to initial condition, $\rho(0, x) = \rho_0(x)$. The hyperbolicity of the system (1.24) is understood in the sense that its Jacobian, $A'(\rho)$, has a complete *real* eigensystem, $(a_k(\rho), r_k(\rho)), k = 1, \ldots, m$. `For example, the existence of a convex entropy function guarantees the symmetry of $A'(\rho)$ ($-$ w.r.t. $\eta''(\rho)$), and hence the complete real eigensystem. For most of our discussion we shall assume the stronger *strict hyperbolicity*, i.e, distinct real eigenvalues, $a_k(\rho) \neq a_j(\rho)$.

A fundamental building block for the construction of approximate solutions in the one-dimensional case is the solution of *Riemann's problem*.

Riemann's problem

Here one seeks a weak solution of (1.24) subject to the piecewise constant initial data

$$\rho(x, 0) = \begin{cases} \rho_\ell, & x < 0 \\ \rho_r, & x > 0. \end{cases} \tag{1.25}$$

The solution is composed of m simple waves, each of which is associated with one (right-)eigenpair, $(a_k(\rho), r_k(\rho)), 1 \leq k \leq m$. There are three types of such waves: if the k-th field is *genuinely nonlinear* in the sense that $r_k \cdot \nabla_\rho a_k \neq 0$, these are either k-shock or k-rarefaction waves; or, if the k-th field is *linearly degenerate* in the sense that $r_k \cdot \nabla_\rho a_k \equiv 0$, this is a k-th contact wave.

These three simple waves are *centered*, depending on $\xi = \frac{x}{t}$ (which is to be expected from the dilation invariance of (1.24),(1.25)). The structure of these three centered waves is as follows:

- A k-shock discontinuity of the form

$$\rho(\xi) = \begin{cases} \rho_\ell, & \xi < s \\ \rho_r, & \xi > s; \end{cases}$$

  here s denotes the shock speed which is determined by a Rankine-Hugoniot relation so that $a_k(\rho_\ell) > s > a_k(\rho_r)$.
- A k-rarefaction wave, $\rho(\xi)$, which is directed along the corresponding k-th eigenvector, $\dot{\rho}(\xi) = r_k(\rho(\xi))$. Here r_k is the normalized k-eigenvector, $r_k \cdot \nabla a_k \equiv 1$ so that the gap between $a_k(\rho_\ell) < a_k(\rho_r)$ is filled with a fan of the form

$$a_k(\rho(\xi)) = \begin{cases} a_k(\rho_\ell), & \xi < a_k(\rho_\ell) \\ \xi, & a_k(\rho_\ell) < \xi < a_k(\rho_r) \\ a_k(\rho_r), & a_k(\rho_r) < \xi \end{cases}$$

– A k-contact discontinuity of the form

$$\rho(\xi) = \begin{cases} \rho_\ell, \, \xi < s \\ \rho_r, \, \xi > s \end{cases}$$

where s denotes the shock speed which is determined by a Rankine-Hugoniot relation so that $a_k(\rho_\ell) = s = a_k(\rho_r)$.

We are concerned with *admissible* systems — systems which consist of either genuinely nonlinear or linearly degenerate fields. We refer to [92] for the full story which is summarized in the celebrated

Theorem 1.3 (Lax solution of Riemann's problem) *The strictly hyperbolic admissible system (1.24), subject to Riemann initial data (1.25) with $\rho_\ell - \rho_r$ sufficiently small, admits a weak entropy solution, which consists of shock- rarefaction- and contact-waves.*

For a detailed account on the solution of Riemann problem consult [16]. An extension to a generalized Riemann problem subject to piecewise-linear initial data can be found in [5], [99]. In this context we also mention the *approximate* Riemann solvers, which became useful computational alternatives to Lax's construction. Roe introduced in [139] a linearized Riemann solver, which resolves jumps discontinuities solely in terms of shock waves. Roe's solver has the computational advantage of sharp resolution (at least when there is one dominant wave per computational cell); it may lead, however, to unstable shocks. Osher and Solomon in [131] used, instead, an approximate Riemann solver based solely on rarefaction fans; one then achieves stability at the expense of deteriorated resolution of shock discontinuities.

Godunov, Lax-Friedrichs and Glimm schemes

We let $\rho^{\Delta x}(t, x)$ be the entropy solution in the slab $t^n \leq t < t + \Delta t$, subject to piecewise constant data $\rho^{\Delta x}(t = t^n, x) = \sum \rho_\nu^n \chi_\nu(x)$. Here $\chi_\alpha(x) := 1_{\{|x - \alpha \Delta x| \leq \Delta x/2\}}$ denotes the usual indicator function. Observe that in each slab, $\rho^{\Delta x}(t, x)$ consists of successive *noninteracting* Riemann solutions, at least for a sufficiently small time interval Δt, for which the CFL condition, $\Delta t/\Delta x \max |a_k(\rho)| \leq \frac{1}{2}$ is met. In order to realize the solution in the next time level, $t^{n+1} = t^n + \Delta t$, it is extended with a jump discontinuity across the line t^{n+1}, by projecting it back into the finite-dimensional space of piecewise constants. Different projections yield different schemes. We recall the basic three.

Godunov Scheme. Godunov scheme [60] sets

$$\rho^{\Delta x}(t^{n+1}, x) = \sum_\nu \bar{\rho}_\nu^{n+1} \chi_\nu(x),$$

where $\bar{\rho}_\nu^{n+1}$ stands for the cell-average,

$$\bar{\rho}_\nu^{n+1} := \frac{1}{\Delta x} \int_x \rho^{\Delta x}(t^{n+1} - 0, x)\chi_\nu(x)dx,$$

which could be explicitly evaluated in terms of the flux of Riemann solution across the cell interfaces at $x_{\nu\pm\frac{1}{2}}$,

$$\bar{\rho}_\nu^{n+1} = \bar{\rho}_\nu^n - \frac{\Delta t}{\Delta x}\Big\{ A(\rho^{\Delta x}(t^{n+\frac{1}{2}}, x_{\nu+\frac{1}{2}})) - A(\rho^{\Delta x}(t^{n+\frac{1}{2}}, x_{\nu-\frac{1}{2}}))\Big\}. \quad (1.26)$$

Godunov scheme had a profound impact on the field of Computational Fluid Dynamics. His scheme became the forerunner for a large class of upwind finite-volume methods which are evolved in terms of (exact or approximate) Riemann solvers. In my view, the most important aspect of what Richtmyer & Morton describe as Godunov's "ingenious method" ([141, p. 338]), lies in its *global* point of view: one does not simply evolve discrete pointvalues $\{\rho_\nu^n\}$, but instead, one evolves a globally defined solution, $\rho^{\Delta x}(t, x)$, which is realized in terms of its discrete averages, $\{\bar{\rho}_\nu^n\}$.

Lax-Friedrichs Scheme. If the piecewise constant projection is carried out over alternating staggered grids, $\bar{\rho}_{\nu+\frac{1}{2}}^{n+1} := \frac{1}{\Delta x}\int_x \rho^{\Delta x}(t^{n+1}-0, x)\chi_{\nu+\frac{1}{2}}(x)dx$, then one effectively integrates 'over the Riemann fan' which is centered at $(x_{\nu+\frac{1}{2}}, t^n)$. This recovers the Lax-Friedrichs (LxF) scheme, [91], with an explicit recursion formula for the evolution of its cell-averages which reads

$$\bar{\rho}_{\nu+\frac{1}{2}}^{n+1} = \frac{\bar{\rho}_\nu^n + \bar{\rho}_{\nu+1}^n}{2} - \frac{\Delta t}{\Delta x}\Big\{ A(\bar{\rho}_{\nu+1}^n) - A(\bar{\rho}_\nu^n)\Big\}. \quad (1.27)$$

The Lax-Friedrichs scheme had a profound impact on the construction and analysis of approximate methods for time-dependent problems, both linear problems [51] and nonlinear systems [91]. The Lax-Friedrichs scheme was and still is the stable, all purpose benchmark for approximate solution of nonlinear systems.

Both Godunov and Lax-Friedrichs schemes realize the exact solution operator in terms of its finite-dimensional cell-averaging projection. This explains the versatility of these schemes, and at the same time, it indicates their limited resolution due to the fact that waves of different families that are averaged together at each computational cell.

Glimm Scheme. Rather than averaging, Glimm's scheme, [55], keeps its sharp resolution by *randomly sampling* the evolving Riemann waves,

$$\rho^{\Delta x}(t^{n+1}, x) = \sum_\nu \rho^{\Delta x}(t^{n+1} - 0, x_{\nu+\frac{1}{2}} + r^n \Delta x)\chi_{\nu+\frac{1}{2}}(x).$$

This defines the Glimm's approximate solution, $\rho^{\Delta x}(t, x)$, depending on the mesh parameters $\Delta x \equiv \lambda \Delta t$, and on the set of random variable

$\{r^n\}$, uniformly distributed in $[-\frac{1}{2}, \frac{1}{2}]$. In its deterministic version, Liu [114] employs equidistributed rather than a random sequence of numbers $\{r^n\}$.

Glimm solution, $\rho^{\Delta x}(t, x)$, was then used to construct a solution for one-dimensional admissible systems of conservation laws. Glimm's celebrated theorem, [55], is still serving today as the cornerstone for existence theorems which are concerned with general one-dimensional systems, e.g. [114],[20],[146].

Theorem 1.4 *(Existence in the large).* *There exists a weak entropy solution, $\rho(t, \cdot) \in L^\infty[[0, T], BV \cap L^\infty(\mathbb{R}_x)]$, of the strictly hyperbolic system (1.24), subject to initial conditions with sufficiently small variation, $\|\rho_0(\cdot)\|_{BV \cap L^\infty(\mathbb{R}_x)} \leq \epsilon$.*

Glimm's scheme has the advantage of retaining sharp resolution, since in each computational cell, the local Riemann solution is realized by a randomly chosen 'physical' Riemann wave. Glimm's scheme was turned into a computational tool known as the Random Choice Method (RCM) in [22], and it serves as the building block inside the front tracking method of Glimm and his co-workers, [57], [21].

Multidimensional systems ($m > 1, d > 1$)
Very little rigor is known on m conservation laws in d spatial dimensions once $(m - 1)(d - 1)$ becomes positive, i.e., general multidimensional systems. We address few major achievements.

Short time existence. For H^s-initial data ρ_0, with $s > \frac{d}{2}$, an H^s-solution exists for a time interval $[0, T]$, with $T = T(\|\rho_0\|_{H^s})$, consult e.g, [83],[78, §5.3].

Short time existence – piecewise analytic data. An existence resultconjectured by Richtmyer was proved by Harabetian in terms of a Cauchy-Kowalewski type existence result [68].

Short time stability – piecewise smooth shock data. Existence for piecewise smooth initial data where smoothness regions are separated by shock discontinuities was studied in [118],[106].

Riemann invariants. The gradients of *Riemann invariants* enable us to 'diagonalize' one-dimensional systems. More is known about 2×2 systems in one space dimension thanks to the existence of Riemann invariants. Consult [56], [157], [148]. Beyond $m = 2$ equations, only special systems admits a full set of Riemann invariants (consult [148] and the references therein).

Riemann problem. Already in the $d = 2$-dimensional case, the collection of simple waves and their composed interaction in the construction of Riemann solution (– subject to piecewise constant initial data), is considerably more complicated than in the one-dimensional setup. We refer to the recent book [33] for a detailed discussion.

Compressible Euler equations. These system of $m = 5$ equations governing the evolution of density, 3-vector of momentum and Energy in $d = 3$-space variables was – and still is, the prime target for further developments in our understanding of general hyperbolic conservation laws. We refer to Majda, [118], for a definitive summary of this aspect.

1.3 Total Variation Bounds

Finite Difference Methods

We begin by covering the space and time variables with a discrete grid: it consists of time-steps of size Δt and rectangular spatial cells of size $\Delta x := (\Delta x_1, \ldots, \Delta x_d)$. Let C_ν denotes the cell which is centered around the gridpoint $x_\nu = \nu \Delta x := (\nu_1 \Delta x_1, \ldots, \nu_d \Delta x_d)$, and let $\{\rho_\nu^n\}$ denote the gridfunction associated with this cell at time $t^n = n\Delta t$. The gridfunction $\{\rho_\nu^n\}$ may represent approximate gridvalues, $\rho(t^n, x_\nu)$, or approximate cell averages, $\bar\rho(t^n, x_\nu)$ (as in the Godunov and LxF schemes), or a combination of higher moments, e.g., [23].

To construct a finite difference approximation of the conservation law (1.1), one introduce a discrete *numerical flux*, $H(\rho^n) := (H_1(\rho^n), \ldots, H_d(\rho^n))$, where $H_j(\rho^n) = H_j(\rho_{\nu-p}^n, \ldots, \rho_{\nu+q}^n)$ is an approximation to the $A_j(\rho^n)$ flux across the interface separating the cell C_ν and its neighboring cell on the x_j's direction, $C_{\nu+e_j}$. Next, exact derivatives in (1.1) are replaced by divided differences: the time-derivative is replaced with forward time difference, and spatial derivatives are replaced by spatial divided differences expressed in terms of $D_{+x_j}\varphi_\nu := (\varphi_{\nu+e_j} - \varphi_\nu)/\Delta x_j$. We arrive at the finite-difference scheme of the form

$$\rho_\nu^{n+1} = \rho_\nu^n - \Delta t \sum_{j=1}^d D_{+x_j} H_j(\rho_{\nu-p}^n, \ldots, \rho_{\nu+q}^n). \tag{1.28}$$

The essential feature of the difference schemes (1.28) is their *conservation form*: perfect derivatives in (1.1) are replaced here by 'perfect differences'. It implies that the change in mass over any spatial domain Ω, $\sum_{\{\nu | x_\nu \in \Omega\}} \rho_\nu^{n+1} |C_\nu| - \sum_{\{\nu | x_\nu \in \Omega\}} \rho_\nu^n |C_\nu|$, depends solely on the discrete flux across the boundaries of that domain. This is a discrete analogue for the notion of a weak solution of (1.1). In their seminal paper [96], Lax & Wendroff introduced the notion of conservative schemes, and prove that their *strong* limit solutions are indeed weak solutions of (1.1).

Theorem 1.5 *(Lax & Wendroff [96])* Consider the conservative difference scheme (1.28), with consistent numerical flux so that $H_j(\rho, \ldots, \rho) = A_j(\rho)$. Let $\Delta t \downarrow 0$ with fixed grid-ratios $\lambda_j := \frac{\Delta t}{\Delta x_j} \equiv Const_j$, and let $\rho^{\Delta t} = \{\rho_\nu^n\}$ denote the corresponding solution (parameterized w.r.t. the vanishing grid-size). Assume that $\rho^{\Delta t}$ converges strongly, $s \lim \rho^{\Delta t}(t^n, x_\nu) = \rho(t, x)$, then $\rho(x, t)$ is a weak solution of the conservation law (1.1).

The Lax-Wendroff theorem plays a fundamental role in the development of the so called 'shock capturing' methods. Instead of tracking jump discontinuities (– by evolving the smooth pieces of the approximate solution on both sides of such discontinuities), conservative schemes capture a discretized version of shock discontinuities. Equipped with the Lax-Wendroff theorem, it remains to prove strong convergence, which leads us to discuss the compactness of $\{\rho_\nu^n\}$.

TVD schemes ($m = d = 1$)

We deal with scalar gridfunctions, $\{\rho_\nu^n\}$, defined on the one-dimensional Cartesian grid $x_\nu := \nu\Delta x, t^n := n\Delta t$ with fixed mesh ratio $\lambda := \frac{\Delta t}{\Delta x}$. The total variation of such gridfunction at time-level t^n is given by $\sum_\nu |\Delta\rho_{\nu+\frac{1}{2}}^n|$, where $\Delta\rho_{\nu+\frac{1}{2}}^n := \rho_{\nu+1}^n - \rho_\nu^n$. It is said to be total-variation-diminishing (TVD) if

$$\sum_\nu |\Delta\rho_{\nu+\frac{1}{2}}^n| \leq \sum_\nu |\Delta\rho_{\nu+\frac{1}{2}}^0|. \tag{1.29}$$

Clearly, the TVD condition (1.29) is the discrete analogue of the scalar TV-bound (1.23). Approximate solutions of difference schemes which respect the TVD property (1.29), share the following desirable properties:

- *Convergence* – by Helly's compactness argument, the piecewise-constant approximate solution, $\rho^{\Delta x}(t^n, x) = \sum_\nu \rho_\nu^n \chi_\nu(x)$, converges strongly to a limit function, $\rho(t^n, x)$ as we refine the grid, $\Delta x \downarrow 0$. This together with equicontinuity in time and the Lax-Wendroff theorem, yield a weak solution, $\rho(t, x)$, of the conservation law (1.1).
- *Spurious oscillations* – are excluded by the TVD condition (1.23).
- *Accuracy* – is not restricted to the first-order limitation of monotone schemes. To be more precise, let us use $\rho^{\Delta t}(t, x)$ to denote a global realization (say – piecewise polynomial interpolant) of the approximate solution $\rho_\nu^n \sim \rho^{\Delta t}(t^n, x_\nu)$. The *truncation error* of the difference scheme is the amount by which the approximate solution, $\rho^{\Delta t}(t, x)$, fails to satisfy the conservation laws (1.1). The difference scheme is α-order accurate if its truncation error is, namely,

$$\|\partial_t \rho^{\Delta t} + \nabla_x \cdot A(\rho^{\Delta t})\| = \mathcal{O}((\Delta t)^\alpha). \tag{1.30}$$

(Typically, a strong norm $\|\cdot\|$ is used which is appropriate to the problem; in general, however, accuracy is indeed a norm-dependent quantity). Consider for example, monotone difference schemes. Monotone schemes are characterized by the fact that ρ_ν^{n+1} is an increasing function of the preceding gridvalues which participate in its stencil (1.28), $\rho_{\nu-p}^n, \ldots, \rho_{\nu+q}^n$ (— so that the monotonicity property (1.10) holds) . A classical result of Harten, Hyman & Lax [72] states that monotone schemes are at most first-order accurate. TVD schemes, however, are not restricted to this first-order accuracy limitation, at least in the one-dimensional case[1]. We demonstrate this point in the context of second-order TVD difference schemes.

We distinguish between two types of TVD schemes, depending on the size of their stencils.

Three-point schemes

Three-point schemes ($p = q = 1$ in (1.28)) are the simplest ones – their stencil occupies the three neighboring gridvalues, $\rho_{\nu-1}^n, \rho_\nu^n, \rho_{\nu+1}^n$. Three-point conservative schemes take the form

$$\rho_\nu^{n+1} = \rho_\nu^n - \frac{\lambda}{2}\left\{ A(\rho_{\nu+1}^n) - A(\rho_{\nu-1}^n) \right\} + \frac{1}{2}\left\{ Q_{\nu+\frac{1}{2}}^n \Delta\rho_{\nu+\frac{1}{2}}^n - Q_{\nu-\frac{1}{2}}^n \Delta\rho_{\nu-\frac{1}{2}}^n \right\}.$$
(1.31)

Thus, three-point schemes are identified solely by their numerical viscosity coefficient, $Q_{\nu+\frac{1}{2}}^n = Q(\rho_\nu^n, \rho_{\nu+1}^n)$, which characterize the TVD condition

$$\lambda|a_{\nu+\frac{1}{2}}^n| \le Q_{\nu+\frac{1}{2}}^n \le 1, \qquad a_{\nu+\frac{1}{2}}^n := \frac{\Delta A_{\nu+\frac{1}{2}}^n}{\Delta\rho_{\nu+\frac{1}{2}}^n}.$$
(1.32)

The schemes of Roe [139], Godunov [60], and Engquist-Osher (EO) [47], are canonical examples of *upwind* schemes, associated with (increasing amounts of) numerical viscosity coefficients, which are given by,

$$Q_{\nu+\frac{1}{2}}^{\text{Roe}} = \lambda|a_{\nu+\frac{1}{2}}^n|,$$
(1.33)

$$Q_{\nu+\frac{1}{2}}^{\text{Godunov}} = \lambda \max_{\zeta \in \mathcal{C}_{\nu+\frac{1}{2}}} \left[\frac{A(\rho_{\nu+1}^n) - 2A(\zeta) + A(\rho_\nu^n)}{\Delta\rho_{\nu+\frac{1}{2}}^n} \right],$$
(1.34)

$$Q_{\nu+\frac{1}{2}}^{\text{EO}} = \lambda \frac{1}{\Delta\rho_{\nu+\frac{1}{2}}^n} \int_{\rho_\nu^n}^{\rho_{\nu+1}^n} |A'(\zeta)|d\zeta.$$
(1.35)

The viscosity coefficients of the three upwind schemes are the same, $Q_{\nu+\frac{1}{2}}^n = \lambda|a_{\nu+\frac{1}{2}}^n|$, except for their different treatment of sonic points (where $a(\rho_\nu^n) \cdot a(\rho_{\nu+1}^n) < 0$). The Lax-Friedrichs (LxF) scheme (1.27) is the canonical *central* scheme. It has a larger numerical viscosity coefficient,

$$Q_{\nu+\frac{1}{2}}^{LxF} \equiv 1.$$
(1.36)

All the three-point TVD schemes are limited to first-order accuracy. Indeed, condition (1.32) is *necessary* for the TVD property of three-point

---

[1] Consult [65], regarding the first-order accuracy limitation for multidimensional $d > 1$ TVD schemes.

schemes, [162], and hence it excludes numerical viscosity associated with the second-order Lax-Wendroff scheme, [96], $Q_{\nu+\frac{1}{2}}^{LW} = \lambda^2(a_{\nu+\frac{1}{2}}^n)^2$. Therefore, scalar TVD schemes with more than first-order accuracy require at least five-point stencils.

Five-point schemes

Following the influential works of Boris & Book [6], van Leer [98], Harten [70], Osher [130], Roe [139] and others, many authors have constructed second order TVD schemes, using five-point (– or wider) stencils. For a more complete account of these works we refer to the recent books by LeVeque, [102], and Godlewski & Raviart, [58]. A large number of these schemes were constructed as second-order upgraded versions of the basic three-point *upwind* schemes. The FCT scheme of Boris & Book, [6], van Leer's MUSCL scheme [98], and the ULTIMATE scheme of Harten, [70], are prototype for this trend. In particular, in [70], Harten provided a useful sufficient criterion for the scalar TVD property, which led to the development of many non-oscillatory high-resolution schemes in the mid-80's.

Higher order *central* schemes can be constructed by upgrading the staggered LxF scheme (1.27). This will be the subject of our next lecture II. Here we quote a five-point TVD scheme of Nessyahu-Tadmor (NT) [126] – a second-order predictor-corrector upgrade of the staggered LxF scheme,

$$\rho_\nu^{n+\frac{1}{2}} = \bar{\rho}_\nu^n - \frac{\lambda}{2}(A(\bar{\rho}_\nu^n))', \tag{1.37}$$

$$\bar{\rho}_{\nu+\frac{1}{2}}^{n+1} = \frac{\bar{\rho}_\nu^n + \bar{\rho}_{\nu+1}^n}{2} +$$
$$-\frac{(\rho_\nu^n)' - (\rho_{\nu+1}^n)'}{8} - \frac{\Delta t}{\Delta x}\Big\{ A(\rho_{\nu+1}^{n+\frac{1}{2}}) - A(\rho_\nu^{n+\frac{1}{2}}) \Big\}. \tag{1.38}$$

Here, $\{w_\nu'\}$ denotes the *discrete numerical derivative* of an arbitrary gridfunction $\{w_\nu\}$. The choice $w_\nu' \equiv 0$ recovers the original first-order LxF scheme (1.27). Second-order accuracy requires $w_\nu' \sim \Delta x \partial_x w(x_\nu)$. To guarantee the non-oscillatory properties is a key issue in the construction of higher (– than first-order..) resolution schemes; this requires more than just the naive divided differences as discrete numerical derivatives. A prototype example is the so called min-mod limiter,

$$w_\nu' = \frac{1}{2}(s_{\nu-\frac{1}{2}} + s_{\nu+\frac{1}{2}}) \cdot \min\{|\Delta w_{\nu-\frac{1}{2}}|, |\Delta w_{\nu+\frac{1}{2}}|\}, \quad s_{\nu+\frac{1}{2}} := sgn(\Delta w_{\nu+\frac{1}{2}}). \tag{1.39}$$

(We shall say more on (nonlinear) limiters like the min-mod below.) With this choice of a limiter, the central NT scheme (1.37)-(1.38) satisfies the TVD property, and at the same time, it retains formal second order accuracy (at least away from extreme gridvalues, ρ_ν, where $\rho_\nu' = s_{\nu-\frac{1}{2}} + s_{\nu+\frac{1}{2}} = 0$).

We conclude we few additional Remarks.

Limiters A variety of discrete TVD limiters like (1.39) was explored during the '80s, e.g, [161] and the references therein. For example, a generalization of (1.39) is provided by the family of min-mod limiters depending on tuning parameters, $0 < \theta_{\nu \pm \frac{1}{2}} < 1$,

$$
w'_\nu(\theta) = \frac{1}{2}(s_{\nu - \frac{1}{2}} + s_{\nu + \frac{1}{2}}) \times
$$

$$
\times \min\{\theta_{\nu \pm \frac{1}{2}} |\Delta w_{\nu \pm \frac{1}{2}}|, \frac{1}{2}|w_{\nu+1} - w_{\nu-1}|\}. \tag{1.40}
$$

An essential feature of these limiters is *co-monotonicity*: they are 'tailored' to produce piecewise-linear reconstruction of the form $\sum[w_\nu + \frac{1}{\Delta x}w'_\nu(x - x_\nu)]\chi_\nu(x)$, which is co-monotone with (and hence, share the TVD property of –) the underlying piecewise-constant approximation $\sum w_\nu \chi_\nu(x)$. Another feature is the limiting property at extrema grid-values (where $w'_\nu = 0$), which is necessary in order to satisfy the TVD property (1.29). In this context, limiters can viewed as extrema detectors: the detection is *global*, yet they are activated *locally* (at extrema gridvalues). The study of the TVD property along these lines can be found in [164]. In particular, limiters are necessarily *nonlinear* in the sense of their stencils' dependence on the discrete gridfunction.

Systems – one-dimensional Godunov-type schemes The question of convergence for approximate solution of hyperbolic systems is tied to the question of existence of an entropy solution – in both cases there are no general theories with $m > 1$ equations[2]. Nevertheless, the ingredients of scalar high-resolution schemes were successfully integrated in the approximate solution of system of conservation laws.

Many of these high-resolution methods for systems, employ the Godunov approach, where one evolves a globally defined approximate solution, $\rho^{\Delta x}(t, x)$, which is governed by iterating the evolution-projection cycle,

$$
\rho^{\Delta x}(\cdot, t) = \begin{cases} T_{\{t - t^{n-1}\}}\rho(\cdot, t^{n-1}), & t^{n-1} < t < t^n = n\Delta t, \\ \\ P^{\Delta x}\rho(\cdot, t^n - 0), & t = t^n, \end{cases}
$$

Here, T_t denotes the evolution operator (see (1.16), and $P^{\Delta x}$ is an *arbitrary, possibly nonlinear* conservative projection, which which is realized as a piecewise polynomial,

$$
\rho^{\Delta x}(x, t^n) = \sum_j p_j(x)\chi_j(x), \quad \bar{p}_\nu(x_\nu) = \bar{\rho}_\nu^n. \tag{1.41}
$$

Typically, this piecewise polynomial approximate solution is reconstructed from the previously computed cell averages, $\{\bar{\rho}_\nu^n\}$, and in this context we

---

[2] There is a large literature concerning two equations – the 2×2 p-system and related equations are surveyed in [157].

may, again, distinguish between two main classes of methods: *upwind* and *central* methods.

Upwind schemes evaluate cell averages at the center of the piecewise polynomial elements; integration of (1.24) over $C_\nu \times [t^n, t^{n+1}]$ yields

$$\bar{\rho}_\nu^{n+1} = \bar{\rho}_\nu^n - \frac{1}{\Delta x}\left[\int_{\tau=t^n}^{t^{n+1}} f(\rho(\tau, x_{\nu+\frac{1}{2}},))d\tau - \int_{\tau=t^n}^{t^{n+1}} f(\rho(\tau, x_{\nu-\frac{1}{2}},))d\tau\right].$$

This in turn requires the evaluation of fluxes along the discontinuous cell interfaces, $(\tau \times x_{\nu+\frac{1}{2}})$. Consequently, upwind schemes must take into account the characteristic speeds along such interfaces. Special attention is required at those interfaces in which there is a combination of forward- and backward-going waves, where it is necessary to decompose the "Riemann fan" and determine the separate contribution of each component by tracing "the direction of the wind". The original first-order accurate Godunov scheme (1.26) is the forerunner for all other upwind Godunov-type schemes. A variety of second- and higher-order sequels to Godunov upwind scheme were constructed, analyzed and implemented with great success during the seventies and eighties, starting with van-Leer's MUSCL scheme [98], followed by [139,70,130,26]. These methods were subsequently adapted for a variety of nonlinear related systems, ranging from incompressible Euler equations, [4], [46], to reacting flows, semiconductors modeling, ... We refer to [59,102] and the references therein a for a more complete accounts on these developments.

In contrast to upwind schemes, *central* schemes evaluate staggered cell averages at the breakpoints between the piecewise polynomial elements,

$$\bar{\rho}_{\nu+\frac{1}{2}}^{n+1} = \bar{\rho}_{\nu+\frac{1}{2}}^n - \frac{1}{\Delta x}\left[\int_{\tau=t^n}^{t^{n+1}} f(\tau, \rho(x_{\nu+1}))d\tau - \int_{\tau=t^n}^{t^{n+1}} f(\rho(\tau, x_\nu))d\tau\right].$$

Thus, averages are integrated over the entire Riemann fan, so that the corresponding fluxes are now evaluated at the smooth centers of the cells, (τ, x_ν). Consequently, costly Riemann-solvers required in the upwind framework, can be now replaced by straightforward quadrature rules. The first-order Lax-Friedrichs (LxF) scheme (1.27) is the canonical example of such central difference schemes. The LxF scheme (like Godunov's scheme) is based on a piecewise constant approximate solution, $p_\nu(x) = \bar{\rho}_\nu$. Its Riemann-solver-free recipe, however, is considerably simpler. Unfortunately, the LxF scheme introduces excessive numerical viscosity (already in the scalar case outlined in §1.3 we have $Q^{LxF} \equiv 1 > Q^{\text{Godunov}}$), resulting in relatively poor resolution. The central scheme (1.37)-(1.38) is a second-order sequel to LxF scheme, with greatly improved resolution. An attractive feature of the central scheme (1.37)-(1.38) is that it avoids Riemann solvers: instead of characteristic variables, one may use a componentwise extension of the non-oscillatory limiters (1.40).

Multidimensional systems There are basically two approaches.

One approach is to reduce the problem into a series of one-dimensional problems. Alternating Direction (ADI) methods and the closely related dimensional splitting methods, e.g., [141, §8.8-9], are effective, widely used tools to solve multidimensional problems by piecing them from one-dimensional problems – one dimension at a time. Still, in the context of nonlinear conservation laws, dimensional splitting encounters several limitations, [31]. A particular instructive example for the effect of dimensional splitting errors can be found in the approximate solution of the weakly hyperbolic system studied in [49],[81, §4.3].

The other approach is 'genuinely multidimensional'. There is a vast literature in this context. The beginning is with the pioneering multidimensional second-order Lax-Wendroff scheme, [97]. To retain high-resolution of multidimensional schemes without spurious oscillations, requires one or more of several ingredients: a careful treatment of waves propagations ('unwinding'), or alternatively, a correctly tuned numerical dissipation which is free of Riemann-solvers ('central differencing'), or the use of adaptive grids (which are not-necessarily rectangular), ... Waves propagation in the context of multidimensional upwind algorithms were studied in [25,103,140,156] .... Another 'genuinely multidimensional' approach can be found in the positive schemes of [95]. The pointwise formulation of ENO schemes due to Shu & Osher, [153,154], is another approach which avoids dimensional splitting: here, the reconstruction of cell-averages is bypassed by the reconstruction *pointvalues* of the fluxes in each dimension; the semi-discrete fluxed are then integrated in time using non-oscillatory ODEs solvers (which are briefly mentioned in §1.3 below). Multidimensional non-oscillatory *central* scheme was presented in [81], generalizing the one-dimensional (1.37)-(1.38); consult [105],[89] for applications to the multidimensional incompressible Euler equations. Finite volume methods, [85,86,24,29]... , and finite-element methods (the **streamline-diffusion and discontinuous Galerkin schemes, [76,79,80,148,122]...**) have the advantage of a 'built-in' recipe for discretization over general triangular grids (we shall say more on these methods in §1.5 below). Another 'genuinely multidimensional' approach is based on a relaxation approximation was introduced in [82]. It employs a central scheme of the type (1.37)-(1.38) to discretize the relaxation models models, [178], [19], [125], ...

TVD filters
Every discretization method is associated with an appropriate finite-dimensional projection. It is well known that *linear* projections which are monotone (or equivalently, positive), are at most first-order accurate, [60]. The lack of monotonicity for higher order projections is reflected by spurious oscillations in the vicinity of jump discontinuities. These are evident with the second-order (and higher) centered differences, whose dispersive nature is respon-

sible to the formation of binary oscillations [64],[104]. With highly-accurate spectral projections, for example, these $\mathcal{O}(1)$ oscillations reflect the familiar Gibbs phenomena.

TVD schemes avoid spurious oscillations — to this end they use the necessarily *nonlinear* projections (expressed in terms of nonlinear limiters like those in (1.40)). TVD filters, instead, suppress spurious oscillations. At each time-level, one post-process the computed (possibly oscillatory) solution $\{\rho(t^n)\}$. In this context we highlight the following.

• **Linear filters.** Consider linear convection problems with discontinuous initial data. Approximate solutions of such problems suffer from loss of accuracy due to propagation of singularities and their interference over domain of dependence of the numerical scheme. Instead, one can show, by duality argument, that the numerical scheme retains its original order of accuracy when the truncation in (1.30) is measured w.r.t. sufficiently large negative norm, [121]. Linear filters then enable to accurately recover the exact solution in any smoothness region of the exact solution, bounded away from its singular support. These filters amount to finite-order mollifiers [121], or spectrally accurate mollifiers, [119], [67], which accurately recover pointvalues from high-order moments.

• **Artificial compression.** Artificial compression was introduced by Harten [69] as a method to sharpen the poor resolution of contact discontinuities. (Typically, the resolution of contacts by α-order schemes diffuses over a fan of width $(\Delta t)^{(\alpha)/(\alpha+1)}$). The idea is to enhance the focusing of characteristics by adding an anti-diffusion modification to the numerical fluxes: if we let $H_{\nu+\frac{1}{2}}$ denote the numerical flux of a three-point TVD scheme (1.31), then one replaces it with a modified flux, $H_{\nu+\frac{1}{2}} \longrightarrow H_{\nu+\frac{1}{2}} + \tilde{H}_{\nu+\frac{1}{2}}$, which is expressed in terms of the min-mod limiter (1.39)

$$\tilde{H}_{\nu+\frac{1}{2}} := \frac{1}{\lambda}\{\rho'_\nu + \rho'_{\nu+1} - sgn(\Delta\rho_{\nu+\frac{1}{2}})|\rho'_{\nu+1} - \rho'_\nu|\}. \tag{1.42}$$

Artificial compression can be used as a second-order TVD filter as well. Let $Q_{\nu+\frac{1}{2}}$ be the numerical viscosity of a three-point TVD scheme (1.31). Then, by adding an artificial compression modification (1.42) which is based on the θ-limiters (1.40), $\rho'_\nu = \rho'_\nu(\theta)$ with $\theta_{\nu+\frac{1}{2}} := Q_{\nu+\frac{1}{2}} - \lambda^2 a^2_{\nu+\frac{1}{2}}$, one obtains a second-order TVD scheme, [70], [133]. Thus, in this case the artificial compression (1.42) can be viewed as a second-order anti-diffusive TVD filter of first-order TVD schemes

$$\rho^{n+1}_\nu \longleftarrow \rho^{n+1}_\nu - \{\tilde{H}_{\nu+\frac{1}{2}}(\rho^n) - \tilde{H}_{\nu-\frac{1}{2}}(\rho^n)\}. \tag{1.43}$$

• **TVD filters.** A particularly useful and effective, general-purpose TVD filter was introduced by Engquist et. al. in [48]; it proceeds in three steps. {i} (Isolate extrema). First, isolate extrema cells where $\Delta\rho^n_{\nu-\frac{1}{2}} \cdot \Delta\rho^n_{\nu+\frac{1}{2}} < 0$.

{ii} (Measure local oscillation). Second, measure local oscillation, osc_ν, by setting

$$osc_\nu := \min\{m_\nu, \frac{1}{2}M_\nu\}, \qquad \{^{m_\nu}_{M_\nu}\} = \{^{\min}_{\max}\}(\Delta\rho^n_{\nu-\frac{1}{2}}, \Delta\rho^n_{\nu+\frac{1}{2}})$$

{iii} (Filtering). Finally, oscillatory minima (respectively – oscillatory maxima) are increased (and respectively, increased) by updating $\rho^n_\nu \to \rho^n_\nu + sgn(\Delta\rho^n_{\nu+\frac{1}{2}})osc_\nu$, and the corresponding neighboring gridvalues is modified by subtracting the same amount to retain conservation. This postprocessing can be repeated, if necessary, and one may use a local maximum principle, $\min_j\rho^n_j \leq \rho^n_\nu \leq \max_j\rho^n_j$ as a stopping criterion. In this case, the above filter becomes TVD once the binary oscillations are removed, [155].

TVB approximations ($m \geq 1, d = 1$)

One sided stability

As an example for Total variation Bounded (TVB) approximations, we begin with the example of approximate solutions satisfying the one-sided Lip^+ stability condition.

Let $\{\rho^\varepsilon(t, x)\}$ be a family of approximate solutions, tagged by their small-scale parameterization, ε. To upper-bound the convergence rate of such approximations, we shall need the usual two ingredients of stability and consistency.

– Lip^+-stability. The family $\{\rho^\varepsilon\}$ is Lip^+-stable if

$$\|\rho^\varepsilon(t, \cdot)\|_{Lip^+} := \sup_x \partial_x\rho^\varepsilon(t, x) \leq Const. \qquad (1.44)$$

This notion of Lip^+-stability is motivated by Oleinik's One-Sided Lipschitz Condition (OSLC), $\rho_x(t, \cdot) \leq Const$, which uniquely identifies the entropy solution of convex conservation laws, (1.24), with scalar $A'' > 0$ (we refer to [100] for a recent contribution concerning the one-sided stability of one-dimensional systems). Since the Lip^+-(semi)-norm dominates the total-variation,

$$\|\rho^\varepsilon(t, \cdot)\|_{BV} \leq Const.\|\rho^\varepsilon(t, \cdot)\|_{Lip^+} + \|\rho^\varepsilon_0(\cdot)\|_{L^1}, \quad Const = 2|\text{supp}_x\rho^\varepsilon(t, \cdot)|,$$

$\{\rho^\varepsilon\}$ are TVB and by compactness, convergence follows. Equipped with Lip^+-stability, we are able to quantify this convergence statement. To this end, we measure the local truncation error in terms of

– Lip'-consistency. The family $\{\rho^\varepsilon\}$ is Lip'-consistent of order ε if

$$\|\partial_t \rho^\varepsilon + \partial_x A(\rho^\varepsilon)\|_{Lip'(t,x)} \sim \varepsilon. \tag{1.45}$$

It follows that the stability+consistency in the above sense, imply the convergence of $\{\rho^\varepsilon\}$ to the entropy solution, ρ, and that the following error estimates hold [166], [127],

$$\|\rho^\varepsilon(t,\cdot) - \rho(t,\cdot)\|_{W^s(L^p(x))} \sim \varepsilon^{\frac{1-sp}{2p}}, \quad -1 \le s \le 1/p. \tag{1.46}$$

The case $(s,p) = (-1,1)$ corresponds to a sharp Lip'-error estimate of order ε — the Lip'-size of the truncation in (1.45); the case $(s,p) = (0,1)$ yields an L^1-error estimate of order one-half, in agreement with Kuznetsov's general convergence theory, [90].

Multidimensional extensions to convex Hamilton-Jacobi equations are treated in [107]. We note in passing that the requirement of Lip^+ stability restricts our discussion to *convex* problems; at the same time, it yields more than just convergence. Indeed, the above error estimate, as well as additional *local* error estimates will discussed in lecture IV.

Higher resolution schemes (with three letters acronym)

We have already mentioned the essential role played by nonlinear limiters in TVD schemes. The mechanism in these nonlinear limiters is switched on in extrema cells, so that the zero discrete slope $\rho' = 0$ avoids new spurious extrema. This, in turn, leads to deteriorated first-order local accuracy at non-sonic extrema, and global accuracy is therefore limited to second-order[3].

To obtain an improved accuracy, one seeks a more accurate realization of the approximate solution, in terms of higher (than first-order) piecewise polynomials

$$\rho^{\Delta x}(t^n, x) = \sum_\nu p_\nu(x)\chi_\nu(x), \quad p_\nu(x) = \sum_j \rho_\nu^{(j)}(\frac{x - x_\nu}{\Delta x})^j/j!. \tag{1.47}$$

Here, the exact solution is represented in a cell C_ν in terms of an r-order polynomial p_ν, which is reconstructed from the its neighboring cell averages, $\{\bar{\rho}_{\nu_\mu}\}$. If we let $\rho^{\Delta x}(t \ge t^n, \cdot)$ denote the entropy solution subject to the reconstructed data at $t = t^n$, $P^{\Delta x}\rho(t^n, \cdot)$, then the corresponding Godunov-type scheme governs the evolution of cell averages

$$\bar{\rho}_\nu^{n+1} := \frac{1}{\Delta x} \int_x \rho^{\Delta x}(t^{n+1} - 0, x)\chi_\nu(x)dx. \tag{1.48}$$

---

[3] The implicit assumption is that we seek an approximation to *piecewise-smooth* solutions with finitely many oscillations, [169]. The convergence theories apply to general BV solutions. Yet, general BV solutions cannot be resolved in *actual* computations in terms of 'classical' macroscopic discretizations – finite-difference, finite-element, spectral methods, etc. Such methods can faithfully resolve piecewise smooth solutions.

The properties of Godunov-type scheme are determined by the polynomial reconstruction should meet three contradicting requirements:

{i} *Conservation:* $p_\nu(x)$ should be cell conservative in the sense that $\fint_{\mathcal{C}_\nu} p_\nu(x) = \fint_{\mathcal{C}_\nu} \rho_\nu(x)$. This tells us that $P^{\Delta x}$ is a (possibly nonlinear) projection, which in turn makes (1.48) a conservative scheme in the sense of Lax-Wendroff, (1.28).

{ii} *Accuracy:* $\rho_\nu^{(j)} \sim (\Delta x \partial_x)^j \rho(t^n, x_\nu)$.

At this stage, we have to relax the TVD requirement. This brings us to the third requirement of

{iii} *TVB bound:* we seek a bound on the total variation on the computed solution. Of course, a bounded variation, $\|\rho^{\Delta x}(t^n, \cdot)\|_{BV} \leq$ Const. will suffice for convergence by L^1-compactness arguments (Helly's theorem).

The (re-)construction of non-oscillatory polynomials led to new high-resolution schemes. In this context we mention the following methods (which **were popularized by their trade-mark of three-letters acronym ...**): the Piecewise-Parabolic Method (PPM) [26], the Uniformly Non-Oscillatory (UNO) scheme [74], and the Essentially Non-Oscillatory schemes (ENO) of Harten et. al. [71]. The particular topic of ENO schemes is covered in C.-W. Shu's lectures elsewhere in this volume.

There is large numerical evidence that these highly-accurate methods are TVB (and hence convergent), at least for a large class of piecewise-smooth solutions. We should note, however, that the convergence question of these schemes is open. (It is my opinion that new characterizations of the (piece-wise) regularity of solutions to conservation laws, e.g., [38],[169] together with additional tools to analyze their compactness, are necessary in order to address the questions of convergence and stability of these highly-accurate schemes).

There are alternative approach to to construct high-resolution approximations which circumvent the TVD limitations. We conclude by mentioning the following two.

One approach is to evolve more than one-piece of information per cell. This is fundamentally different from standard Godunov-type schemes where only the cell average is evolved (and higher order projections are *reconstructed* from these averages – one per cell). In this context we mention the quasi-monotone TVB schemes introduced in [23]. Here, one use a TVD evolution of cell averages together with additional higher moments. Another instructive example for this approach is found in the third-order TVB scheme, [144]: in fact, Sanders constructed a third-order non-expansive scheme (circumventing the first-order limitation of [72]), by using a 2×2 system which governs the first two moments of the scalar solution. More recently, Bouchut et. al. [8], constructed a second-order MUSCL scheme which respects a discrete version of the entropy inequality (1.3) w.r.t *all* Kružkov's scalar entropy pairs in (1.8); this circumvents the second-order limitation of Osher & Tadmor [133,

Theorem 7.3], by evolving *both* – the cell average and the discrete slope in each computational cell.

Another approach to enforce a TVB bound on higher($>$ 2)-resolution schemes, makes use of gridsize-dependent limiters, $\rho^{(j)} = \rho^{(j)}\{\bar{\rho}^n, \Delta x\}$, such that the following holds, e.g., [151],

$$\|\rho^{\Delta x}(t^{n+1}, \cdot)\|_{BV} \leq \|\rho^{\Delta x}(t^{n+1}, \cdot)\|_{BV} + \text{Const} \cdot \Delta x.$$

Such Δx-dependent limiters fail to satisfy, however, the basic dilation invariance of (1.24)-(1.25), $(t, x) \rightarrow (ct, cx)$.

Time discretizations

One may consider separately the discretization of time and spatial variables. Let P_N denote a (possibly nonlinear) finite-dimensional spatial discretization of (1.1); this yields an N-dimensional approximate solution, $\rho_N(t)$, which is governed by the system of N nonlinear ODEs

$$\frac{d}{dt}\rho_N(t) = P_N(\rho_N(t)). \tag{1.49}$$

System (1.49) is a *semi-discrete* approximation of (1.1). For example, if we let $P_N = P^{\Delta x}$, $N \sim (\Delta x)^{-d}$, to be one of the piecewise-polynomial reconstructions associated with Godunov-type methods in (1.47), then one ends up with a semi-discrete finite-difference method, the so called method of lines. In fact, our discussion on streamline-diffusion and spectral approximations in §1.4 and §1.4 below will be primarily concerned with such semi-discrete approximations.

An explicit time discretization of (1.49) proceeds by either a multi-level or a Runge-Kutta method. A CFL condition should be met, unless one accounts for wave interactions, consult [101]. For the construction of non-oscillatory schemes, one seeks time discretizations which retain the non-oscillatory properties of the spatial discretization, P_N. In this context we mention the TVB time-discretizations of Shu & Osher, [152],[153,154]. Here, one obtains high-order multi-level and Runge-Kutta time discretizations as *convex combinations* of the standard forward time differencing, which amounts to the first-order accurate forward Euler method. Consequently, the time discretizations [153,154] retain the nonoscillatory properties of the low-order forward Euler time differencing — in particular, TVD/TVB bounds, and at the same time, they enable to match the time accuracy with the high-order spatial accuracy.

Cell entropy inequality

Approximate solutions with bounded variation (obtained by TVD/TVB schemes) converge to *a weak* solution; the question of uniqueness is addressed by an entropy condition. In the context of finite-difference scheme, one seeks a cell entropy inequality – a conservative discrete analogue of the entropy inequality (1.3),

$$\eta(\rho_\nu^{n+1}) \le \eta(\rho_\nu^n) - \Delta t \sum_{j=1}^{d} D_{+x_j} G_j(\rho_{\nu-p}^n, \dots, \rho_{\nu+q}^n). \tag{1.50}$$

By arguments à la Lax & Wendroff (Theorem 1.5), any approximate solution which satisfies (1.50) with a consistent numerical entropy flux, $G_j(\rho, \dots, \rho) = F_j(\rho)$, its strong limit satisfies (1.3), which in turn yields uniqueness, at least in the scalar case. Crandall & Majda, [30], following Harten, Hyman & Lax in [72], were the first to implement this approach in the context of monotone difference schemes (in fact, the abstract setup of Theorem 1.2 directly applies in this case). Osher [130] introduced the so-called numerical E-fluxes to guarantee the cell entropy inequality. In [163] we prove the entropy inequality for general fully-discrete E-schemes: the proof is based on the key observation that the numerical viscosity (— quantified in terms of the numerical viscosity coefficient Q in (1.31)), associated with any E-flux, is a convex combination of the Godunov and Lax-Friedrichs viscosities, given in (1.34) and (1.36), respectively. Applications to the question of multidimensional convergence can be found in [85],[86],[24],[128].... E-fluxes are restricted to first-order accuracy, since they are consistent with *all* Kružkov's entropy pairs. A systematic study of the cell entropy inequality for second-order resolution scheme can be found in [133] (for upwind schemes) and in [126] (for central schemes). The above discussion is restricted to scalar problems. Of course, general Godunov and LxF schemes ($m > 1, d = 1$), satisfy a cell entropy inequality because the Riemann solutions do. (For the LxF scheme, we refer to Lax, [93], who proved the cell entropy inequity *independently* of the Riemann solution.

1.4 Entropy Production Bounds

Compensated compactness ($m \le 2, d = 1$)
We deal with a family of approximate solutions, $\{\rho^\varepsilon\}$, such that

(i) *It is uniformly bounded,* $\rho^\varepsilon \in L^\infty$, *with a weak* limit,* $\rho^\varepsilon \rightharpoonup \rho$;
(ii) *The entropy production,* for all convex entropies η, lies in a compact subset of $W_{loc}^{-1}(L^2(t, x))$,

$$\forall \eta'' > 0: \qquad \partial_t \eta(\rho^\varepsilon) + \partial_x F(\rho^\varepsilon) \hookrightarrow W_{loc}^{-1}(L^2(t, x)). \tag{1.51}$$

The conclusion is that $A(\rho^\varepsilon) \rightharpoonup A(\rho)$, and hence ρ is a weak solution; in fact, there is a strong convergence, $\rho^\varepsilon \to \rho$, on any nonaffine interval of $A(\cdot)$. For a complete account on the theory of compensated compactness we refer to the innovative works of Tartar [171] and Murat [124]. In the present context, compensated compactness argument is based on a clever application of the div-curl lemma. First scalar applications are due to Murat-Tartar, [123],[171], followed by extensions to certain $m = 2$ systems by DiPerna [40] and Chen [17].

The current framework has the advantage of dealing with L^2-type estimates rather than the more intricate BV framework. How does one verify the $W_{loc}^{-1}(L^2)$-condition (1.51)? we illustrate this point with canonical viscosity approximation (1.2). Multiplication by η' shows that its entropy production amounts to $\varepsilon(\eta' Q \rho_x^\varepsilon)_x - \varepsilon \eta'' Q (\rho_x^\varepsilon)^2$. By entropy convexity, $\varepsilon \eta'' Q > 0^4$, and space-time integration yields

– *An entropy production bound*

$$\sqrt{\varepsilon} \| \frac{\partial \rho^\varepsilon}{\partial x} \|_{L_{loc}^2(t,x)} \leq Const. \tag{1.52}$$

Though this bound is too weak for strong compactness, it is the key estimate behind the $W_{loc}^{-1}(L^2)$-compactness condition (1.51). We continue with the specific examples of streamline-diffusion in §1.4 and spectral viscosity methods in §1.4.

The streamline diffusion finite-element method
The Streamline Diffusion (SD) finite element scheme, due to Hughes, Johnson, Szepessy and their co-workers [76], [79], [80], was one of the first methods whose convergence was analyzed by compensated compactness arguments. (Of course, finite-element methods fit into L^2-type Hilbert-space arguments). In the SD method, formulated here in several space dimensions, one seeks a piecewise polynomial, $\{\rho^{\Delta x}\}$, which is uniquely determined by requiring for all piecewise polynomial test functions $\psi^{\Delta x}$,

$$\langle \partial_t \rho^{\Delta x} + \nabla_x \cdot A(\rho^{\Delta x}), \ \psi^{\Delta x} + |\Delta x| \cdot \boxed{(\psi_t^{\Delta x} + A'(\rho^{\Delta x}) \psi_x^{\Delta x})} \rangle = 0. \tag{1.53}$$

Here, Δx denotes the spatial grid size (for simplicity we ignore time discretization). The expression inside the framed box on the left represents a diffusion term along the streamlines, $\dot{x} = A'(\rho^{\Delta x})$. Setting the test function, $\psi^{\Delta x} = \rho^{\Delta x}$, (1.53) yields the desired entropy production bound

$$\sqrt{\Delta x} \| \partial_t \rho^{\Delta x} + \nabla_x \cdot A(\rho^{\Delta x}) \|_{L_{loc}^2(t,x)} \leq Const. \tag{1.54}$$

Thus, the spatial derivative in (1.52) is replaced here by a streamline-directional gradient. This together with an L^∞-bound imply $W_{loc}^{-1}(L^2)$-compact entropy production, (1.51), and convergence follows [79],[80],[160]. We note in passing that the extension of the SD method for systems of equations is carried out by projection into entropy variables, [120], which in turn provide the correct interpretation of (1.54) as an entropy production bound.
The lectures of C. Johnson in this volume will present a comprehensive discussion of the streamline diffusion method and its related extensions.

---

[4] Observe that the viscosity matrix is therefore required to be positive w.r.t. the Hessian η''.

The spectral viscosity method

Since spectral projections are inherently oscillatory, they do not lend themselves to a priori TVB bound. Spectral methods provide another example for
a family of approximate solutions whose convergence could be better dealt,
therefore, by compensated compactness arguments. Spurious Gibbs oscillations violate the strict TVD condition in this case. Instead, an entropy production bound, analogous to (1.52) is sought. Indeed, such bound could be
secured by spectrally accurate hyper-viscosity which is expressed in terms
of the computed Fourier coefficients. This leads us to a discussion on the
Spectral Viscosity (SV) method.

Let P_N denote an appropriate spatial projection into the space of N-
degree polynomials,

$$P_N \rho(t, x) = \sum_{|k| \leq N} \hat{\rho}_k(t) \varphi_k(x);$$

here $\{\varphi_k\}$ stands for a given family of orthogonal polynomials, either trigonometric or algebraic ones, e.g., $\{e^{ikx}\}$, $\{L_k(x)\}$, $\{T_k(x)\}$, etc. The corresponding N-degree approximate solution, $\rho_N(t, x)$, is governed by the spectral viscosity (SV) approximation

$$\partial_t \rho_N + \partial_x P_N A(\rho_N) = \frac{(-1)^{s+1}}{N^{2s-1}} \partial_x^s (Q * \partial_x^s \rho_N). \tag{1.55}$$

The left hand side of (1.55) is the standard spectral approximation of the
conservation law (1.1). The expression on the right

$$\frac{(-1)^{s+1}}{N^{2s-1}} \partial_x^s (Q * \partial_x^s \rho_N) := \frac{(-1)^{s+1}}{N^{2s-1}} \sum_{|k| > N^\theta} \hat{Q}_k \hat{\rho}_k(t) \varphi_k^{(2s)}(x), \tag{1.56}$$

represents the so called spectral viscosity introduced in [165]. It contains a
minimal amount of high-modes regularization which retains the underlying
spectral accuracy of the overall approximation. The case $s = 1$ corresponds
to a truncated second-order viscosity

$$\frac{1}{N} \partial_x (Q * \partial_x \rho_N) := \frac{1}{N} \sum_{|k| > N^\theta} \hat{Q}_k \hat{\rho}_k(t) \varphi_k''(x).$$

It involves a viscous-free zone for the first N^θ modes, $0 < \theta < \frac{1}{2}$. High modes
diffusion is tuned by the viscosity coefficients $\hat{Q}_k$.

Larger s's corresponds to truncated hyper-diffusion of order $2s$. This allows for even a larger viscosity-free zone of size N^θ, with $0 < \theta < \frac{2s-1}{2s}$
(with possibly $s = s_N \leq \sqrt{N}$), consult [167]. The underlying hyper-viscosity
approximation (for say $s = 2$) reads

$$\partial_t \rho^\varepsilon + \partial_x A(\rho^\varepsilon) + \varepsilon^3 \partial_x^4 \rho^\varepsilon = 0. \tag{1.57}$$

We note that already the solution operator associated with (1.57) is not monotone, hence L^1-contraction and the TVD condition fail in this case.

Instead, an L^2-type entropy production estimate analogous to (1.52)

$$\frac{1}{\sqrt{N}}\|\frac{\partial \rho_N}{\partial x}\|_{L^2_{loc}(t,x)} \leq Const.$$

together with an L^∞-bound, carry out the convergence analysis by compensated compactness arguments, [165], [117]. Extensions to certain $m = 2$ systems can be found in [145]. We shall return to a detailed discussion on the SV method in our lecture III.

1.5   Measure-valued solutions($m = 1, d \geq 1$)

We turn our attention to the multidimensional scalar case, dealing with a families of uniformly bounded approximate solutions, $\{\rho^\varepsilon\}$, with weak* limit, $\rho^\varepsilon \rightharpoonup \rho$. DiPerna's result [42] states that if the entropy production of such a family tends weakly to a negative measure, $m \leq 0$,

$$\forall \eta'' > 0: \qquad \partial_t \eta(\rho^\varepsilon) + \nabla_x \cdot F(\rho^\varepsilon) \rightharpoonup m \leq 0, \qquad (1.58)$$

then the measure-valued solution ρ coincides with the entropy solution, and convergence follows. This framework was used to prove the convergence of multidimensional finite-difference schemes [27], streamline diffusion method [79],[80], spectral-viscosity approximations [18] and finite-volume schemes [24], [86],[85]. We focus our attention on the latter.

Finite volume schemes ($d \geq 1$)

We are concerned with finite-volume schemes based on possibly *unstructured* triangulation grid $\{T_\nu\}$ (for simplicity we restrict attention to the $d = 2$ case). The spatial domain is covered by a triangulation, $\{T_\nu\}$, and we compute approximate averages over these triangles, $\bar{\rho}_\nu^n \sim \frac{1}{|T_\nu|} \int_{T_\nu} \rho(t^n, x)dx$, governed by the finite volume (FV) scheme

$$\bar{\rho}_\nu^{n+1} = \bar{\rho}_\nu^n - \frac{\Delta t}{|T_\nu|} \sum_\mu \tilde{A}_{\nu_\mu}(\rho_\nu^n, \rho_{\nu_\mu}^n). \qquad (1.59)$$

Here $\tilde{A}_{\nu_\mu}$ stand for approximate fluxes across the interfaces of T_ν and its neighboring triangles (identified by a secondary index μ).

Typically, the approximate fluxes, $\tilde{A}_{\nu_\mu}$ are derived on the basis of approximate Riemann solvers across these interfaces, which yield a monotone scheme. That is, the right hand side of (1.59) is a monotone function of its arguments $(\rho_\nu^n, \rho_{\nu_\mu}^n)$, and hence the corresponding FV scheme is L^1-contractive. However, at this stage one cannot proceed with the previous compactness arguments which apply to TVD schemes over fixed Cartesian grid: since the grid is unstructured, the discrete solution operator is not translation invariant and L^1-contraction need not imply a TV bound. Instead, an entropy dissipation estimate yields

$$\sum_n \Delta t \sum_{\nu,\mu} |\rho_\nu^n - \rho_{\nu_\mu}^n|(\Delta x)^\theta \le Const, \quad 0 < \theta < 1. \tag{1.60}$$

Observe that (1.60) is weaker than a TV bound (corresponding to $\theta = 0$), yet it suffices for convergence to a measure-valued solution, consult [24], [85]. These questions will be addressed in B. Cockburn's lectures, later in this volume.

1.6 Kinetic Approximations

By a *kinetic formulation* of (1.1) we mean a representation of the solution $\rho(t, x)$ as the average of a 'microscopic' density function, $f(t, x, v)$. The formulation is a kinetic one by its analogy with the classical kinetic models such as Boltzmann or Vlasov models - see for instance [15],[44]. In particular, we add a real-valued variable called velocity, v, and the unknown becomes a 'density-like' function, $f(t, x, v)$, which is governed by an appropriate transport equation.

A useful tool in this context is the *velocity averaging lemma*, dealing with the regularity of the moments for such transport solutions.

Velocity averaging lemmas ($m \ge 1, d \ge 1$)
We deal with solutions to transport equations

$$a(v) \cdot \nabla_x f(x, v) = \partial_v^s g(x, v). \tag{1.61}$$

The averaging lemmas, [62], [53], [45], state that in the generic non-degenerate case, averaging over the velocity space, $\bar{f}(x) := \int_v f(x, v) dv$, yields a gain of *spatial* regularity. The prototype statement reads

Lemma 1.2 *([62],[45],[111]). Let $f \in L^p(x, v)$ be a solution of the transport equation (1.61) with $g \in L^q(x, v), 1 \le q < p \le 2$. Assume the following non-degeneracy condition holds*

$$meas_v\{v| \ |a(v) \cdot \xi| < \delta\}_{|\xi|=1} \le Const \cdot \delta^\alpha, \quad \alpha \in (0, 1). \tag{1.62}$$

Then $\bar{f}(x) := \int_v f(x, v) dv$ belongs to Sobolev space $W^\theta(L^r(x))$,

$$\bar{f}(x) \in W^\theta(L^r(x)), \qquad \theta < \frac{\alpha}{\alpha(1 - \frac{p'}{q'}) + (s + 1)p'}, \quad \frac{1}{r} = \frac{\theta}{q} + \frac{1 - \theta}{p}. \tag{1.63}$$

Variants of the averaging lemmas were used by DiPerna and Lions to construct global weak (renormalized) solutions of Boltzmann, Vlasov-Maxwell and related kinetic systems, [43], [44]; in Bardos et. al., [2], averaging lemmas were used to construct solutions of the incompressible Navier-Stokes equations. We turn our attention to their use in the context of nonlinear conservation laws and related equations.

Nonlinear conservation laws

As a prototype example we begin with a Boltzmann-like – or more precisely, a BGK-like model proposed in [136]. Its 'hydrodynamical limit' describes both the scalar conservation law (1.1) together with its entropy inequalities, (1.20). It consists in solving the transport equation

$$\frac{\partial f^{\varepsilon}}{\partial t} + a(v) \cdot \nabla_x f^{\varepsilon} = \frac{1}{\varepsilon}\left(\chi_{\rho^{\varepsilon}}(v) - f^{\varepsilon}\right), \quad (t, x, v) \in \mathbb{R}_t^+ \times \mathbb{R}_x^d \times \mathbb{R}_v, \quad (1.64)$$

$$f^{\varepsilon}|_{t=0} = \chi_{\rho_0(x)}(v), \quad (x, v) \in \mathbb{R}_x^d \times \mathbb{R}_v. \quad (1.65)$$

Here, $\chi_{\rho^{\varepsilon}(t,x)}(v)$ denotes the 'pseudo-Maxwellian',

$$\chi_{\rho^{\varepsilon}}(v) := \begin{cases} +1 & 0 < v < \rho^{\varepsilon} \\ -1 & \rho^{\varepsilon} < v < 0 \\ 0 & |v| > \rho^{\varepsilon} \end{cases}, \quad (1.66)$$

which is associated with the average of f^{ε},

$$\rho^{\varepsilon}(t, x) = \bar{f}^{\varepsilon} := \int_{\mathbb{R}} f^{\varepsilon}(t, x, v) dv, \quad (t, x) \in \mathbb{R}_t^+ \times \mathbb{R}_x^d. \quad (1.67)$$

Notice that the BGK-like model in (1.64)-(1.67) is a semilinear, nonlocal, hyperbolic (first-order) equation which is rather simple to solve for fixed $\varepsilon > 0$. This kinetic model was introduced in [136], following the earlier works [9],[54]. It follows that that if $\rho_0 \in L^1(\mathbb{R}^d) \cap L^{\infty}(\mathbb{R}^d)$, then ρ^{ε} converges in $L^1((0, T) \times \mathbb{R}^d)$ to the unique entropy solution (1.1), (1.20). In fact, there is a convergence on the underlying microscopic level, to a kinetic formulation of (1.1), (1.20). The latter is described by a limiting 'density-function', $f(t, x, v)$, which is governed by the transport equation

$$\frac{\partial f}{\partial t} + a(v) \cdot \nabla_x f = \frac{\partial m}{\partial v} \quad (t, x, v) \in \mathcal{D}'(\mathbb{R}_t^+ \times \mathbb{R}_x^d \times \mathbb{R}_v) \quad (1.68)$$

subject to initial conditions

$$f = \chi_{\rho_0(t,x)}(v). \quad (1.69)$$

Here, m is a nonnegative bounded measure on $\mathbb{R}_t^+ \times \mathbb{R}_x^d \times \mathbb{R}_v$.

In what sense does the kinetic formulation (1.68-1.69) 'describe' the conservation law (1.1-1.20)? observe that by averaging of (1.68) one recovers the conservation law (1.1), and taking its higher moments by integration against $\eta'(v)$, one recovers Kružkov entropy inequalities (1.20)) for *all convex* entropies η.

Theorem 1.6 *Consider the BGK-like model (1.64)-(1.65).*

{i} There exists a nonnegative measure, $m^{\varepsilon}(t, x, v)$, which is bounded independently of ε, such that the relaxation term on the right of (1.64) admits

$$\frac{1}{\varepsilon}(\chi_{\rho^\varepsilon} - f^\varepsilon) = \frac{\partial m^\varepsilon}{\partial v}, \quad m^\varepsilon \geq 0. \tag{1.70}$$

{ii} *The solution f^ε of the kinetic model (1.64)-(1.65) converges in $L^1((0,T) \times \mathbb{R}_x^d \times \mathbb{R}_v)$ ($\forall\, T < \infty$) to the solution of (1.68)-(1.69). In addition, its associated measure, m^ε, converges weakly to the measure, m, uniquely determined by the kinetic formulation (1.68)-(1.69) with $f = \chi_\rho$.*

Remark 1.4 One may deduce from the above result and from [136] that m vanishes on open sets of the form $\{(x,v,t) \,/\, (x,t) \in \mathcal{O}\, v \in \mathbb{R}\}$ where $\mathcal{O}$ is an open set on which ρ is locally Lipschitz. In other words, m is 'supported by the shocks".

Proof. Several proof are available, each highlights the related aspects of this issue.

One approach makes use of the simple H-functions, à la Boltzmann, constructed in [136], $H_c(f^\varepsilon) := |f^\varepsilon - \chi_c|$.

Lemma 1.3 *([136, Corollary 3.2]) For any real c the following functions*

$$H_c(f^\varepsilon) := |f^\varepsilon - \chi_c|$$

are kinetic entropy functions, i.e., we have

$$\int_v [\partial_t + a(v) \cdot \nabla_x] |f^\varepsilon - \chi_c| dv \leq 0. \tag{1.71}$$

Let us Remark that our kinetic entropy functions, $H_c(f^\varepsilon)$, are intimately related to Kružkov entropy functions, (1.8). Indeed, in [136] we prove that as $\varepsilon \downarrow 0$, f_ε approaches χ_ρ. With this in mind, the inequality (1.71) turns into Kružkov's entropy inequality (1.20). The entropy (or H-)inequality, (1.71), then yields macroscopic convergence by compensated compactness arguments in the one-dimensional case, and by BV+entropy production bounds in the multidimensional case. Earlier works on kinetic models related to (1.68) can be found in [9],[54],[84].

An alternative proof, presented in [111], makes use of the averaging lemma, 1.2. In view of the results recalled above, we just have to verify that (1.70) holds, $\frac{1}{\varepsilon}(\chi_{\rho^\varepsilon} - f^\varepsilon) = \frac{\partial m^\varepsilon}{\partial v}$. This fact can be shown in several ways.

One way is to observe that if $g(v)$ is an $L^1(\mathbb{R})$ function which satisfy (– as f^ε does),

$$0 \leq \text{sign}(v)g \leq 1, \quad \int_\mathbb{R} g(v)\, dv = \alpha \tag{1.72}$$

then there exists a nonnegative, bounded, continuous q such that

$$\chi_\alpha(v) - g(v) = q'(v), \quad q \in \mathcal{C}_+. \tag{1.73}$$

Indeed, set $q(v) = \int_{-\infty}^{v}(\chi_\alpha(w) - g(w))dw$: in the case $\alpha > 0$ (– the other case being treated similarly), we see that q is nondecreasing on $(-\infty, \alpha)$ and nonincreasing on $(\alpha, +\infty)$ and we conclude since $q(-\infty) = 0$ and $q(+\infty) = \alpha - \int_{\mathbb{R}} g \, dv = 0$.

The characterization in (1.73) of g's satisfying (1.72), is in fact equivalent with the following elementary lemma due to Brenier [9], which in turn yields still another possible proof for the desired representation of the relaxation term in (1.70).

Lemma 1.4 *([9]) Let $\alpha \in \mathbb{R}$ and let φ be a C^1 convex function on $\mathbb{R}$ such that φ' is bounded. Then, $\chi_\alpha(v)$ is a minimizer of $\inf_{\mathcal{G}} \left\{ \int_{\mathbb{R}} \varphi'(v)g(v) \, dv \right\}$ where the infimum is taken over all $g \in \mathcal{G} := \left\{ g \in L^1(\mathbb{R}), \quad \int_{\mathbb{R}} g \, dv = \alpha, \quad 0 \le g \operatorname{sign}(v) \le 1 \right\}$. In addition, $\chi_\alpha(v)$ is the unique minimizer if φ' is strictly increasing on $\mathbb{R}$.*

Granted that (1.70) holds, i.e., the relaxation term on the right of (1.64) belongs to $W_v^{-1}(\mathcal{M}_{t,x})$, then the averaging lemma 1.2 applies with $s = q = 1$, $p = 2$ (here we identify, $t \leftrightarrow x_0$, $\tau \leftrightarrow \xi_0$, $a_0(v) \equiv 1$). It follows that if the conservation law is linearly non-degenerate in the sense that (1.62) holds, that is, if $\exists \alpha \in (0, 1)$ such that

$$meas\{v| \; |\tau + A'(v) \cdot \xi| < \delta\} \le Const \cdot \delta^\alpha, \quad \forall \, \tau^2 + |\xi|^2 = 1, \qquad (1.74)$$

then, $\{\rho^\varepsilon\}$ is compact – in fact $\{\rho^\varepsilon(t > 0, \cdot)\}$ gains Sobolev regularity of order $s = \frac{\alpha}{\alpha+4}$. ∎

We conclude this section with several remarks.

Regularizing effect

We have shown above how the averaging lemma implies convergence under the non-degeneracy condition (1.74). Moreover, in this case we quantified the Sobolev W^s-regularity of the approximate solutions, $\{\rho^\varepsilon\}$. In fact, even more can be said if the solution operator associated with $\{\rho^\varepsilon\}$ is translation invariant: a bootstrap argument presented in [111] yields the improved regularity of order $s = \frac{\alpha}{\alpha+2}$,

$$\rho^\varepsilon(t > 0, \cdot) \in W^{\frac{\alpha}{\alpha+2}}(L^1(x)). \qquad (1.75)$$

This shows that due to nonlinearity, (1.74), the corresponding solution operator, T_t, has a *regularization* effect, as it maps $L_c^\infty \longrightarrow W^s(L^1)$ with $s, t > 0$.

In particular, this framework provides an alternative route to analyze the convergence of general entropy stable multidimensional schemes, *independent* of the underlying kinetic formulations. Here we refer to finite-difference, finite-volume, streamline-diffusion and spectral approximations ..., which were studied in [29,24,85] and [86,79,80,18], for example. Indeed, the key feature in the convergence proof for all of these methods is the $W_{loc}^{-1}(L^2)$-compact entropy production,

$$\partial_t \eta(\rho^\varepsilon) + \nabla_x \cdot F(\rho^\varepsilon) \hookrightarrow W_{loc}^{-1}(L^2(t,x)), \quad \forall \eta'' > 0. \tag{1.76}$$

Hence, if the underlying conservation law satisfies the non-linear degeneracy condition (1.74), then the corresponding family of approximate solutions, $\{\rho^\varepsilon(t > 0, \cdot)\}$ becomes compact. Moreover, if the entropy production is in fact a bounded measure, (– and here positive measures are included compared with the *nonpositive* entropy production required from measure-valued solutions in (1.58)), then there is actually a *gain* of Sobolev regularity of order $\frac{\alpha}{\alpha+4}$, and of order $\frac{\alpha}{\alpha+2}$ for the translation invariant case. (The expected optimal order is α). We shall outline this general framework for studying the regularizing effect of approximate solutions to multidimensional scalar equations in Lecture V.

Kinetic schemes

There is more than one way to convert microscopic kinetic formulations of nonlinear equations, into macroscopic algorithms for the approximate solution of such equations. We mention the following three examples (in the context of conservation laws).

- **Brenier's transport collapse method, [9], is a macroscopic projection method** which preceded the BGK-like model (1.64), see also [54]. Here one alternates between transporting microscopic 'pseudo-Maxwellians' which start with $f(t^n, \cdot, v) := \chi_{\rho(t^n, \cdot)}(v)$, and projecting their macroscopic averaging, $\rho(t^{n+1}, \cdot) = \bar{f}(t^{n+1}, \cdot, v)$. A convergence analysis of this method by the velocity averaging lemma was recently worked out in [176].

- Another approach is based on Chapman-Enskog asymptotic expansions, [15]. We refer to [147], for an example of macroscopic approximation other than the usual Navier-Stokes-like viscosity regularization (– the scalar version of this regularized Chapman-Enskog expansion is studied in Lecture IV).

- Still another approach is offered by Godunov-type schemes, (1.3), based on projections of the Maxwellians associated with the specific kinetic formulations. These amount to specific Riemann solvers which were studied in [39], [135], [137].

We conclude by noting that kinetic formulations like those mentioned above in the context of scalar conservation laws apply in more general situations. **For extensions consult [111] for degenerate parabolic equations, [112],[110]** for the system of 2×2 isentropic equations, [77] for the system of chromatographic equations, ... We shall say more on these issues in Lecture V.

References

1. L. ALVAREZ AND J.-M. MOREL *Formulation and computational aspects of image analysis*, Acta Numerica (1994), 1–59.

2. C. BARDOS, F. GOLSE AND D. LEVERMORE, *Fluid dynamic limits of kinetic equations II: convergence proofs of the Boltzmann equations*, Comm. Pure Appl. Math. XLVI (1993), 667–754.

3. G. BARLES AND P.E. SOUGANIDIS, *Convergence of approximation schemes for fully nonlinear second order equations*, Asympt. Anal. 4 (1991), 271–283.

4. J.B. BELL, P. COLELLA, AND H.M. GLAZ, *A Second-Order Projection Method for the Incompressible Navier-Stokes Equations*, J. Comp. Phys. 85 (1989), 257–283.

5. M. BEN-ARTZI AND J. FALCOVITZ, *Recent developments of the GRP method*, JSME (Ser.b) 38 (1995), 497–517.

6. J.P. BORIS AND D. L. BOOK, *Flux corrected transport: I. SHASTA, a fluid transport algorithm that works*, J. Comput. Phys. 11 (1973), 38–69.

7. F. BOUCHUT AND B. PERTHAME *Kružkov's estimates for scalar conservation laws revisited*, Universite D'Orleans, preprint, 1996.

8. F. BOUCHOT, CH. BOURDARIAS AND B. PERTHAME, *A MUSCL method satisfying all the numerical entropy inequalities*, Math. Comp. 65 (1996) 1439–1461.

9. Y. BRENIER, *Résolution d'équations d'évolution quasilinéaires en dimension N d'espace à l'aide d'équations linéaires en dimension N + 1*, J. Diff. Eq. 50 (1983), 375–390.

10. Y. BRENIER AND S.J. OSHER, *The discrete one-sided Lipschitz condition for convex scalar conservation laws*, SIAM J. Numer. Anal. 25 (1988), 8–23.

11. A. BRESSAN, *The semigroup approach to systems of conservation laws*, 4th Workshop on PDEs, Part I (Rio de Janeiro, 1995). Mat. Contemp. 10 (1996), 21–74.

12. A. BRESSAN, *Decay and structural stability for solutions of nonlinear systems of conservation laws*, 1st Euro-Conference on Hyperbolic Conservation Laws, Lyon, Feb. 1997.

13. A. BRESSAN AND R. COLOMBO *The semigroup generated by 2×2 conservation laws*, Arch. Rational Mech. Anal. 133 (1995), no. 1, 1–75.

14. A. BRESSAN AND R. COLOMBO *Unique solutions of 2×2 conservation laws with large data*, Indiana Univ. Math. J. 44 (1995), no. 3, 677–725.

15. C. CERCIGNANI, The Boltzmann Equation and its Applications, Appl. Mathematical Sci. 67, Springer, New-York, 1988.

16. T. CHANG AND L. HSIAO, *The Riemann Problem and Interaction of Waves in Gasdynamics*, Pitman monographs and surveys in pure appl. math, 41, John Wiley, 1989.

17. G.-Q. CHEN, *The theory of compensated compactness and the system of isentropic gas dynamics*, Preprint MCS-P154-0590, Univ. of Chicago, 1990.

18. G.-Q. CHEN, Q. DU AND E. TADMOR, *Spectral viscosity approximation to multidimensional scalar conservation laws*, Math. of Comp. 57 (1993).

19. G.-Q. CHEN, D. LEVERMORE AND T. P. LIU, *Hyperbolic conservation laws with stiff relaxation terms and entropy* Comm. Pure Appl. Math. 47 (1994) 787–830.

20. I. L. CHERN, *Stability theorem and truncation error analysis for the Glimm scheme and for a front tracking method for flows with strong discontinuities*, Comm. Pure Appl. Math. XLII (1989), 815–844.
21. I.L. CHERN, J. GLIMM, O. McBRYAN, B. PLOHR AND S. YANIV, *Front Tracking for gas dynamics* J. Comput. Phys. 62 (1986) 83–110.
22. A. J. CHORIN, *Random choice solution of hyperbolic systems*, J. Comp. Phys. 22 (1976), 517–533.
23. B. COCKBURN, *Quasimonotone schemes for scalar conservation laws. I. II, III.* SIAM J. Numer. Anal. 26 (1989) 1325–1341, 27 (1990) 247–258, 259–276.
24. B. COCKBURN, F. COQUEL AND P. LEFLOCH, *Convergence of finite volume methods for multidimensional conservation laws*, SIAM J. Numer. Anal. 32 (1995), 687–705.
25. P. COLELLA, *Multidimensional upwind methods for hyperbolic conservation laws*, J. Comput. Phys. 87 (1990), 87–171.
26. P. COLELLA AND P. WOODWARD, *The piecewise parabolic method (PPM) for gas-dynamical simulations*, JCP 54 (1984), 174–201.
27. F. COQUEL AND P. LEFLOCH, *Convergence of finite difference schemes for conservation laws in several space dimensions: a general theory*, SIAM J. Numer. Anal. (1993).
28. M. G. CRANDALL, *The semigroup approach to first order quasilinear equations in several space dimensions*, Israel J. Math. 12 (1972), 108–132.
29. M. G. CRANDALL AND P. L. LIONS, *Viscosity solutions of Hamilton-Jacobi equations*, Trans. Amer. Math. Soc. 277 (1983), 1–42.
30. M. G. CRANDALL AND A. MAJDA, *Monotone difference approximations for scalar conservation laws*, Math. of Comp. 34 (1980), 1–21.
31. M. G. CRANDALL AND A. MAJDA, *The method of fractional steps for conservation laws*, Numer. Math. 34 (1980), 285–314.
32. M. G. CRANDALL AND L. TARTAR, *Some relations between non expansive and order preserving mapping*, Proc. Amer. Math. Soc. 78 (1980), 385–390.
33. T. CHANG AND S. YANG, Two-Dimensional Riemann Problems for Systems of Conservation Laws, Pitman Monographs and Surveys in Pure and Appl. Math., 1995.
34. C. DAFERMOS, *Polygonal approximations of solutions of initial-value problem for a conservation law*, J. Math. Anal. Appl. 38 (1972) 33–41.
35. C. DAFERMOS, *Hyperbolic systems of conservation laws*, in "Systems of Nonlinear PDEs", J. M. Ball, ed, NATO ASI Series C, No. 111, Dordrecht, D. Reidel (1983), 25–70.
36. C. DAFERMOS, private communication.
37. R. DEVORE & G. LORENTZ, Constructive Approximation, Springer-Verlag, 1993.
38. R. DEVORE AND B. LUCIER, *High order regularity for conservation laws*, Indiana Univ. Math. J. 39 (1990), 413–430.
39. S. M. DESHPANDE, *A second order accurate, kinetic-theory based, method for inviscid compressible flows*, NASA Langley Tech. paper No. 2613, 1986.
40. R. DiPERNA, *Convergence of approximate solutions to conservation laws*, Arch. Rat. Mech. Anal. 82 (1983), 27–70.
41. R. DiPERNA, *Convergence of the viscosity method for isentropic gas dynamics*, Comm. Math. Phys. 91 (1983), 1–30.
42. R. DiPERNA, *Measure-valued solutions to conservation laws*, Arch. Rat. Mech. Anal. 88 (1985), 223-270.

43. R. DiPERNA AND P. L. LIONS, *On the Cauchy problem for Boltzmann equations: Global existence and weak stability*, Ann. Math. 130 (1989), 321–366.

44. R. DiPERNA AND P.L. LIONS, *Global weak solutions of Vlasov-Maxwell systems*, Comm. Pure Appl. Math. 42 (1989), 729–757.

45. R. DiPERNA, P.L. LIONS AND Y. MEYER, *L^p regularity of velocity averages*, Ann. I.H.P. Anal. Non Lin. 8(3-4) (1991), 271–287.

46. W. E. AND C.-W.SHU, *A numerical resolution study of high order essentially non-oscillatory schemes applied to incompressible flow* J. Comp. Phys. 110, (1993) 39-46.

47. B. ENGQUIST AND S.J. OSHER *One-sided difference approximations for non-linear conservation laws*, Math. Comp. 36 (1981) 321–351.

48. B. ENGQUIST, P. LOTSTEDT AND B. SJOGREEN, *Nonlinear filters for efficient shock computation*, Math. Comp. 52 (1989), 509–537.

49. B. ENGQUIST AND O. RUNBORG, *Multi-phase computations in geometrical optics*, J. Comp. Appl. Math. 74 (1996) 175–192.

50. C. EVANS, Weak Convergence Methods for Nonlinear Partial Differential equations, AMS Regional Conference Series in Math. 74, Providence R.I. 1990.

51. K. FRIEDRICHS *Symmetric hyperbolic linear differential equations*, CPAM 7 (1954) 345–.

52. K. O. FRIEDRICHS AND P. D. LAX, *Systems of conservation laws with a convex extension*, Proc. Nat. Acad. Sci. USA 68 (1971), 1686–1688.

53. P. GÉRARD, *Microlocal defect measures*, Comm. PDE 16 (1991), 1761–1794.

54. Y. GIGA AND T. MIYAKAWA, *A kinetic construction of global solutions of first-order quasilinear equations*, Duke Math. J. 50 (1983), 505–515.

55. J. GLIMM, *Solutions in the large for nonlinear hyperbolic systems of equations*, Comm. Pure Appl. Math. 18 (1965), 697–715.

56. J. GLIMM AND P. D. LAX, Decay of solutions of systems of nonlinear hyperbolic conservation laws, Amer. Math. Soc. Memoir 101, AMS Providence, 1970.

57. J. GLIMM, B. LINDQUIST AND Q. ZHANG, *Front tracking, oil reservoirs, engineering scale problems and mass conservation* in *Multidimensional Hyperbolic Problems and Computations*, Proc. IMA workshop (1989) IMA vol. Math Appl 29 (J. Glimm and A. Majda Eds.), Springer-Verlag, New-York (1991), 123-139.

58. E. GODLEWSKI AND P.-A. RAVIART, Hyperbolic Systems of Conservation Laws, Ellipses, Paris, 1991.

59. E. GODLEWSKI AND P.-A. RAVIART, Numerical Approximation of Hyperbolic Systems of Conservation Laws, Springer, 1996.

60. S. K. GODUNOV, *A difference scheme for numerical computation of discontinuous solutions of fluid dynamics*, Mat. Sb. 47 (1959), 271–306.

61. S. K. GODUNOV, *An interesting class of quasilinear systems*, Dokl. Akad. Nauk. SSSR 139(1961), 521–523.

62. F. GOLSE, P. L. LIONS, B. PERTHAME AND R. SENTIS, *Regularity of the moments of the solution of a transport equation*, J. of Funct. Anal. 76 (1988), 110–125.

63. J. GOODMAN, private communication.

64. J. GOODMAN AND P. D. LAX, *On dispersive difference schemes. I*, Comm. Pure Appl. Math. 41 (1988), 591–613.

65. J. GOODMAN AND R. LeVEQUE, *On the accuracy of stable schemes for 2D scalar conservation laws*, Math. of Comp. 45 (1985), 15–21.

66. J. GOODMAN AND XIN, *Viscous limits for piecewise smooth solutions to systems of conservation laws*, Arch. Rat. Mech. Anal. 121 (1992), 235–265.

67. D. GOTTLIEB AND E. TADMOR, *Recovering Pointwise Values of Discontinuous Data within Spectral Accuracy*, in "Progress and Supercomputing in Computational Fluid Dynamics", Progress in Scientific Computing, Vol. 6 (E. M. Murman and S. S. Abarbanel, eds.), Birkhauser, Boston, 1985, 357–375.

68. E. HARABETIAN, *A convergent series expansion for hyperbolic systems of conservation laws*, Trans. Amer. Math. Soc. 294 (1986), no. 2, 383–424.

69. A. HARTEN, *The artificial compression method for the computation of shocks and contact discontinuities:I. single conservation laws*, CPAM 39 (1977), 611–638.

70. A. HARTEN, *High resolution schemes for hyperbolic conservation laws*, J. Comput. Phys. 49 (1983), 357–393.

71. A. HARTEN, B. ENGQUIST, S. OSHER AND S.R. CHAKRAVARTHY, *Uniformly high order accurate essentially non-oscillatory schemes. III*, JCP 71, 1982, 231–303.

72. A. HARTEN M. HYMAN AND P. LAX, *On finite-difference approximations and entropy conditions for shocks*, Comm. Pure Appl. Math. 29 (1976), 297–322.

73. A. HARTEN P.D. LAX AND B. VAN LEER, *On upstream differencing and Godunov-type schemes for hyperbolic conservation laws*, SIAM Rev. 25 (1983), 35–61.

74. A. HARTEN AND S. OSHER, *Uniformly high order accurate non-oscillatory scheme. I*, SIAM J. Numer. Anal. 24 (1982) 229–309.

75. C. HIRSCH, Numerical Computation of Internal and External Flows, Wiley, 1988.

76. T. J. R. HUGHES AND M. MALLET, *A new finite element formulation for the computational fluid dynamics: III. The general streamline operator for multidimensional advective-diffusive systems*, Comput. Methods Appl. Mech. Engrg. 58 (1986), 305–328.

77. F. JAMES, Y.-J PENG AND B. PERTHAME, *Kinetic formulation for chromatography and some other hyperbolic systems*, J. Math. Pures Appl. 74 (1995), 367–385.

78. F. JOHN, Partial Differential Equations, 4th ed. Springer, New-York, 1982.

79. C. JOHNSON AND A. SZEPESSY, *Convergence of a finite element methods for a nonlinear hyperbolic conservation law*, Math. of Comp. 49 (1988), 427–444.

80. C. JOHNSON, A. SZEPESSY AND P. HANSBO, *On the convergence of shock-capturing streamline diffusion finite element methods for hyperbolic conservation laws*, Math. of Comp. 54 (1990), 107–129.

81. G.-S. JIANG AND E. TADMOR, *Nonoscillatory Central Schemes for Multidimensional Hyperbolic Conservation Laws*, SIAM J. Sci. Compt., in press.

82. S. JIN AND Z. XIN, *The relaxing schemes for systems of conservation laws in arbitrary space dimensions*, Comm. Pure Appl. Math. 48 (1995) 235–277.

83. T. KATO, *The Cauchy problem for quasi-linear symmetric hyperbolic systems*, Arch. Rat. Mech. Anal. 58 (1975), 181–205.

84. Y. KOBAYASHI, *An operator theoretic method for solving $u_t = \Delta\psi(u)$*, Hiroshima Math. J. 17 (1987) 79–89.

85. D. KRÖNER, S. NOELLE AND M. ROKYTA, *Convergence of higher order upwind finite volume schemes on unstructured grids for scalar conservation laws in several space dimensions*, Numer. Math. 71 (1995) 527–560.

86. D. KRÖNER AND M. ROKYTA, *Convergence of Upwind Finite Volume Schemes for Scalar Conservation Laws in two space dimensions*, SINUM 31 (1994) 324–343.

87. S.N. KRUŽKOV, *The method of finite difference for a first order non-linear equation with many independent variables*, USSR comput Math. and Math. Phys. 6 (1966), 136–151. (English Trans.)

88. S.N. KRUŽKOV, *First order quasilinear equations in several independent variables*, Math. USSR Sbornik 10 (1970), 217–243.

89. R. KUPFERMAN AND E. TADMOR, *A fast high-resolution second-order central scheme for incompressible flows*, Proc. Nat. Acad. Sci. 94 (1997), 4848–4852

90. N.N. KUZNETSOV, *Accuracy of some approximate methods for computing the weak solutions of a first-order quasi-linear equation*, USSR Comp. Math. and Math. Phys. 16 (1976), 105–119.

91. P.D. LAX, *Weak solutions of non-linear hyperbolic equations and their numerical computations*, Comm. Pure Appl. Math. 7 (1954), 159–193.

92. P. D. LAX, *Hyperbolic systems of conservation laws II*, Comm. Pure Appl. Math. 10 (1957), 537–566.

93. P.D. LAX, *Shock waves and entropy*, in *Contributions to nonlinear functional analysis*, E.A. Zarantonello Ed., Academic Press, New-York (1971), 603–634.

94. P.D. LAX, Hyperbolic Systems of Conservation Laws and the Mathematical Theory of Shock Waves (SIAM, Philadelphia, 1973).

95. P.D. LAX AND X.-D. LIU, *Positive schemes for solving multi-dimensional hyperbolic systems of conservation laws*, Courant Mathematics and Computing Laboratory Report NYU, 95-003 (1995), Comm. Pure Appl. Math.

96. P. LAX AND B. WENDROFF, *Systems of conservation laws*, Comm. Pure Appl. Math. 13 (1960), 217–237.

97. P. LAX AND B. WENDROFF, *Difference schemes for hyperbolic equations with high order of accuracy*, Comm. Pure Appl. Math. 17 (1964), 381–.

98. B. VAN LEER, *Towards the ultimate conservative difference scheme. V. A second-order sequel to Godunov's method*, J. Comput. Phys. 32 (1979), 101–136.

99. F. LEFLOCH AND P.A. RAVIART, *An asymptotic expansion for the solution of the generalized Riemann problem, Part I: General theory*, Ann. Inst. H. Poincare, Nonlinear Analysis 5 (1988), 179–

100. P. LEFLOCH AND Z. XIN, *Uniqueness via the adjoint problem for systems of conservation laws*, CIMS Preprint.

101. R. LEVEQUE, *A large time step generalization of Godunov's method for systems of conservation laws*, SIAM J. Numer. Anal. 22(6) (1985), 1051–1073.

102. R. LEVEQUE, Numerical Methods for Conservation Laws, Lectures in Mathematics, Birkhäuser, Basel 1992.

103. R. LEVEQUE, *Wave propagation algorithms for multi-dimensional hyperbolic systems*, Preprint.

104. D. LEVERMORE AND J.-G. LIU, *Oscillations arising in numerical experiments*, NATO ARW seies, Plenum, New-York (1993), To appear.

105. D. LEVY AND E. TADMOR, *Non-oscillatory central schemes for the incompressible 2-D Euler equations*, Math. Res. Lett., 4 (1997) 1-20.

106. D. LI, *Riemann problem for multi-dimensional hyperbolic conservation laws*, Free boundary problems in fluid flow with applications (Montreal, PQ, 1990), 64–69, Pitman Res. Notes Math. Ser., 282.

107. C.-T. LIN AND E. TADMOR, L^1-stability and error estimates for approximate *Hamilton-Jacobi solutions*, in preparation.

108. P. L. LIONS, Generalized Solutions of Hamilton-Jacobi Equations, Pittman, London 1982.

109. P.L. LIONS, *On kinetic equations*, in Proc. Int'l Congress of Math., Kyoto, 1990, Vol. II, Math. Soc. Japan, Springer (1991), 1173-1185.

110. P. L. LIONS, B. PERTHAME AND P. SOUGANIDIS, *Existence and stability of entropy solutions for the hyperbolic systems of isentropic gas dynamics in Eulerian and Lagrangian coordinates*, Comm. Pure and Appl. Math. 49 (1996), 599-638.

111. P. L. LIONS, B. PERTHAME AND E. TADMOR, *Kinetic formulation of scalar conservation laws and related equations*, J. Amer. Math. Soc. 7(1) (1994), 169–191

112. P. L. LIONS, B. PERTHAME AND E. TADMOR, *Kinetic formulation of the isentropic gas-dynamics equations and p-systems*, Comm. Math. Phys. 163(2) (1994), 415–431.

113. T.-P. LIU, *The entropy condition and the admissibility of shocks*, J. Math. Anal. Appl. 53 (1976), 78–88.

114. T. P. LIU, *The deterministic version of the Glimm scheme*, Comm. Math. Phys. 57 (1977), 135–148.

115. B. LUCIER, *Error bounds for the methods of Glimm, Godunov and LeVeque*, SIAM J. Numer. Anal. 22 (1985), 1074–1081.

116. B. Lucier, *Lecture Notes*, 1993.

117. Y. MADAY, S. M. OULD-KABER AND E. TADMOR, *Legendre pseudospectral viscosity method for nonlinear conservation laws*, SIAM J. Numer. Anal. 30 (1993), 321–342.

118. A. MAJDA, Compressible Fluid Flow and Systems of Conservation Laws in Several Space Variables, Springer-Verlag New-York, 1984.

119. A. MAJDA, J. MCDONOUGH AND S. OSHER, *The Fourier method for nonsmooth initial data*, Math. Comp. 30 (1978), 1041–1081.

120. M.S. MOCK, *Systems of conservation laws of mixed type*, J. Diff. Eq. 37 (1980), 70–88.

121. M. S. MOCK AND P. D. LAX, The computation of discontinuous solutions of linear hyperbolic equations, Comm. Pure Appl. Math. 31 (1978), 423–430.

122. K.W. MORTON, *Lagrange-Galerkin and characteristic-Galerkin methods and their applications*, 3rd Int'l Conf. Hyperbolic Problems (B. Engquist & B. Gustafsson, eds.), Studentlitteratur, (1991), 742–755.

123. F. MURAT, *Compacité par compensation*, Ann. Scuola Norm. Sup. Pisa 5 (1978), 489–507.

124. F. MURAT, *A survey on compensated compactness*, in 'Contributions to Modern calculus of variations' (L. Cesari, ed), Pitman Research Notes in Mathematics Series, John Wiley New-York, 1987, 145–183.

125. R. NATALINI, *Convergence to equilibrium for the relaxation approximations of conservation laws*, Comm. Pure Appl. Math. 49 (1996), 1–30.

126. H. NESSYAHU AND E. TADMOR, Non-oscillatory central differencing for hyperbolic conservation laws. J. Comp. Phys. 87 (1990), 408–463.

127. H. NESSYAHU AND E. TADMOR, *The convergence rate of approximate solutions for nonlinear scalar conservation laws*, SIAM J. Numer. Anal. 29 (1992), 1–15.

128. S. NOELLE, *A note on entropy inequalities and error estimates for higher-order accurate finite volume schemes on irregular grids*, Math. Comp. to appear.

129. O. A. OLĚINIK *Discontinuous solutions of nonlinear differential equations*, Amer. Math. Soc. Transl. (2), 26 (1963), 95–172.

130. S. OSHER, *Riemann solvers, the entropy condition, and difference approximations*, SIAM J. Numer. Anal. 21 (1984), 217-235.

131. S. OSHER AND F. SOLOMON, *Upwind difference schemes for hyperbolic systems of conservation laws*, Math. Comp. 38 (1982), 339–374.

132. S. OSHER AND J. SETHIAN, *Fronts propagating with curvature dependent speed: Algorithms based on Hamilton-Jacobi formulations*, J. Comp. Phys. 79 (1988), 12–49.

133. S. OSHER AND E. TADMOR, *On the convergence of difference approximations to scalar conservation laws*, Math. of Comp. 50 (1988), 19–51.

134. B. PERTHAME, *Global existence of solutions to the BGK model of Boltzmann equations*, J. Diff. Eq. 81 (1989), 191-205.

135. B. PERTHAME, *Second-order Boltzmann schemes for compressible Euler equations*, SIAM J. Num. Anal. 29, (1992), 1–29.

136. B. PERTHAME AND E. TADMOR, *A kinetic equation with kinetic entropy functions for scalar conservation laws*, Comm. Math. Phys.136 (1991), 501–517.

137. K. H. PRENDERGAST AND K. XU, *Numerical hydrodynamics from gas-kinetic theory*, J. Comput. Phys. 109(1) (1993), 53–66.

138. A. RIZZI AND B. ENGQUIST, Selected topics in the theory and practice of computational fluid dynamics, J. Comp. Phys. 72 (1987), 1–69.

139. P. ROE, *Approximate Riemann solvers,parameter vectors, and difference schemes*, J. Comput. Phys. 43 (1981), 357–372.

140. P. ROE, *Discrete models for the numerical analysis of time-dependent multidimensional gas dynamics*, J. Comput. Phys. 63 (1986), 458–476.

141. R. RICHTMYER AND K.W. MORTON, Difference Methods for Initial-Value Problems, Interscience, 2nd ed., 1967.

142. F. SABAC, PhD Thesis, Univ. S. Carolina, 1994.

143. R. SANDERS, *On Convergence of monotone finite difference schemes with variable spatial differencing*, Math. of Comp. 40 (1983), 91–106.

144. R. SANDERS, *A third-order accurate variation nonexpansive difference scheme for single conservation laws*, Math. Comp. 51 (1988), 535–558.

145. S. SCHOCHET, *The rate of convergence of spectral viscosity methods for periodic scalar conservation laws*, SIAM J. Numer. Anal. 27 (1990), 1142–1159.

146. S. SCHOCHET, *Glimm's scheme for systems with almost-planar interactions*, Comm. Partial Differential Equations 16(8-9) (1991), 1423–1440.

147. S. SCHOCHET AND E. TADMOR, *Regularized Chapman-Enskog expansion for scalar conservation laws*, Archive Rat. Mech. Anal. 119 (1992), 95–107.

148. D. SERRE, *Richness and the classification of quasilinear hyperbolic systems*, in "Multidimensional Hyperbolic Problems and Computations", Minneapolis MN 1989, IMA Vol. Math. Appl. 29, Springer NY (1991), 315–333.

149. D. SERRE, Systemés de Lois de Conservation, Diderot, Paris 1996.

150. D. SERRE, private communication.

151. C.W. SHU, *TVB uniformly high-order schemes for conservation laws*, Math. Comp. 49 (1987) 105–121.

152. C. W. SHU, *Total-variation-diminishing time discretizations*, SIAM J. Sci. Comput. 6 (1988), 1073–1084.

153. C. W. SHU AND S. OSHER, *Efficient implementation of essentially non-oscillatory shock-capturing schemes*, J. Comp. Phys. 77 (1988), 439–471.

154. C. W. SHU AND S. OSHER, *Efficient implementation of essentially non-oscillatory shock-capturing schemes. II*, J. Comp. Phys. 83 (1989), 32–78.

155. W. SHYY, M.-H CHEN, R. MITTAL AND H.S. UDAYKUMAR, *On the suppression on numerical oscillations using a non-linear filter*, J. Comput. Phys. 102 (1992), 49–62.

156. D. SIDILKOVER, *Multidimensional upwinding: unfolding the mystery*, Barriers and Challenges in CFD, ICASE workshop, ICASE, Aug, 1996.

157. J. SMOLLER, Shock Waves and Reaction-Diffusion Equations, Springer-Verlag, New York, 1983.

158. G. SOD, *A survey of several finite difference methods for systems of nonlinear hyperbolic conservation laws*, JCP 22 (1978) 1–31.

159. G. SOD, Numerical Methods for Fluid Dynamics, Cambridge University Press, 1985.

160. A. SZEPESSY, *Convergence of a shock-capturing streamline diffusion finite element method for scalar conservation laws in two space dimensions*, Math. of Comp. (1989), 527–545.

161. P. R. SWEBY, *High resolution schemes using flux limiters for hyperbolic conservation laws*, SIAM J. Num. Anal. 21 (1984), 995–1011.

162. E. TADMOR, *The large-time behavior of the scalar, genuinely nonlinear Lax-Friedrichs scheme* Math. Comp. 43 (1984), no. 168, 353–368.

163. E. TADMOR, *Numerical viscosity and the entropy condition for conservative difference schemes* Math. Comp. 43 (1984), no. 168, 369–381.

164. E. TADMOR, *Convenient total variation diminishing conditions for nonlinear difference schemes*, SIAM J. on Numer. Anal. 25 (1988), 1002-1014.

165. E. TADMOR, *Convergence of Spectral Methods for Nonlinear Conservation Laws*, SIAM J. Numer. Anal. 26 (1989), 30–44.

166. E. TADMOR, *Local error estimates for discontinuous solutions of nonlinear hyperbolic equations*, SIAM J. Numer. Anal. 28 (1991), 891–906.

167. E. TADMOR, *Super viscosity and spectral approximations of nonlinear conservation laws*, in "Numerical Methods for Fluid Dynamics", Proceedings of the 1992 Conference on Numerical Methods for Fluid Dynamics (M. J. Baines and K. W. Morton, eds.), Clarendon Press, Oxford, 1993, 69–82.

168. E. TADMOR, *Approximate Solution of Nonlinear Conservation Laws and Related Equations*, in "Recent Advances in Partial Differential Equations and Applications" Proceedings of the 1996 Venice Conference in honor of Peter D. Lax and Louis Nirenberg on their 70th Birthday (R. Spigler and S. Venakides eds.), AMS Proceedings of Symposia in Applied Mathematics 54, 1997, 321–368

169. E. TADMOR AND T. TASSA, *On the piecewise regularity of entropy solutions to scalar conservation laws*, Com. PDEs 18 91993), 1631-1652.

170. T. TANG & Z. H. TENG, *Viscosity methods for piecewise smooth solutions to scalar conservation laws*, Math. Comp., 66 (1997), pp. 495–526.

171. L. TARTAR, *Compensated compactness and applications to partial differential equations*, in *Research Notes in Mathematics 39*, Nonlinear Analysis and Mechanics, Heriott-Watt Symposium, Vol. 4 (R.J. Knopps, ed.) Pittman Press, (1975), 136–211.

172. L. TARTAR, *Discontinuities and oscillations*, in Directions in PDEs, Math Res. Ctr Symposium (M.G. Crandall, P.H. Rabinowitz and R.E. Turner eds.) Academic Press (1987), 211-233.

173. T. TASSA, *Applications of compensated compactness to scalar conservation laws*, M.Sc. thesis (1987), Tel-Aviv University (in Hebrew).

174. B. TEMPLE, *No L^1 contractive metrics for systems of conservation laws*, Trans. AMS 288(2) (1985), 471–480.

175. Z.-H. TENG *Particle method and its convergence for scalar conservation laws*, SIAM J. Num. Anal. 29 (1992) 1020–1042.

176. VASSEUR, *Kinetic semi-discretization of scalar conservation laws and convergence using averaging lemmas*, SIAM J. Numer. Anal.

177. A. I. VOL'PERT, *The spaces BV and quasilinear equations*, Math. USSR-Sb. 2 (1967), 225–267.

178. G.B. WHITHAM, Linear and Nonlinear Waves, Wiley-Interscience, 1974.

179. K. YOSIDA, Functional Analysis, Springer-Verlag, 1980.

2 Non-oscillatory central schemes

Abstract. We discuss a new class of high-resolution approximations for hyperbolic systems of conservation laws, which are based on *central* differencing. Its two main ingredients include:

#1. A non-oscillatory reconstruction of pointvalues from their given cell averages; and

#2. A central differencing based on *staggered* evolution of the reconstructed averages.

Many of the modern high-resolution schemes for such systems, are based on Godunov-type *upwind* differencing; their intricate and time consuming part involves the field-by-field characteristic decomposition, which is required in order to identify the "direction of the wind". Instead, our proposed central (staggered) stencils enjoy the main advantage is simplicity: no Riemann problems are solved, and hence field-by-field decompositions are avoided. This could be viewed as the high-order sequel to the celebrated Lax-Friedrichs (staggered) scheme. Typically, staggering suffers from excessive numerical dissipation. Here, excessive dissipation is compensated by using modern, high-resolution, non-oscillatory reconstructions.

We highlight several features of this new class of central schemes.

Scalar equations. For both the second- and third-order schemes we prove variation bounds (– which in turn yield convergence with precise error estimates), as well as entropy and multidimensional L^∞-stability estimates.

Systems of equations. Extension to systems is carried out by *componentwise* application of the scalar framework. It is in this context that our central schemes offer a remarkable advantage over the corresponding upwind framework.

Multidimensional problems. Since we bypass the need for (approximate) Riemann solvers, multidimensional problems are solved *without* dimensional splitting. In fact, the proposed class of central schemes is utilized for a variety of nonlinear transport equations.

A variety of numerical experiments confirm the high-resolution content of the proposed central schemes. They include second- and third-order approximations for one- and two-dimensional Euler, MHD, as well as other compressible and incompressible equations. These numerical experiments demonstrate that the proposed central schemes offer *simple, robust, Riemann-solver-free* approximations, while at the same time, they retain the high-resolution content of the more expensive upwind schemes.

2.1 Introduction

In recent years, central schemes for approximating solutions of hyperbolic conservation laws, received a considerable amount of renewed attention. A family of high-resolution, non-oscillatory, *central* schemes, was developed to handle such problems. Compared with the 'classical' *upwind* schemes, these *central* schemes were shown to be both simple and stable for a large variety of problems ranging from one-dimensional scalar problems to multi-dimensional systems of conservation laws. They were successfully implemented for a variety of other related problems, such as, e.g., the incompressible Euler equations [25],[22],[20], [21], the magneto-hydrodynamics equations [45], viscoelastic flows—[20] hyperbolic systems with relaxation source terms [4],[37],[38] non-linear optics [36],[7], and slow moving shocks [17].

The family of high-order *central* schemes we deal with, can be viewed as a direct extension to the first-order, Lax-Friedrichs (LxF) scheme [9], which on one hand is robust and stable, but on the other hand suffers from excessive dissipation. To address this problematic property of the LxF scheme, a Godunov-like second-order central scheme was developed by Nessyahu and Tadmor (NT) in [31] (see also [41]). It was extended to higher-order of accuracy as well as for more space dimensions (consult [1], [16], [2], [3] and [21], for the two-dimensional case, and [40], [14], [29] and [24] for the third-order schemes).

The NT scheme is based on reconstructing, in each time step, a piecewise-polynomial interpolant from the cell-averages computed in the previous time step. This interpolant is then (exactly) evolved in time, and finally, it is projected on its staggered averages, resulting with the staggered cell-averages at the next time-step. The one- and two-dimensional second-order schemes, are based on a piecewise-linear MUSCL-type reconstruction, whereas the third-order schemes are based on the non-oscillatory piecewise-parabolic reconstruction [28],[29]. Higher orders are treated in [39].

Like *upwind* schemes, the reconstructed piecewise-polynomials used by the central schemes, also make use of non-linear limiters which guarantee the overall non-oscillatory nature of the approximate solution. But unlike the upwind schemes, central schemes avoid the intricate and time consuming Riemann solvers; this advantage is particularly important in the multi-dimensional setup, where no such Riemann solvers exist.

2.2 A Short guide to Godunov-Type schemes

We want to solve the hyperbolic system of conservation laws

$$u_t + f(u)_x = 0 \qquad\qquad (2.77)$$

by Godunov-type schemes. To this end we proceed in two steps. First, we introduce a small spatial scale, Δx, and we consider the corresponding (Steklov) sliding average of $u(\cdot, t)$,

$$\bar{u}(x,t) := \frac{1}{|I_x|} \int_{I_x} u(\xi, t) d\xi, \qquad I_x = \left\{ \xi \ \Big| \ |\xi - x| \le \frac{\Delta x}{2} \right\}.$$

The sliding average of (2.77) then yields

$$\bar{u}_t(x,t) + \frac{1}{\Delta x} \left[f(u(x + \frac{\Delta x}{2}, t)) - f(u(x - \frac{\Delta x}{2}, t)) \right] = 0. \qquad (2.78)$$

Next, we introduce a small time-step, Δt, and integrate over the slab $t \leq \tau \leq t + \Delta t$,

$$\bar{u}(x, t + \Delta t) = \bar{u}(x, t) \tag{2.79}$$
$$-\frac{1}{\Delta x}\left[\int_{\tau=t}^{t+\Delta t} f(u(x + \frac{\Delta x}{2}, \tau))d\tau\right.$$
$$\left. - \int_{\tau=t}^{t+\Delta t} f(u(x - \frac{\Delta x}{2}, \tau))d\tau\right].$$

We end up with an equivalent reformulation of the conservation law (2.77): it expresses the precise relation between the sliding averages, $\bar{u}(\cdot, t)$, and their underlying pointvalues, $u(\cdot, t)$. We shall use this reformulation, (2.79), as the starting point for the construction of Godunov-type schemes.

We construct an approximate solution, $w(\cdot, t^n)$, at the discrete time-levels, $t^n = n\Delta t$. Here, $w(x, t^n)$ is a piecewise polynomial written in the form

$$w(x, t^n) = \sum p_j(x)\chi_j(x), \quad \chi_j(x) := 1_{I_j},$$

where $p_j(x)$ are algebraic polynomials supported at the discrete cells, $I_j = I_{x_j}$, centered around the midpoints, $x_j := j\Delta x$. An *exact* evolution of $w(\cdot, t^n)$ based on (2.79), reads

$$\bar{w}(x, t^{n+1}) = \bar{w}(x, t^n) \tag{2.80}$$
$$-\frac{1}{\Delta x}\left[\int_{t^n}^{t^{n+1}} f(w(x + \frac{\Delta x}{2}, \tau))d\tau\right.$$
$$\left. - \int_{t^n}^{t^{n+1}} f(w(x - \frac{\Delta x}{2}, \tau))d\tau\right].$$

To construct a Godunov-type scheme, we *realize* (2.80) — or at least an accurate approximation of it, at discrete gridpoints. Here, we distinguish between the main methods, according to their way of *sampling* (2.80): these two main sampling methods correspond to upwind schemes and central schemes.

Upwind schemes

Let $\bar{w}_j^n$ abbreviates the cell averages, $\bar{w}_j^n := \frac{1}{\Delta x}\int_{I_j} w(\xi, t^n)d\xi$. By sampling (2.80) at the *mid-cells*, $x = x_j$, we obtain an evolution scheme for these averages, which reads

$$\bar{w}_j^{n+1} = \bar{w}_j^n - \frac{1}{\Delta x}\left[\int_{\tau=t^n}^{t^{n+1}} f(w(x_{j+\frac{1}{2}}, \tau))d\tau - \int_{\tau=t^n}^{t^{n+1}} f(w(x_{j-\frac{1}{2}}, \tau))d\tau\right].$$
$$\tag{2.81}$$

Here, it remains to recover the *pointvalues*, $\{w(x_{j+\frac{1}{2}}, \tau)\}_j$, $t^n \leq \tau \leq t^{n+1}$, in terms of their known cell averages, $\{\bar{w}_j^n\}_j$, and to this end we proceed in two steps:

- First, the *reconstruction* – we recover the pointwise values of $w(\cdot, \tau)$ at $\tau = t^n$, by a reconstruction of a piecewise polynomial approximation

$$w(x, t^n) = \sum_j p_j(x)\chi_j(x), \quad \bar{p}_j(x_j) = \bar{w}_j^n. \tag{2.82}$$

- Second, the *evolution* — $w(x_{j+\frac{1}{2}}, \tau \geq t^n)$ are determined as the solutions of the generalized Riemann problems

$$w_t + f(w)_x = 0, \quad t \geq t^n; \quad w(x, t^n) = \begin{cases} p_j(x) & x < x_{j+\frac{1}{2}}, \\ p_{j+1}(x) & x > x_{j+\frac{1}{2}}. \end{cases} \tag{2.83}$$

The solution of (2.83) is composed of a family of nonlinear waves – left-going and right-going waves. An exact Riemann solver, or at least an approximate one is used to distribute these nonlinear waves between the two neighboring cells, I_j and I_{j+1}. It is this distribution of waves according to their direction which is responsible for *upwind differencing*, consult Figure 2.2. We briefly recall few canonical examples for this category of upwind Godunov-type schemes.

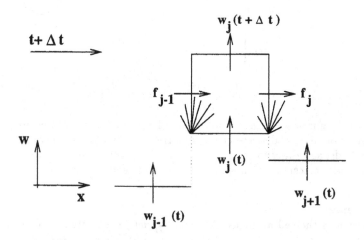

Fig. 2.1: Upwind differencing by Godunov-type scheme.

The original Godunov scheme is based on piecewise-constant reconstruction, $w(x, t^n) = \Sigma \bar{w}_j^n \chi_j$, followed by an exact Riemann solver. This results in a first-order accurate upwind method [11], which is the forerunner for all other Godunov-type schemes. A second-order extension was introduced by van Leer [19]: his MUSCL scheme reconstructs a piecewise linear approximation, $w(x, t^n) = \Sigma p_j(x)\chi_j(x)$, with linear pieces of the form $p_j(x) = \bar{w}_j^n + w_j' \left(\frac{x-x_j}{\Delta x}\right)$ so that $\bar{p}_j(x_j) = \bar{w}_j^n$. Here the w_j'-s are possibly limited slopes which are reconstructed from the known cell-averages, $w_j' = \{(w_j^n)'\} = \{w'(\bar{w}_k^n)_{k=j-1}^{j+1}\}$. (Throughout this lecture we use primes, $w_j', w_j'', \ldots$, to denote *discrete* derivatives, which approximate the corresponding differential ones). A whole library of limiters is available in this context, so that the co-monotonicity of $w(x, t^n)$ with $\Sigma \bar{w}_j \chi_j$ is guaranteed, e.g., [42]. The Piecewise-Parabolic Method (PPM) of Colella-Woodward [6] and respectively, ENO schemes of Harten et.al. [13], offer, respectively, third- and higher-order Godunov-type upwind schemes. (A detailed account of ENO schemes can be found in lectures of C.W. Shu in this volume). Finally, we should not give the impression that limiters are used exclusively in conjunction with Godunov-type schemes. The *positive schemes* of Liu and Lax, [27], offer simple and fast upwind schemes for multidimensional systems, based on an alternative positivity principle.

Central schemes

As before, we seek a piecewise-polynomial, $w(x, t^n) = \Sigma p_j(x)\chi_j(x)$, which serves as an approximate solution to the *exact* evolution of sliding averages in (2.80),

$$\bar{w}(x, t^{n+1}) = \bar{w}(x, t^n) - \frac{1}{\Delta x} \left[\int_{t^n}^{t^{n+1}} f(w(x + \frac{\Delta x}{2}, \tau))d\tau \right. \tag{2.84}$$

$$\left. - \int_{t^n}^{t^{n+1}} f(w(x - \frac{\Delta x}{2}, \tau))d\tau \right].$$

Note that the polynomial pieces of $w(x, t^n)$ are supported in the cells, $I_j = \left\{\xi \,\middle|\, |\xi - x_j| \leq \frac{\Delta x}{2}\right\}$, with interfacing breakpoints at the half-integers gridpoints, $x_{j+\frac{1}{2}} = (j + \frac{1}{2})\Delta x$.

We recall that upwind schemes (2.81) were based on sampling (2.80) in the *midcells*, $x = x_j$. In contrast, central schemes are based on sampling (2.84) at the *interfacing breakpoints*, $x = x_{j+\frac{1}{2}}$, which yields

$$\bar{w}_{j+\frac{1}{2}}^{n+1} = \bar{w}_{j+\frac{1}{2}}^n - \frac{1}{\Delta x} \left[\int_{\tau=t^n}^{t^{n+1}} f(w(x_{j+1}, \tau))d\tau - \int_{\tau=t^n}^{t^{n+1}} f(w(x_j, \tau))d\tau \right].$$
$$\tag{2.85}$$

We want to utilize (2.85) in terms of the known cell averages at time level $\tau = t^n$, $\{\bar{w}_j^n\}_j$. The remaining task is therefore to recover the *pointvalues* $\{w(\cdot, \tau)|\ t^n \leq \tau \leq t^{n+1}\}$, and in particular, the *staggered averages*, $\{\bar{w}_{j+\frac{1}{2}}^n\}$. As before, this task is accomplished in two main steps:

– First, we use the given cell averages $\{\bar{w}_j^n\}_j$, to *reconstruct* the pointvalues of $w(\cdot, \tau = t^n)$ as piecewise polynomial approximation

$$w(x, t^n) = \sum_j p_j(x)\chi_j(x), \quad \bar{p}_j(x_j) = \bar{w}_j^n. \qquad (2.86)$$

In particular, the staggered averages on the right of (2.85) are given by

$$\bar{w}_{j+\frac{1}{2}}^n = \frac{1}{\Delta x}\left[\int_{x_j}^{x_{j+\frac{1}{2}}} p_j(x)dx + \int_{x_{j+\frac{1}{2}}}^{x_{j+1}} p_{j+1}(x)dx\right]. \qquad (2.87)$$

The resulting central scheme (2.85) then reads

$$\bar{w}_{j+\frac{1}{2}}^{n+1} = \frac{1}{\Delta x}\left[\int_{x_j}^{x_{j+\frac{1}{2}}} p_j(x)dx + \int_{x_{j+\frac{1}{2}}}^{x_{j+1}} p_{j+1}(x)dx\right] + \qquad (2.88)$$

$$-\frac{1}{\Delta x}\left[\int_{\tau=t^n}^{t^{n+1}} f(w(x_{j+1}, \tau))d\tau - \int_{\tau=t^n}^{t^{n+1}} f(w(x_j, \tau))d\tau\right].$$

– Second, we follow the *evolution* of the pointvalues along the mid-cells, $x = x_j$, $\{w(x_j, \tau \geq t^n)\}_j$, which are governed by

$$w_t + f(w)_x = 0, \quad \tau \geq t^n; \quad w(x, t^n) = p_j(x) \quad x \in I_j. \qquad (2.89)$$

Let $\{a_k(u)\}_k$ denote the eigenvalues of the Jacobian $A(u) := \frac{\partial f}{\partial u}$. By hyperbolicity, information regarding the interfacing discontinuities at $(x_{j\pm\frac{1}{2}}, t^n)$ propagates no faster than $\max_k|a_k(u)|$. Hence, the mid-cells values governed by (2.89), $\{w(x_j, \tau \geq t^n)\}_j$, remain free of discontinuities, at least for sufficiently small time step dictated by the CFL condition $\Delta t \leq \frac{1}{2}\Delta x \cdot \max_k|a_k(u)|$. Consequently, since the numerical fluxes on the right of (2.88), $\int_{\tau=t^n}^{t^{n+1}} f(w(x_j, \tau))d\tau$, involve only smooth integrands, they can be computed within any degree of desired accuracy by an appropriate quadrature rule.

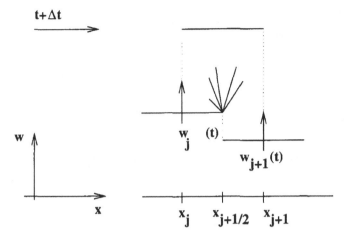

Fig. 2.2: Central differencing by Godunov-type scheme.

It is the *staggered* averaging over the fan of left-going and right-going waves centered at the half-integered interfaces, $(x_{j+\frac{1}{2}}, t^n)$, which characterizes the *central* differencing, consult Figure 2.2. A main feature of these central schemes – in contrast to upwind ones, is the computation of *smooth* numerical fluxes along the mid-cells, $(x = x_j, \tau \geq t^n)$, which avoids the costly (approximate) Riemann solvers. A couple of examples of central Godunov-type schemes is in order.

The first-order Lax-Friedrichs (LxF) approximation is the forerunner for such central schemes — it is based on piecewise constant reconstruction, $w(x, t^n) = \Sigma p_j(x) \chi_j(x)$ with $p_j(x) = \bar{w}_j^n$. The resulting central scheme, (2.88), then reads (with the usual fixed mesh ratio $\lambda := \frac{\Delta t}{\Delta x}$)

$$\bar{w}_{j+\frac{1}{2}}^{n+1} = \frac{1}{2}(\bar{w}_j + \bar{w}_{j+1}) - \lambda\Big[f(\bar{w}_{j+1}) - f(\bar{w}_j)\Big]. \qquad (2.90)$$

Our main focus in the rest of this chapter is on non-oscillatory higher-order extensions of the LxF schemes.

2.3 Central schemes in one-space dimension

The second-order Nessyahu-Tadmor scheme

In this section we overview the construction of high-resolution central schemes in one-space dimension. We begin with the reconstruction of the second-order, non-oscillatory Nessyahu and Tadmor (NT) scheme, [31]. To approximate solutions of (2.77), we introduce a piecewise-linear approximate solution at the discrete time levels, $t^n = n\Delta t$, based on linear functions $p_j(x, t^n)$ which are supported at the cells I_j (see Figure 2.3),

$$w(x,t)|_{t=t^n} = \sum_j p_j(x,t^n)\chi_j(x) \qquad (2.91)$$

$$:= \sum_j \left[\bar{w}_j^n + w_j'\left(\frac{x-x_j}{\Delta x}\right)\right]\chi_j(x), \qquad \chi_j(x) := 1_{I_j}.$$

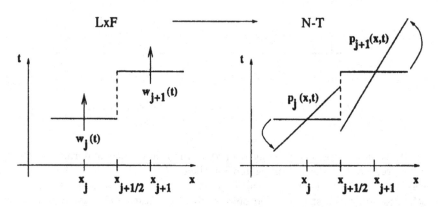

Fig. 2.3: The second-order reconstruction

Second-order of accuracy is guaranteed if the discrete slopes approximate the corresponding derivatives, $w_j' \sim \Delta x \cdot \partial_x w(x_j, t^n) + O(\Delta x)^2$. Such a non-oscillatory approximation of the derivatives is possible, e.g., by using built-in non-linear limiters of the form

$$w_j' = MM\{\theta(\bar{w}_{j+1}^n - \bar{w}_j^n), \frac{1}{2}(\bar{w}_{j+1}^n - \bar{w}_{j-1}^n), \theta(\bar{w}_j^n - \bar{w}_{j-1}^n)\}. \qquad (2.92)$$

Here and below, $\theta \in (0, 2)$ is a non-oscillatory limiter and MM denotes the Min-Mod function

$$MM\{x_1, x_2, ...\} = \begin{cases} \min_i\{x_i\} & \text{if } x_i > 0, \forall i \\ \max_i\{x_i\} & \text{if } x_i < 0, \forall i \\ 0 & \text{otherwise.} \end{cases}$$

An *exact* evolution of w, based on integration of the conservation law over the staggered cell, $I_{j+\frac{1}{2}}$, then reads, (2.85)

$$\bar{w}_{j+\frac{1}{2}}^{n+1} = \frac{1}{\Delta x}\int_{I_{j+\frac{1}{2}}} w(x,t^n)dx - \frac{1}{\Delta x}\int_{\tau=t^n}^{t^{n+1}} [f(w(x_{j+1},\tau)) - f(w(x_j,\tau))]\, d\tau.$$

The first integral is the staggered cell-average at time t^n, $\bar{w}_{j+\frac{1}{2}}^n$, which can be computed directly from the above reconstruction,

$$\bar{w}_{j+\frac{1}{2}}^n := \frac{1}{\Delta x} \int_{x_j}^{x_{n+1}} w(x, t^n) dx = \frac{1}{2}(\bar{w}_j^n + \bar{w}_{j+1}^n) + \frac{1}{8}(w_j' - w_{j+1}'). \quad (2.93)$$

The time integrals of the flux are computed by the second-order accurate mid-point quadrature rule

$$\int_{\tau=t^n}^{t^{n+1}} f(w(x_j, \tau)) d\tau \sim \Delta t \cdot f(w(x_j, t^{n+\frac{1}{2}})).$$

Here, the Taylor expansion is being used to predict the required mid-values of w

$$w(x_j, t^{n+\frac{1}{2}}) \sim w(x_j, t) + \frac{\Delta t}{2} w_t(x_j, t^n)$$

$$= \bar{w}_j^n - \frac{\Delta t}{2} A(\bar{w}_j^n)(p_j(x_j, t^n))_x = \bar{w}_j^n - \frac{\lambda}{2} A_j^n w_j'.$$

In summary, we end up with the central scheme, [31], which consists of a first-order *predictor step*,

$$w_j^{n+\frac{1}{2}} = \bar{w}_j^n - \frac{\lambda}{2} A_j^n w_j', \quad A_j^n := A(\bar{w}_j^n), \quad (2.94)$$

followed by the second-order *corrector step*, (2.88),

$$\bar{w}_{j+\frac{1}{2}}^{n+1} = \frac{1}{2}(\bar{w}_j^n + \bar{w}_{j+1}^n) + \frac{1}{8}(w_j' - w_{j+1}') - \lambda \left[f(w_{j+1}^{n+\frac{1}{2}}) - f(w_j^{n+\frac{1}{2}}) \right]. \quad (2.95)$$

The *scalar* non-oscillatory properties of (2.94)-(2.95) were proved in [31], [32], including the TVD property, cell entropy inequality, L_{loc}^1- error estimates, etc. Moreover, the numerical experiments, reported in [30], [31], [2], [3], [45], [37], [38], [39], with one-dimensional *systems* of conservation laws, show that such second-order central schemes enjoy the same high-resolution as the corresponding second-order upwind schemes do. Thus, the excessive smearing typical to the first-order LxF central scheme is compensated here by the second-order accurate MUSCL reconstruction.

In figure 2.4 we compare, side by side, the upwind ULT scheme of Harten, [12], with our central scheme (2.94)-(2.95). The comparable high-resolution of this so called Lax's Riemann problem is evident.

At the same time, the central scheme (2.94)-(2.95) has the advantage over the corresponding upwind schemes, in that no (approximate) Riemann solvers, as in (2.83), are required. Hence, these Riemann-free central schemes provide an efficient high-resolution alternative in the one-dimensional case, and a particularly advantageous framework for multidimensional computations, e.g., [3], [2], [16]. This advantage in the multidimensional case will be explored in the next section. Also, *staggered* central differencing, along the lines of the Riemann-free Nessyahu-Tadmor scheme (2.94)-(2.95), admits simple efficient extensions in the presence of general source terms, [8], and in particular, stiff source terms, [4]. Indeed, it is a key ingredient behind the relaxation schemes studied in [18].

It should be noted, however, that the component-wise version of these central schemes might result in deterioration of resolution at the computed extrema. The second-order computation presented in figure 2.3 below demonstrates this point. (this will be corrected by higher order central methods). Of course, this – so called extrema clipping, is typical to high-resolution upwind schemes as well; but it is more pronounced with our central schemes due to the built-in extrema-switching to the dissipative LxF scheme. Indeed, once an extrema cell, I_j, is detected (by the limiter), it sets a zero slope, $w'_j = 0$, in which case the second-order scheme (2.94)-(2.95) is reduced back to the first-order LxF, (2.90).

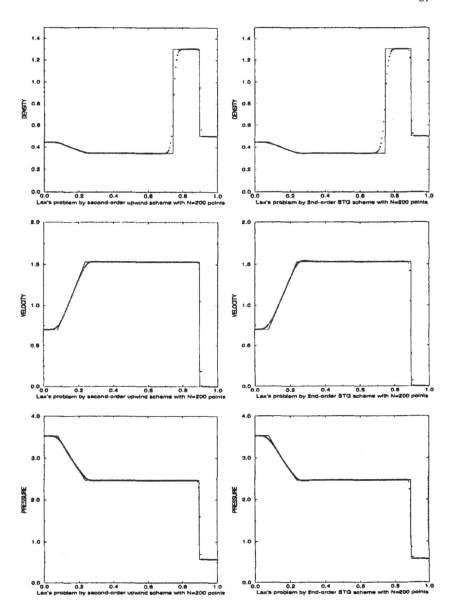

Fig. 2.4: 2nd order: central (STG) vs. upwind (ULT) — Lax's Riemann problem

The third-order central scheme

Following the framework outlined in §2.3, the upgrade to third-order central scheme consists of two main ingredients:

(i) A third-order accurate, piecewise-quadratic polynomial reconstruction which enjoys desirable non-oscillatory properties;

(ii) An appropriate quadrature rule to approximate the numerical fluxes along cells' interfaces.

Following [29], we proceed as follows. The piecewise-parabolic reconstruction takes the form

$$p_j(x) = w_j^n + w_j' \left(\frac{x - x_j}{\Delta x}\right) + \frac{1}{2}w_j'' \left(\frac{x - x_j}{\Delta x}\right)^2. \qquad (2.96)$$

Here, w_j'' are the (pointvalues of) the *reconstructed second derivatives*

$$w_j'' := \theta_j \Delta_+ \Delta_- \bar{w}_j^n; \qquad (2.97)$$

w_j' are the (pointvalues of) the *reconstructed slopes*,

$$w_j' := \theta_j \Delta_0 \bar{w}_j^n; \qquad (2.98)$$

and w_j^n are the *reconstructed pointvalues*

$$w_j^n := \bar{w}_j^n - \frac{w_j''}{24}. \qquad (2.99)$$

Observe that, starting with third- (and higher-) order accurate methods, pointwise values *cannot* be interchanged with cell averages, $w_j^n \neq \bar{w}_j^n$.

Here, θ_j are appropriate nonlinear limiters which guarantee the non-oscillatory behavior of the third-order reconstruction; its precise form can be found in [28], [29]. They guarantee that the reconstruction (2.96) is non-oscillatory in the sense that $N(w(\cdot, t^n))$ — the number of extrema of $w(x, t^n)$, does not exceed that of its piecewise-constant projection, $N(\Sigma \bar{w}_j^n \chi_j(\cdot))$,

$$N(w(\cdot, t^n)) \leq N(\Sigma \bar{w}_j^n \chi_j(\cdot)). \qquad (2.100)$$

Next we turn to the evolution of the piecewise-parabolic reconstructed solution. To this end we need to evaluate the staggered averages, $\{\bar{w}_{j+\frac{1}{2}}^n\}$, and to approximate the interface fluxes, $\left\{\int_{\tau=t^n}^{t^{n+1}} f(w(x_j, \tau)) d\tau\right\}$.

With $p_j(x) = w_j^n + w_j' \left(\frac{x - x_j}{\Delta x}\right) + \frac{1}{2}w_j'' \left(\frac{x - x_j}{\Delta x}\right)^2$ specified in (2.96)-(2.99), one evaluates the staggered averages of the third-order reconstruction $w(x, t^n) = \Sigma p_j(x) \chi_j(x)$

$$\bar{w}_{j+\frac{1}{2}}^n = \frac{1}{\Delta x} \int_{x_j}^{x_{j+1}} w(x, t^n) dx = \frac{1}{2}(\bar{w}_j + \bar{w}_{j+1}) + \frac{1}{8}(w_j' - w_{j+1}'). \quad (2.101)$$

Remarkably, we obtain here the same formula for the staggered averages as in the second-order cases, consult (2.93); the only difference is the use of the new limited slopes in (2.98), $w'_j = \theta_j \Delta_0 \bar{w}^n_j$.

Next, we approximate the (exact) numerical fluxes by Simpson's quadrature rule, which is (more than) sufficient for retaining the overall third-order accuracy,

$$\frac{1}{\Delta x} \int_{\tau = t^n}^{t^{n+1}} f(w(x_j, \tau)) d\tau \sim \frac{\lambda}{6} \left[f(w^n_j) + 4f(w^{n+\frac{1}{2}}_j) + f(w^{n+1}_j) \right]. \quad (2.102)$$

This in turn, requires the three approximate *pointvalues* on the right, $w^{n+\beta}_j \sim w(x_j, t^{n+\beta})$ for $\beta = 0, \frac{1}{2}, 1$. Following our approach in the second-order case, [31], we use Taylor expansion to *predict*

$$w^n_j = \bar{w}^n_j - \frac{w''_j}{24}; \quad (2.103)$$

$$\dot{w}^n_j \equiv (\Delta x \cdot \partial_t) w(x_j, t^n) = -\Delta x \cdot \partial_x f(w(x_j, t^n)) =$$
$$= -a(w^n_j) \cdot w'_j,; \quad (2.104)$$

$$\ddot{w}^n_j \equiv (\Delta x \cdot \partial_t)^2 w(x_j, t^n) =$$
$$= \Delta x \cdot \partial_x \left[a(w^n_j) \Delta x \cdot \partial_x f(w(x_j, t^n)) \right] =$$
$$= a^2(w^n_j) w''_j + 2a(w^n_j) a'(w^n_j)(w'_j)^2. \quad (2.105)$$

In summary of the scalar setup, we end up with a two step scheme where, starting with the reconstructed pointvalues

$$w^n_j = \bar{w}^n_j - \frac{w''_j}{24}, \quad (2.106)$$

we *predict* the pointvalues $w^{n+\beta}_j$ by, e.g. Taylor expansions,

$$w^{n+\beta}_j = w^n_j + \lambda \beta \dot{w}^n_j + \frac{(\lambda \beta)^2}{2} \ddot{w}^n_j, \qquad \beta = \frac{1}{2}, 1; \quad (2.107)$$

this is followed by the *corrector* step

$$\bar{w}^n_{j+\frac{1}{2}} = \frac{1}{2}(\bar{w}^n_j + \bar{w}^n_{j+1}) + \frac{1}{8}(w'_j - w'_{j+1}) + \quad (2.108)$$

$$- \frac{\lambda}{6} \left\{ \left[f(w^n_{j+1}) + 4f(w^{n+\frac{1}{2}}_{j+1}) + f(w^{n+1}_{j+1}) \right] \right.$$

$$\left. - \left[f(w^n_j) + 4f(w^{n+\frac{1}{2}}_j) + f(w^{n+1}_j) \right] \right\}.$$

In figure 2.3 we revisit the so called Woodward-Colella problem, [46], where we compare the second vs. the third-order results. The improvement in resolving the density field is evident.

We conclude this section with several remarks.

Remarks.

1. Stability.

   We briefly mention the stability results for the scalar central schemes. In the second order case, the NT scheme was shown to be both TVD and entropy stable in the sense of satisfying a cell entropy inequality – consult [31]. The third-order scalar central scheme is stable in the sense of satisfying the NED property, (2.100), namely

   Theorem 2.1 *([29]) Consider the central scheme (2.106),(2.107), (2.108), based on the third-order accurate quadratic reconstruction, (2.96)-(2.99). Then it satisfies the so-called Number of Extrema Diminishing (NED) property, in the sense that*

   $$N\left(\sum_{\nu}\bar{w}^{n+1}_{v+\frac{1}{2}}\chi_{\nu+\frac{1}{2}}(x)\right) \leq N\left(\sum_{\nu}\bar{w}^{n}_{\nu}\chi_{\nu}(x)\right). \qquad (2.109)$$

2. Source terms, radial coordinates, ...

   Extensions of the central framework which deal with both, stiff and non-stiff source terms can be found in [37],[38], [8], [4]. In particular, Kupferman in [20],[21] developed the central framework within the radial coordinates which require to handle both – variable coefficients + source terms.

3. Higher order central schemes.

   We refer to [39], where a high-order ENO reconstruction is realized by a staggered cell averaging. Here, intricate Riemann solvers are replaced by high order quadrature rules. and for this purpose, one can effectively use the RK method (rather than the Taylor expansion outlined above):

4. Taylor vs. Runge-Kutta.

   The evaluations of Taylor expansions could be substituted by the more economical Runge-Kutta integrations; the simplicity becomes more pronounced with *systems*. A particular useful approach in this context was proposed in [39], using the natural continuous extensions of RK schemes.

5. Systems.

   One of the main advantages of our central-staggered framework over that of the upwind schemes, is that expensive and time-consuming characteristic decompositions can be avoided. Specifically, all the non-oscillatory computations can be carried out with diagonal limiters, based on a *component-wise* extension of the scalar limiters outlined above.

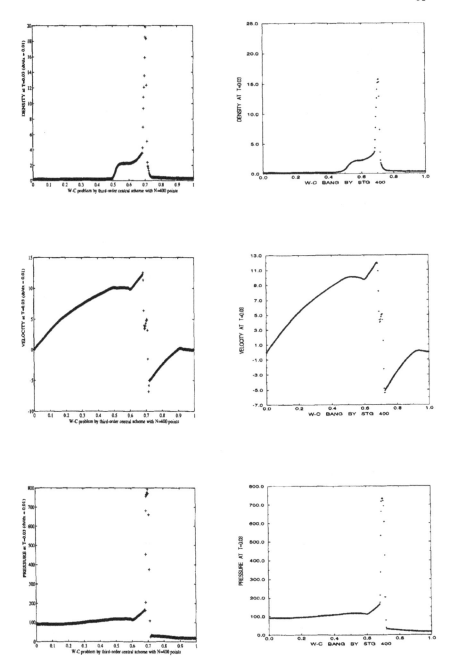

Fig. 2.5: 3rd vs. 2nd order central schemes — Woodward-Colella problem at $t = 0.03$

2.4 Central schemes in two space dimensions

Following the one dimensional setup, one can derive a non-oscillatory, two-dimensional central scheme. Here we sketch the construction of the second-order two-dimensional scheme following [16] (see also [2],[1]). For the two-dimensional third-order accurate scheme, we refer to [24].

We consider the two-dimensional hyperbolic system of conservation laws

$$u_t + f(u)_x + g(u)_y = 0. \tag{2.110}$$

To approximate a solution to (2.110), we start with a two-dimensional linear reconstruction

$$w(x,y,t^n) = \sum_{j,k} p_{j,k}(x,y)\chi_{j,k}(x,y), \tag{2.111}$$

$$p_{j,k}(x,y) = \bar{w}_{j,k}^n + w'_{j,k}\left(\frac{x-x_j}{\Delta x}\right) + w^{\backslash}_{j,k}\left(\frac{y-y_k}{\Delta y}\right).$$

Here, the discrete slopes in the x and in the y direction approximate the corresponding derivatives, $w'_{j,k} \sim \Delta x \cdot w_x(x_j,y_k,t^n) + O(\Delta x)^2$, $w^{\backslash}_{j,k} \sim \Delta y \cdot w_y(x_j,y_k,t^n) + O(\Delta y)^2$, and $\chi_{j,k}(x,y)$ is the characteristic function of the cell $C_{j,k} := \{(\xi,\eta)|\ i\,|\xi - x_j| \leq \frac{\Delta x}{2}, |\eta - y_k| \leq \frac{\Delta y}{2}\} = I_j \otimes J_k$. Of course, it is essential to reconstruct the discrete slopes, w' and $w^{\backslash}$, with built in *limiters*, which guarantee the non-oscillatory character of the reconstruction; the family of min-mod limiters is a prototype example

$$w'_{jk} = MM\{\theta(\bar{w}_{j+1,k}^n - \bar{w}_{j,k}^n), \frac{1}{2}(\bar{w}_{j+1,k}^n - \bar{w}_{j-1,k}^n), \theta(\bar{w}_{j,k}^n - \bar{w}_{j-1,k}^n)\} \tag{2.112'}$$

$$w^{\backslash}_{jk} = MM\{\theta(\bar{w}_{j,k+1}^n - \bar{w}_{j,k}^n), \frac{1}{2}(\bar{w}_{j,k+1}^n - \bar{w}_{j,k-1}^n), \theta(\bar{w}_{j,k}^n - \bar{w}_{j,k-1}^n)\}. \tag{2.112^{\backslash}}$$

An exact evolution of this reconstruction, which is based on integration of the conservation law over the staggered volume yields

$$\bar{w}_{j+\frac{1}{2},k+\frac{1}{2}}^{n+1} = \fint_{C_{j+\frac{1}{2},k+\frac{1}{2}}} w(x,y,t^n)dxdy + \tag{2.113}$$

$$- \lambda \left\{ \fint_{\tau=t^n}^{t^{n+1}} \fint_{y \in J_{k+\frac{1}{2}}} [f(w(x_{j+1},y,\tau)) - f(w(x_j,y,\tau))]\,dyd\tau \right\} +$$

$$- \mu \left\{ \fint_{\tau=t^n}^{t^{n+1}} \fint_{x \in I_{j+\frac{1}{2}}} [g(w(x,y_{k+1},\tau)) - g(w(x,y_k,\tau))]\,dxd\tau \right\}.$$

The exact averages at t^n – consult the floor plan in Figure 2.6 yields

$$\bar{w}^n_{j+\frac{1}{2},k+\frac{1}{2}} := \fint_{C_{j+\frac{1}{2},k+\frac{1}{2}}} w(x,y,t^n)dxdy = \tag{2.114}$$

$$= \frac{1}{4}(\bar{w}^n_{jk} + \bar{w}^n_{j+1,k} + \bar{w}^n_{j,k+1} + \bar{w}^n_{j+1,k+1}) +$$

$$+ \frac{1}{16}\Big\{(w'_{jk} - w'_{j+1,k}) + (w'_{j,k+1} - w'_{j+1,k+1}) +$$

$$+ (w^{\backslash}_{jk} - w^{\backslash}_{j,k+1}) + (w^{\backslash}_{j+1,k} - w^{\backslash}_{j+1,k+1})\Big\}.$$

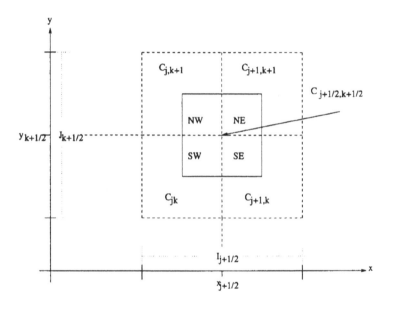

Fig. 2.6: Floor plan of the staggered grid.

So far everything is *exact*. We now turn to *approximate* the four fluxes on the right of (2.113), starting with the one along the East face, consult figure 2.7, $\int_{t^n}^{t^{n+1}} \fint_{J_{k+\frac{1}{2}}} f(w(x_{j+1},y,\tau))dyd\tau$. We use the midpoint quadrature rule for second-order approximation of the temporal integral, $\fint_{y \in J_{k+\frac{1}{2}}} f(w(x_{j+1},y,t^{n+\frac{1}{2}}))dy$; and, for reasons to be clarified below, we use the second-order rectangular quadrature rule for the spatial integration across the y-axis, yielding

$$\int_{t^n}^{t^{n+1}} \int_{y \in J_{k+\frac{1}{2}}} f(w(x_{j+1}, y, \tau)) dy d\tau \sim \frac{1}{2} \left[f(w_{j+1,k}^{n+\frac{1}{2}}) + f(w_{j+1,k+1}^{n+\frac{1}{2}}) \right].$$

(2.115)

In a similar manner we approximate the remaining fluxes.

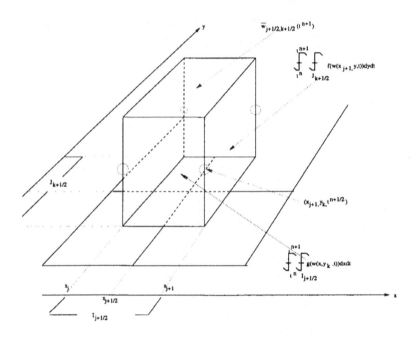

Fig. 2.7: The central, staggered stencil.

These approximate fluxes make use of the midpoint values, $w_{jk}^{n+\frac{1}{2}} \equiv w(x_j, y_k, t^{n+\frac{1}{2}})$, and it is here that we take advantage of utilizing these midvalues for the spatial integration by the rectangular rule. Namely, since these midvalues are secured at the smooth center of their cells, C_{jk}, bounded away from the jump discontinuities along the edges, we may use Taylor expansion, $w(x_j, y_k, t^{n+\frac{1}{2}}) = \bar{w}_{jk}^n + \frac{\Delta t}{2} w_t(x_j, y_k, t^n) + \mathcal{O}(\Delta t)^2$. Finally, we use the conservation law (2.110) to express the time derivative, w_t, in terms of the spatial derivatives, $f(w)'$ and $g(w)\backslash$,

$$w_{jk}^{n+\frac{1}{2}} = \bar{w}_{jk}^n - \frac{\lambda}{2}f(w)'_{jk} - \frac{\mu}{2}g(w)\backslash_{jk}. \tag{2.116}$$

Here, $f(w)'_{jk} \sim \Delta x \cdot f(w(x_j, y_k, t^n))_x$ and $g(w)\backslash_{jk} \sim \Delta y \cdot g(w(x_j, y_k, t^n))_y$, are one-dimensional discrete slopes in the x- and y-directions, of the type reconstructed in (2.112')-(2.112\`); for example, multiplication by the corresponding Jacobians A and B yields

$$f(w)'_{jk} = A(\bar{w}_{jk}^n)w'_{jk}, \qquad g(w)\backslash_{jk} = B(\bar{w}_{jk}^n)w\backslash_{jk}.$$

Equipped with the midvalues (2.116), we can now evaluate the approximate fluxes, e,g., (2.115). Inserting these values, together with the staggered average computed in (2.115), into (2.113), we conclude with new staggered averages at $t = t^{n+1}$, given by

$$\bar{w}_{j+\frac{1}{2},k+\frac{1}{2}}^{n+1} = \frac{1}{4}(\bar{w}_{jk}^n + \bar{w}_{j+1,k}^n + \bar{w}_{j,k+1}^n + \bar{w}_{j+1,k+1}^n) + \tag{2.117}$$

$$+ \frac{1}{16}(w'_{jk} - w'_{j+1,k}) - \frac{\lambda}{2}\left[f(w_{j+1,k}^{n+\frac{1}{2}}) - f(w_{j,k}^{n+\frac{1}{2}})\right]$$

$$+ \frac{1}{16}(w'_{j,k+1} - w'_{j+1,k+1}) - \frac{\lambda}{2}\left[f(w_{j+1,k+1}^{n+\frac{1}{2}}) - f(w_{j,k+1}^{n+\frac{1}{2}})\right]$$

$$+ \frac{1}{16}(w\backslash_{jk} - w\backslash_{j,k+1}) - \frac{\mu}{2}\left[g(w_{j,k+1}^{n+\frac{1}{2}}) - g(w_{j,k}^{n+\frac{1}{2}})\right]$$

$$+ \frac{1}{16}(w\backslash_{j+1,k} - w\backslash_{j+1,k+1}) - \frac{\mu}{2}\left[g(w_{j+1,k+1}^{n+\frac{1}{2}}) - g(w_{j+1,k}^{n+\frac{1}{2}})\right].$$

In summary, we end up with a simple two-step predictor-corrector scheme which could be conveniently expressed in terms on the one-dimensional staggered averaging notations

$$< w_{j,.} >_{k+\frac{1}{2}} := \frac{1}{2}(w_{j,k} + w_{j,k+1}), \quad < w_{.,k} >_{j+\frac{1}{2}} := \frac{1}{2}(w_{j,k} + w_{j+1,k}).$$

Our scheme consists of a *predictor step*

$$w_{j,k}^{n+\frac{1}{2}} = w_{j,k}^n - \frac{\lambda}{2}f'_{j,k} - \frac{\mu}{2}g\backslash_{j,k}, \tag{2.118}$$

followed by the *corrector step*

$$\bar{w}_{j+\frac{1}{2},k+\frac{1}{2}}^{n+1} = < \frac{1}{4}(\bar{w}_{j,.}^n + \bar{w}_{j+1,.}^n) + \frac{1}{8}(w'_{j,.} - w'_{j+1,.}) - \lambda(f_{j+1,.}^{n+\frac{1}{2}} - f_{j,.}^{n+\frac{1}{2}}) >_{k+\frac{1}{2}}$$

$$+ < \frac{1}{4}(\bar{w}_{.,k}^n + \bar{w}_{.,k+1}^n) + \frac{1}{8}(w\backslash_{.,k} - w\backslash_{.,k+1}) - \mu(g_{.,k+1}^{n+\frac{1}{2}} - g_{.,k}^{n+\frac{1}{2}}) >_{j+\frac{1}{2}}.$$

In figures 2.8 taken from [16], we present the two-dimensional computation of a double-Mach reflection problem; in figure 2.9 we quote from [45] the two-dimensional computation of MHD solution of Kelvin-Helmholtz instability due to shear flow. The computations are based on our second-order central scheme. It is remarkable that such a simple 'two-lines' algorithm, with no characteristic decompositions and no dimensional splitting, approximates the rather complicated double Mach reflection problem with such high resolution. Couple of remarks are in order.

- The two-dimensional computation is more sensitive to the type of limiter than in the one-dimensional framework [31]. In the context of the double Mach reflection problem, the MM_2 (consult (2.92) with $\theta = 2$) seems to yield the sharper results.

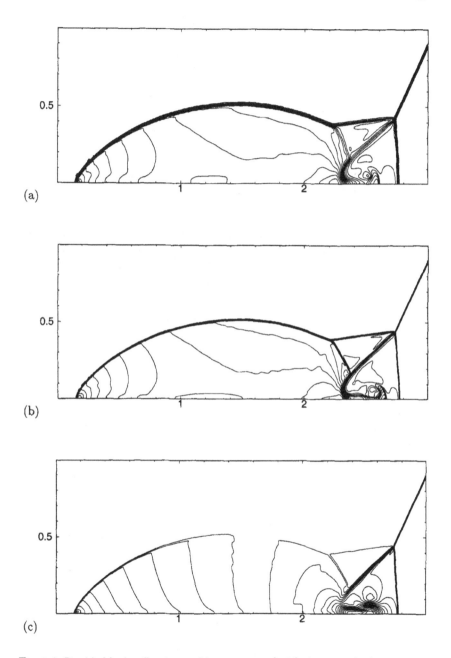

Fig. 2.8: Double Mach reflection problem computed with the central scheme using MM_2 limiter with CFL=0.475 at $t = 0.2$ (a) density computed with 480×120 cells (b) density computed with 960×240 cells (c) x-velocity computed with 960×240 cells

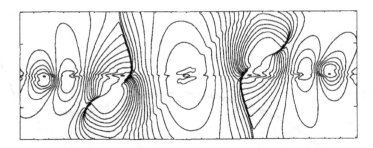

Fig. 2.9: Kelvin-Helmholtz instability due to shear flow. Transverse configuration (B perpendicular to v). Pressure contours at $t = 140$

– No effort was made to optimize the boundary treatment. The staggered stencils require a different treatment for even-odd cells intersecting with the boundaries. A more careful treatment following [26] is presented in §2.4. The lack of boundary resolution could be observed at the bottom of the two Mach stems.

We conclude this section with brief remarks on further results related to central schemes.

Remarks.

1. Simplicity.
   Again, we would like to highlight the simplicity of the central schemes, which is particularly evident in the multidimensional setup: no characteristic information is required – in fact, even the exact Jacobians of the fluxes are not required; also, since no (approximate) Riemann solvers are involved, the central schemes require no dimensional splitting; as an example we refer to the approximation of the incompressible equations by central schemes, §2.5; the results in [7] provide another example of a *weakly* hyperbolic multidimensional system which could be efficiently solved in term of central schemes, by avoiding dimensional splitting.
2. Non-staggering. We refer to [15] for a non-staggered version of the central schemes.
3. Stability.
   The following maximum principle holds for the nonoscillatory scalar central schemes:

   Theorem 2.2 *[16] Consider the two-dimensional scalar scheme (2.116-2.117), with minmod slopes, w' and $w^`$, in (2.112'-2.112'')). Then for any $\theta < 2$ there exists a sufficiently small CFL number, C_θ (– e.g. $C_1 = (\sqrt{7} - 2)/6 \sim 0.1$), such that if the CFL condition is fulfilled,*

   $$\max(\lambda \cdot \max_u |f_u(u)|, \mu \cdot \max_u |g_u(u)|) \leq C_\theta,$$

then the following local maximum principle holds

$$\min_{\substack{|p-(j+\frac{1}{2})|=\frac{1}{2} \\ |q-(k+\frac{1}{2})|=\frac{1}{2}}} \{\bar{w}_{p,q}^n\} \le \bar{w}_{j+\frac{1}{2},k+\frac{1}{2}}^{n+1} \le \max_{\substack{|p-(j+\frac{1}{2})|=\frac{1}{2} \\ |q-(k+\frac{1}{2})|=\frac{1}{2}}} \{\bar{w}_{p,q}^n\}. \qquad (2.119)$$

4. Third-order accuracy. Extensions to third-order accuracy in two space dimensions can be found in [24].

Boundary conditions

Following [25], we demonstrate our boundary treatment in the case of the left-boundary (see Figure 2.10).

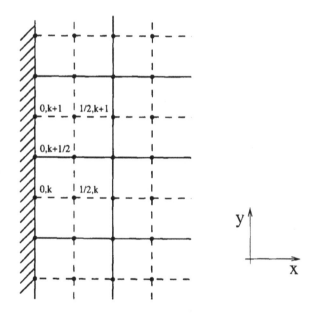

Fig. 2.10: Two dimensions - left boundary

We distinguish between *inflow* $(f'(w_{1/2,k}^n) > 0)$, and *outflow* $(f'(w_{1/2,k}^n) < 0)$, boundary cells.

In *inflow* boundary cells, we reconstruct a constant interpolant from the prescribed point-values at these boundaries,

$$p_{1/2,k}(x,y,t^n) \equiv w_{0,k}^n, \qquad w_{1/2,k}' = 0. \qquad (2.120)$$

This reconstruction is then used to build the approximate solution at time t^{n+1} in the interior cells. At the next-time step, t^{n+1}, the cell-averages at these boundary cells are defined according to the prescribed point-values as

$$\bar{w}_{1/4,k+1/2}^{n+1} := w_{0,k+1/2}^{n+1}.$$

We now turn to the *outflow* boundary cells. Here, we extrapolate the data from the interior of the domain, up to the boundary. First, we determine the discrete slope in the x-direction, $w'_{1/2,k}$. This slope is then used to extrapolate the cell-average up to the boundary,

$$w_{0,k}^n = w_{1/2,k}^n - \frac{\Delta x}{2} w'_{1/2,k},$$

which is then used to predict the mid-value, $w_{0,k}^{n+1/2} = w_{0,k}^n - \frac{\lambda}{2} f'_{0,k} - \frac{\mu}{2} g'_{0,k}$. Here

$$f'_{0,k} = a(w_{0,k}^n) w'_{1/2,k}, \qquad g'_{0,k} = b(w_{0,k}^n) w'_{1/2,k}.$$

The discrete slope in the y-direction, $w'_{0,k}$, is computed in that boundary cell in an analogous way to the interior computation. In summary, the staggered average at time t^{n+1} is given by

$$\bar{w}_{1/4,k+1/2}^{n+1} = \frac{\bar{w}_{1/2,k}^n + \bar{w}_{1/2,k+1}^n}{2} + \tag{2.121}$$
$$+ \frac{1}{8}(-w'_{1/2,k} - w'_{1/2,k+1} + w'_{1/2,k} - w'_{1/2,k+1})$$
$$- \frac{\lambda}{2}(f(w_{1/2,k+1}^n) + f(w_{1/2,k}^n) - f(w_{0,k+1}^n) - f(w_{0,k}^n))$$
$$- \mu(g(w_{1/2,k+1}^n) + g(w_{0,k+1}^n) - g(w_{1/2,k}^n) - g(w_{0,k+1}^n)).$$

This concludes the boundary treatment of the left boundary. Similar expressions hold for the other three boundaries.

We now turn to the corners and as a prototype, consider the upper-left corner (see Figure 2.11). In the corner we repeat the previous boundary treatment with one simple modification. The main difference regarding the boundary scheme in the corner is based on the number of different possible inflow/outflow configurations in that corner.

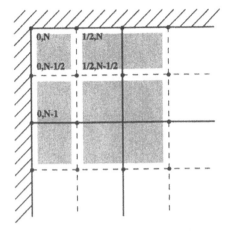

Fig. 2.11: Upper-left corner

Computationally, the most complicated case is when the flow in that upper-left corner is outflow in both directions. In this case, the staggered average at time t^{n+1}, $\bar{w}^{n+1}_{1/4,N-1/4}$, is computed according to

$$\begin{cases} w'_{1/2,N-1/2} = \cdots \\[2mm] w^{\backslash}_{1/2,N-1/2} = \cdots \end{cases} \qquad \text{Limited slopes}$$

$$\begin{cases} w^n_{0,N-1/2} = w^n_{1/2,N-1/2} - \frac{\Delta x}{2} w'_{1/2,N-1/2} \\[2mm] w^{n+1/2}_{0,N-1/2} = w^n_{0,N-1/2} - \frac{\lambda}{2} f'_{0,N-1/2} - \frac{\mu}{2} g^{\backslash}_{0,N-1/2} \end{cases} \qquad \text{Predictor (west)}$$

$$\begin{cases} w^n_{1/2,N} = w^n_{1/2,N-1/2} + \frac{\Delta y}{2} w^{\backslash}_{1/2,N-1/2} \\[2mm] w^{n+1/2}_{1/2,N} = w^n_{1/2,N} - \frac{\lambda}{2} f'_{1/2,N} - \frac{\mu}{2} g^{\backslash}_{1/2,N} \end{cases} \qquad \text{Predictor (north)}$$

$$\begin{cases} w^n_{0,N} = w^n_{1/2,N-1/2} - \frac{\Delta x}{2} w'_{1/2,N-1/2} + \frac{\Delta y}{2} w^{\backslash}_{1/2,N-1/2} \\[2mm] w^{n+1/2}_{0,N} = w^n_{0,N} - \frac{\lambda}{2} f'_{0,N} - \frac{\mu}{2} g^{\backslash}_{0,N} \end{cases} \qquad \text{Predictor (north-west)}$$

The cell-average in the north-west edge of Figure 2.11 in time t^{n+1}, is given in this outflow-outflow case by the *corrector step*

$$\begin{aligned} \bar{w}^{n+1}_{1/4,N-1/4} = \bar{w}^n_{1/2,N-1/2} &+ \frac{-w'_{1/2,N-1/2} + w^{\backslash}_{1/2,N-1/2}}{4} \\ &- \lambda(f(w^n_{1/2,N}) + f(w^n_{1/2,N-1/2}) - f(w^n_{0,N}) - f(w^n_{0,N-1/2})) \\ &- \mu(g(w^n_{1/2,N}) + g(w^n_{0,N}) - g(w^n_{1/2,N-1/2}) - g(w_{0,N-1/2}^n)). \end{aligned} \qquad (2.122)$$

When one of the boundaries is inflow, we have $w'_{1/2,N-1/2} = w\grave{}_{1/2,N-1/2} = 0$, and $\bar{w}^{n+1}_{1/4,N-1/4} = w^{n+1}_{0,N}$ (– the prescribed pointvalues at the corner).

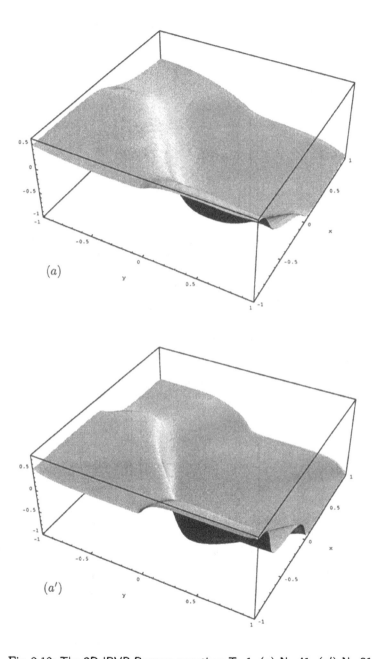

Fig. 2.12: The 2D IBVP Burgers equation: T=1. (a) N=41, (a') N=81

As an example, we we approximate a solution to the two-dimensional Burgers equation

$$u_t + uu_x + uu_y = 0, \tag{2.123}$$

subject to the initial conditions,

$$u_0(x, y) = \begin{cases} 0.5 & -1 \leq x < 0, -1 \leq y < 0 \\ 0 & 0 \leq x \leq 1, -1 \leq y < 0 \\ -1 & 0 \leq x \leq 1, 0 \leq y \leq 1 \\ -0.2 & -1 \leq x < 0, 0 \leq y \leq 1. \end{cases}$$

and augmented with boundary conditions at the inflow boundaries which are equal to the initial values at these same boundaries. Figures 2.12 show the evolution of the solution in time for mesh sizes $41 * 41$ and $81 * 81$. Again, we note that there are no spurious oscillations at the boundaries, oscillations that are inherent with a naive treatment of inflow boundaries.

2.5 Incompressible Euler equations

The vorticity formulation
We are concerned with the approximate solution of the 2D Euler (– and respectively – NS) equations, expressed in terms of the vorticity, $\omega := \nabla \times \boldsymbol{u}$,

$$\omega_t + (u\omega)_x + (v\omega)_y = 0 \ (+\nu\Delta\omega). \tag{2.124}$$

Here, $\boldsymbol{u} = (u, v)$, is the two-component divergence-free velocity field,

$$u_x + v_y = 0. \tag{2.125}$$

Equation (2.124) can be viewed as a nonlinear (viscous) conservation law,

$$\omega_t + f(\omega)_x + g(\omega)_y = 0 \ (+\nu\Delta\omega), \tag{2.126}$$

with a global flux, $(f, g) := (u\omega, v\omega)$. At the same time, the incompressibility (2.125) enables us to rewrite (2.124) in the equivalent convective form

$$\omega_t + u\omega_x + v\omega_y = 0. \tag{2.127}$$

Equation (2.127) guarantees that the vorticity, ω, propagates with finite speed, at least for uniformly bounded velocity field, $\boldsymbol{u} \in L^\infty$. This duality between the conservative and convective forms of the equations plays an essential role in our discussion.

To approximate (2.124) by a second-order central scheme (following [16,31]) we introduce a piecewise-linear polynomial MUSCL approximate solution, $\omega(\cdot,\cdot,t)$, at the discrete time levels, $t^n = n\Delta t$,

$$\omega(x,y,t^n) = \sum_{j,k}\left\{\bar{\omega}_{j,k}^n + \omega_{j,k}'\left(\frac{x-x_j}{\Delta x}\right) + \omega_{j,k}^{\backslash}\left(\frac{y-y_k}{\Delta y}\right)\right\}1_{C_{j,k}}. \quad (2.128)$$

with pieces supported in the cells, $C_{j,k} := \left\{(\xi,\zeta)\,\big|\,|\xi-x_j| \le \frac{\Delta x}{2}, |\zeta - y_k| \le \frac{\Delta y}{2}\right\}$.

As before, we use the *exact* staggered averages at t^n, followed by the mid-point rule to approximate the corresponding flux. For example, the averaged flux, $f = u\omega$ is approximated by Analogous expressions hold for the remaining fluxes. Note that finite speed of propagation (of ω – which is due to the discrete incompressibility relation (2.132) below), guarantees that these values are 'secured' inside a region of local smoothness of the flow. The missing midvalues, $\omega_{j,k}^{n+\frac{1}{2}}$, are predicted using a first-order Taylor expansion (where $\lambda := \frac{\Delta t}{\Delta x}$ and $\mu := \frac{\Delta t}{\Delta y}$, are the usual fixed mesh-ratios),

$$\omega_{j,k}^{n+\frac{1}{2}} = \bar{\omega}_{j,k}^n - \frac{\lambda}{2}f_{j,k}' - \frac{\mu}{2}g_{j,k}^{\backslash} \left(+\Delta t\nu\nabla^2\bar{\omega}_{j,k}^n\right). \quad (2.129)$$

Equipped with these midvalues, we are now able to use the approximate fluxes which yield a second-order corrector step outlined in (2.134) below. Finally, we have to recover the velocity field from the computed values of vorticity. We end up with the following algorithm.

1. **Reconstruct**
   (a) An exact discrete divergence-free reconstruction of the velocity field. We define the discrete vorticity at the mid-cells as the average of the four corners of each cell, i.e.

   $$\omega_{j+\frac{1}{2},k+\frac{1}{2}} := \frac{1}{4}(\omega_{j+1,k+1} + \omega_{j,k+1} + \omega_{j,k} + \omega_{j+1,k}). \quad (2.130)$$

   We then use a streamfunction, ψ, such that $\Delta\psi = -\omega$, which is obtained in the min-cells, e.g., by solving the five-points Laplacian, $\Delta\psi_{j+\frac{1}{2},k+\frac{1}{2}} = -\omega_{j+\frac{1}{2},k+\frac{1}{2}}$. Then, its gradient, $\nabla\psi$ recovers the velocity field

   $$u_{j,k} = \mu_x\nabla_y\psi, \qquad v_{j,k} = -\mu_y\nabla_x\psi. \quad (2.131)$$

   Here, μ_x and μ_y denote averaging in the x-direction and in the y-direction, respectively, such that, e.g.,

   $$u_{j,k} = \frac{1}{2}\left(\psi_{j+\frac{1}{2},k+\frac{1}{2}} - \psi_{j+\frac{1}{2},k-\frac{1}{2}} + \psi_{j-\frac{1}{2},k+\frac{1}{2}} - \psi_{j-\frac{1}{2},k-\frac{1}{2}}\right).$$

Observe that with this integer indexed velocity field, we retain a discrete incompressibility relation, centered around $(j + \frac{1}{2}, k + \frac{1}{2})$,

$$\frac{< u_{j+1,\cdot} - u_{j,\cdot} >_{k+\frac{1}{2}}}{\Delta x} + \frac{< v_{\cdot,k+1} - v_{\cdot,k} >_{j+\frac{1}{2}}}{\Delta y} = 0, \qquad (2.132)$$

which is essential for the maximum principle in (2.5).

2. **Predict**

   (a) Prepare the pointvalues of the divergence-free velocity field , $u(\cdot, \cdot, t^n)$, from the reconstructed vorticity pointvalues, $w_{j,k}^n$. To this end, use the Biot-Savart solver (2.131);

   (b) Predict the midvalues of the vorticity, $w_{j,k}^{n+\frac{1}{2}}$,

$$\omega_{j,k}^{n+\frac{1}{2}} = \bar{\omega}_{j,k}^n - \frac{\lambda}{2} u_{j,k}^n \omega'_{j,k} - \frac{\mu}{2} v_{j,k}^n \omega_{j,k}^{\backprime}. \qquad (2.133)$$

Note: Observe that here we use the predictor step (2.129) in its convective formulation (2.127), that is, $(f', g^{\backprime}) = (u\omega', v\omega^{\backprime})$.

3. **Correct**

   (a) As in step (2a), use the previously calculated values of the vorticity to compute the divergence-free pointvalues of the velocity, at time $t^{n+\frac{1}{2}}$, $u(\cdot, \cdot, t^{n+\frac{1}{2}})$.

   (b) Finally, the previously calculated pointvalues of the velocities and vorticity are plugged into the second-order corrector step in order to compute the staggered cell-averages of the vorticity at time t^{n+1},

$$
\begin{aligned}
\bar{\omega}_{j+\frac{1}{2},k+\frac{1}{2}}^{n+1} = &< \frac{1}{4}(\bar{\omega}_{j,\cdot}^n + \bar{\omega}_{j+1,\cdot}^n) + \frac{1}{8}(\omega'_{j,\cdot} - \omega'_{j+1,\cdot}) >_{k+\frac{1}{2}} + \\
&- < \lambda((u\omega)_{j+1,\cdot}^{n+\frac{1}{2}} - (u\omega)_{j,\cdot}^{n+\frac{1}{2}}) >_{k+\frac{1}{2}} + \\
&+ < \frac{1}{4}(\bar{\omega}_{\cdot,k}^n + \bar{\omega}_{\cdot,k+1}^n) + \frac{1}{8}(\omega_{\cdot,k}^{\backprime} - \omega_{\cdot,k+1}^{\backprime}) >_{j+\frac{1}{2}} + \\
&- < \mu((v\omega)_{\cdot,k+1}^{n+\frac{1}{2}} - (v\omega)_{\cdot,k}^{n+\frac{1}{2}}) >_{j+\frac{1}{2}} .
\end{aligned}
\qquad (2.134)
$$

The specific recovery of the velocity field outlined above, retains the dual convective-conservative form of the vorticity variable, which in turn leads to the maximum principle [25].

$$\min_{\substack{|p-(j+\frac{1}{2})|=\frac{1}{2} \\ |q-(k+\frac{1}{2})|=\frac{1}{2}}} \{\bar{w}_{p,q}^n\} \leq \bar{w}_{j+\frac{1}{2},k+\frac{1}{2}}^{n+1} \leq \max_{\substack{|p-(j+\frac{1}{2})|=\frac{1}{2} \\ |q-(k+\frac{1}{2})|=\frac{1}{2}}} \{\bar{w}_{p,q}^n\}. \qquad (2.135)$$

As in the compressible case – compare (2.119), the main idea in [25] is to rewrite $\bar{\omega}_{j+\frac{1}{2},k+\frac{1}{2}}^{n+1}$ as a *convex* combination of the cell averages at t^n, $\bar{\omega}_{j,k}^n, \bar{\omega}_{j+1,k}^n, \bar{\omega}_{j,k+1}^n, \bar{\omega}_{j+1,k+1}^n$.

In figure 2.14 we show the central computation of a 'thin' shear-layer problem, [5]. For details, consult [25].

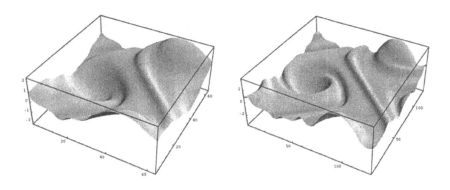

Fig. 2.13: t = 8 , 64*64   Fig. 2.14: t = 8 , 128*128

The "thin" shear-layer problem, solved by the second-order central scheme (2.129),(2.134) with spectral reconstruction of the velocity field.

The velocity formulation

Following [22] our goal is introduce a second-order central difference scheme for incompressible flows, based on *velocity variables*. The use of the velocity formulation yields a more versatile algorithm. The advantage of our proposed central scheme in its velocity formulation is two-fold: generalization to the three dimensional case is straightforward, and the treatment of boundary conditions associated with general geometries becomes simpler. The result is a simple fast high-resolution method, whose accuracy is comparable to that of an upwind scheme. In addition, numerical experiments show the new scheme to be immune to some of the well-known deleterious consequences of under-resolution.

We consider a two-dimensional incompressible flow field, $\mathbf{u} = (u, v)$, so that $\nabla \cdot \mathbf{u} = 0$. The equations of motion for a Newtonian fluid in conservation form are

$$u_t = \left[-u^2 + \nu u_x - p\right]_x + \left[-uv + \nu u_y\right]_y \equiv f^u(u, v, u_x, \dots)_x + g^u(u, v, u_x, \dots)_y$$

$$v_t = \left[-uv + \nu v_x\right]_x + \left[-v^2 + \nu v_y - p\right]_y \equiv f^v(u, v, u_x, \dots)_x + g^v(u, v, u_x, \dots)_y$$

$$\tag{2.136}$$

where p is the pressure, ν is the kinematic viscosity, and subscripts denote partial derivatives. The functions $f^{u,v}(\cdot)$ and $g^{u,v}(\cdot)$ are components of the fluxes of the conserved quantities u and v.

The computational grid consists of rectangular cells of sizes Δx and Δy; at time level $t^n = n\Delta t$, these cells, $C_{i,j}$, are centered at $(x_i = i\Delta x, y_j = j\Delta y)$. Starting with the corresponding cell averages, $\mathbf{u}^n = (u_{i,j}^n, v_{i,j}^n)$, we first reconstruct a piecewise linear polynomial approximation which recovers the point values of the velocity field, $\mathbf{u}^n(x,y) = (u^n(x,y), v^n(x,y))$. For second-order accuracy, the piecewise linear reconstructed velocities take the form,

$$\mathbf{u}^n(x,y) = \mathbf{u}_{i,j}^n + \frac{\mathbf{u}_{i,j}'}{\Delta x}(x - x_i) + \frac{\mathbf{u}_{i,j}^{\backslash}}{\Delta y}(y - y_j), \qquad x,y \in C_{i,j}. \tag{2.137}$$

As before, exact averaging over a staggered control volume yields

$$\tilde{u}_{i+\frac{1}{2},j+\frac{1}{2}}(t^{n+1}) = \fint_{C_{i+\frac{1}{2},j+\frac{1}{2}}} u(x,y,t^n)dxdy \tag{2.138}$$

$$+ \Delta t \left\{ D_x^+ \int_{\tau=t^n}^{t^{n+1}} \fint_{y \in J_{j+\frac{1}{2}}} f^u(x_i,y,\tau)dyd\tau \right\}$$

$$+ \Delta t \left\{ D_y^+ \int_{\tau=t^n}^{t^{n+1}} \fint_{x \in I_{i+\frac{1}{2}}} g^u(x,y_j,\tau)dxd\tau \right\},$$

and a similar averaging applies for $\tilde{v}_{i+\frac{1}{2},j+\frac{1}{2}}^{n+1}$.

An exact computation yields

$$\fint_{C_{i+\frac{1}{2},j+\frac{1}{2}}} u(x,y,t^n)dxdy = \mu_x^+ \mu_y^+ u_{i,j}^n - \frac{\Delta x}{8} D_x^+ \mu_y^+ w_{i,j}' - \frac{\Delta y}{8} D_y^+ \mu_x^+ w_{i,j}^{\backslash}. \tag{2.139}$$

The incompressible fluxes, e.g., $f^u = -u^2 + \nu u_x - p_x$, are approximated in terms of the midpoint rule , which in turn employs predicted midvalues which are obtained from half-step Taylor expansion. Thus our scheme starts with a *predictor step* of the form

$$u_{i,j}^{n+\frac{1}{2}} = u_{i,j}^n - \frac{\Delta t}{2}\left[2u_{i,j}^n \frac{w_{i,j}'}{\Delta x} + u_{i,j}^n \frac{v_{i,j}^{\backslash}}{\Delta y} + v_{i,j}^n \frac{w_{i,j}^{\backslash}}{\Delta y} + G_x p_{i,j}^n - \nu \nabla_h^2 u_{i,j}\right] \tag{2.140}$$

$$v_{i,j}^{n+\frac{1}{2}} = v_{i,j}^n - \frac{\Delta t}{2}\left[v_{i,j}^n \frac{w_{i,j}'}{\Delta x} + u_{i,j}^n \frac{v_{i,j}'}{\Delta x} + 2v_{i,j}^n \frac{v_{i,j}^{\backslash}}{\Delta y} + G_y p_{i,j}^n - \nu \nabla_h^2 v_{i,j}\right].$$

Note that the predictor step is nothing but a forward Euler scheme; conservation form is not essential for the spatial discretization at this stage.

This is followed by a *corrector step*

$$\left(1 - \frac{\nu \Delta t}{2} \nabla_h^2\right) \tilde{\mathbf{u}}_{i+\frac{1}{2},j+\frac{1}{2}}^{n+1} = \mu_x^+ \mu_y^+ \mathbf{u}_{i,j}^n - \frac{\Delta x}{8} D_x^+ \mu_y^+ \mathbf{u}'_{i,j} - \frac{\Delta y}{8} D_y^+ \mu_x^+ \mathbf{u}^{\backslash}_{i,j} +$$

$$(2.141)$$

$$- \Delta t D_x^+ \mu_y^+ \left[u_{i,j}^{n+\frac{1}{2}} u_{i,j}^{n+\frac{1}{2}} - \frac{\nu \mathbf{u}'_{i,j}}{2 \Delta x}\right] +$$

$$- \Delta t D_y^+ \mu_x^+ \left[v_{i,j}^{n+\frac{1}{2}} u_{i,j}^{n+\frac{1}{2}} - \frac{\nu \mathbf{u}^{\backslash}_{i,j}}{2 \Delta y}\right].$$

Note that the viscous terms are handled here by the implicit Crank-Nicholson discretization which is favored due to its preferable stability properties. Here, we ignore the pressure terms; instead, the contribution of the pressure will be integrated by enforcing zero-divergence fluxes at the last *projection step*.

Compute the potential $\varphi_{i,j}$ solving the Poisson equation

$$\left[D_x^+ D_x^- \mu_y^+ \mu_y^- + D_y^+ D_y^- \mu_x^+ \mu_x^-\right]\varphi_{i,j} = \frac{1}{\Delta t}\left[D_x^- \mu_y^- \tilde{u}_{i+\frac{1}{2},j+\frac{1}{2}}^{n+1} + D_y^- \mu_x^- \tilde{v}_{i+\frac{1}{2},j+\frac{1}{2}}^{n+1}\right].$$

$$(2.142)$$

Then, the pressure gradient at t^{n+1} is being updated,

$$G_x p_{i+\frac{1}{2},j+\frac{1}{2}}^{n+1} := D_x^+ \mu_y^+ \varphi_{i,j}, \qquad G_y p_{i+\frac{1}{2},j+\frac{1}{2}}^{n+1} := D_y^+ \mu_x^+ \varphi_{i,j}, \qquad (2.143)$$

and finally, it is used to evaluate the divergence-free velocity field, $\mathbf{u}^{n+1}$

$$\mathbf{u}_{i+\frac{1}{2},j+\frac{1}{2}}^{n+1} = \tilde{\mathbf{u}}_{i+\frac{1}{2},j+\frac{1}{2}}^{n+1} - \Delta t \mathbf{G}_\times p_{i+\frac{1}{2},j+\frac{1}{2}}^{n+1}. \qquad (2.144)$$

In Figure 2.15, we plot vorticity contours for two shear layer problems studied in [5]: the inviscid "thick" shear layer problem corresponding to (u_0^ρ, v_0^δ) with $\rho = 30$, and a viscous "thin" shear layer problem (with $\nu = 5 \cdot 10^{-5}$), corresponding to (u_0^ρ, v_0^δ) with $\rho = 100$. As in [5], both plots in Figures 2.15a and 2.15b are recorded at time $t = 1.2$, and are subject to an initial perturbation v_0^δ, with $\delta = 0.05$.
Further applications of the central schemes for more complex incompressible flows (with 'variable' axisymmetric coefficients, forcing source/viscous terms, ...), can be found in [20],[21].

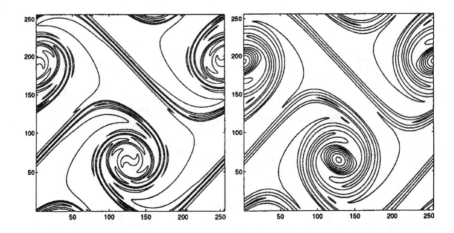

Fig. 2.15: Contour lines of the vorticity, $\omega = v_x - u_y$, at $t = 1.2$ with initial (u^ρ, v^δ), $\delta = 0.05$, using a 256×256 grid. (a) A "thick" shear layer with $\rho = 30$, and $\nu = 0$. The contour levels range from -36 to 36 (cf. Figure 3c in Ref. [5]). (b) A "thin" shear layer with $\rho = 100$, and $\nu = 5 \cdot 10^{-5}$. The contour levels range from -70 to 70 (cf. Figure 9b in Ref. [5]).

References

1. P. ARMINJON, D. STANESCU & M.-C. VIALLON, *A Two-Dimensional Finite Volume Extension of the Lax-Friedrichs and Nessyahu-Tadmor Schemes for Compressible Flow*, (1995), preprint.

2. P. ARMINJON, D. STANESCU & M.-C. VIALLON, *A two-dimensional finite volume extension of the Lax-Friedrichs and Nessyahu-Tadmor schemes for compressible flows*, Preprint.

3. P. ARMINJON & M.-C. VIALLON, *Généralisation du Schéma de Nessyahu-Tadmor pour Une Équation Hyperbolique à Deux Dimensions D'espace*, C.R. Acad. Sci. Paris, **t. 320** , série I. (1995), pp. 85-88.

4. F. BEREUX & L. SAINSAULIEU, *A Roe-type Riemann Solver for Hyperbolic Systems with Relaxation Based on Time-Dependent Wave Decomposition*, Numer. Math,. **77**, (1997), pp. 143-185.

5. D. L. BROWN & M. L. MINION *Performance of under-resolved two-dimensional incompressible flow simulations*, J. Comp. Phys. **122**, (1985) 165-183.

6. P. COLELLA & P. WOODWARD, *The piecewise parabolic method (PPM) for gas-dynamical simulations*, JCP **54**, 1984, pp. 174-201.

7. B. ENGQUIST & O. RUNBORG, *Multi-phase computations in geometrical optics*, J. Comp. Appl. Math., 1996, in press.

8. ERBES, *A high-resolution Lax-Friedrichs scheme for Hyperbolic conservation laws with source term. Application to the Shallow Water equations.* Preprint.

9. K.O. FRIEDRICHS & P.D. LAX, *Systems of Conservation Equations with a Convex Extension*, Proc. Nat. Acad. Sci., **68**, (1971), pp.1686-1688.

10. E. GODLEWSKI & P.-A. RAVIART, *Hyperbolic Systems of Conservation Laws*, Mathematics & Applications, Ellipses, Paris, 1991.

11. S.K. GODUNOV, *A finite difference method for the numerical computation of discontinuous solutions of the equations of fluid dynamics*, Mat. Sb. **47**, 1959, pp. 271-290.

12. A. HARTEN, *High Resolution Schemes for Hyperbolic Conservation Laws*, JCP, **49**, (1983), pp.357-393.

13. A. HARTEN, B. ENGQUIST, S. OSHER & S.R. CHAKRAVARTHY, *Uniformly high order accurate essentially non-oscillatory schemes. III*, JCP **71**, 1982, pp. 231-303.

14. H.T. HUYNH, *A piecewise-parabolic dual-mesh method for the Euler equations*, AIAA-95-1739-CP, The 12th AIAA CFD Conf., 1995.

15. G.-S. JIANG, D. LEVY, C.-T. LIN, S. OSHER & E. TADMOR, *High-resolution Non-Oscillatory Central Schemes with Non-Staggered Grids for Hyperbolic Conservation Laws*, SIAM Journal on Num. Anal., to appear.

16. G.-S JIANG & E. TADMOR, *Nonoscillatory Central Schemes for Multidimensional Hyperbolic Conservation Laws*, SIAM J. Scie. Comp., to appear.

17. S. JIN, private communication.

18. S. JIN AND Z. XIN, *The relaxing schemes for systems of conservation laws in arbitrary space dimensions*, Comm. Pure Appl. Math. 48 (1995) 235–277.

19. B. VAN LEER, *Towards the Ultimate Conservative Difference Scheme, V. A Second-Order Sequel to Godunov's Method*, JCP, **32**, (1979), pp.101-136.

20. R. KUPFERMAN, *Simulation of viscoelastic fluids: Couette-Taylor flow*, J. Comp. Phys., to appear.

21. R. KUPFERMAN, *A numerical study of the axisymmetric Couette-Taylor problem using a fast high-resolution second-order central scheme*, SIAM. J. Sci. Comp., to appear.

22. R. KUPFERMAN & E. TADMOR, *A Fast High-Resolution Second-Order Central Scheme for Incompressible Flow s*, Proc. Nat. Acad. Sci. 94 (1997), 4848–4852

23. R.J. LEVEQUE, *Numerical Methods for Conservation Laws*, Lectures in Mathematics, Birkhauser Verlag, Basel, 1992.

24. D. LEVY, *Third-order 2D Central Schemes for Hyperbolic Conservation Laws*, in preparation.

25. D. LEVY & E. TADMOR, *Non-oscillatory Central Schemes for the Incompressible 2-D Euler Equations*, Math. Res. Let., **4**, (1997), pp.321-340.

26. D. LEVY & E. TADMOR, *Non-oscillatory boundary treatment for staggered central schemes*, preprint.

27. X.-D. LIU & P. D. LAX, *Positive Schemes for Solving Multi-dimensional Hyperbolic Systems of Conservation Laws, Courant Mathematics and Computing Laboratory Report*, Comm. Pure Appl. Math.

28. X.-D. LIU & S. OSHER, *Nonoscillatory High Order Accurate Self-Similar Maximum Principle Satisfying Shock Capturing Schemes I*, SINUM, **33**, no. 2 (1996), pp.760-779.

29. X.-D. LIU & E. TADMOR, *Third Order Nonoscillatory Central Scheme for Hyperbolic Conservation Laws*, Numer. Math., to appear.

30. H. NESSYAHU, *Non-oscillatory second order central type schemes for systems of nonlinear hyperbolic conservation laws*, M.Sc. Thesis, Tel-Aviv University, 1987.

31. H. NESSYAHU & E. TADMOR, *Non-oscillatory Central Differencing for Hyperbolic Conservation Laws*, JCP, **87**, no. 2 (1990), pp.408-463.

32. H. NESSYAHU, E. TADMOR & T. TASSA, *On the convergence rate of Godunov-type schemes*, SINUM **31**, 1994, pp. 1-16.

33. S. OSHER & E. TADMOR, *On the Convergence of Difference Approximations to Scalar Conservation Laws*, Math. Comp., **50**, no. 181 (1988), pp.19-51.

34. P. L. ROE, *Approximate Riemann Solvers, Parameter Vectors, and Difference Schemes*, JCP, **43**, (1981), pp.357-372.

35. A. ROGERSON & E. MEIBURG, *A numerical study of the convergence properties of ENO schemes*, J. Sci. Comput., **5**, 1990, pp. 127-149.

36. O. RUNBORG, *Multiphase Computations in Geometrical Optics*, UCLA CAM report no. 96-52 (1996).

37. V. ROMANO & G. RUSSO, *Numerical solution for hydrodynamical models of semiconductors*, IEEE, to appear.

38. A.M. ANILE, V. ROMANO & G. RUSSO, *Extended hydrodymnamical model of carrier transport in semiconductors*, Phys. Rev. B., to appear.

39. F. BIANCO, G. PUPPO & G. RUSSO, *High order central schemes for hyperbolic systems of conservation laws*, SIAM J. Sci. Comp., to appear.

40. R. SANDERS, *A Third-order Accurate Variation Nonexpansive Difference Scheme for Single Conservation Laws*, Math. Comp., **41** (1988), pp.535-558.

41. R. SANDERS R. & A. WEISER, *A High Resolution Staggered Mesh Approach for Nonlinear Hyperbolic Systems of Conservation Laws*, JCP, **1010** (1992), pp.314-329.

42. P. K. SWEBY, *High Resolution Schemes Using Flux Limiters for Hyperbolic Conservation Laws*, SINUM, **21**, no. 5 (1984), pp.995-1011.

43. C.-W. SHU, *Numerical experiments on the accuracy of ENO and modified ENO schemes*, JCP **5**, 1990, pp. 127-149.

44. G. SOD, *A survey of several finite difference methods for systems of nonlinear hyperbolic conservation laws*, JCP **22**, 1978, pp. 1-31.

45. E. TADMOR & C.C. WU, *Central Scheme for the Multidimensional MHD Equations*, in preparation.

46. P. WOODWARD & P. COLELLA, *The numerical simulation of two-dimensional fluid flow with strong shocks*, JCP **54**, 1988, pp. 115-173.

3 The Spectral Viscosity Method

3.1 Introduction

Let P_N stands for one of the standard spectral projections — Fourier, Chebyshev, Legendre .... It is well known that such spectral projections, $P_N u$, provide highly accurate approximations for sufficiently smooth u's. This superior accuracy is destroyed if u contains discontinuities. Indeed, $P_N u$ produces $\mathcal{O}(1)$ Gibbs' oscillations in the *local* neighborhoods of the discontinuities, and moreover, their *global* accuracy deteriorates to first-order.

We are interested in spectral approximations of nonlinear conservation laws

$$\frac{\partial u}{\partial t} + \frac{\partial}{\partial x} f(u) = 0, \tag{3.145}$$

subject to initial conditions, $u(x, 0) = u_0$, and augmented with appropriate boundary conditions. The purpose of a spectral method is to compute an approximation to the *projection* of $u(\cdot, t)$ rather than $u(\cdot, t)$ itself. Consequently, since nonlinear conservation laws exhibit spontaneous shock discontinuities, the spectral approximation faces two difficulties:

Stability. Numerical tests indicate that the convergence of spectral approximations to nonlinear conservation laws fails. In [26]–[28] we prove[5] that this failure is related to the fact that spurious Gibbs oscillations pollute the entire computational domain, and that the lack of entropy dissipation then renders these spectral approximations unstable.

Accuracy. The accuracy of the spectral computation is limited by the first order convergence rate of $P_N u(\cdot, t)$.

With this in mind we turn to discuss the Spectral Viscosity (SV) method introduced in [26]. Our discussion focuses on three aspects: the periodic Fourier SV method in both – one and several space dimensions and the nonperiodic Legendre SV method.

In §3.2 we begin with the one-dimensional periodic problems. The purpose of the SV method is to stabilize the nonlinear spectral approximation without sacrificing its underlying spectral accuracy. This is achieved by augmenting the standard spectral approximation with high frequency regularization. In §3.2 we briefly review the convergence results of the periodic Fourier SV method, [26]–[29], [17], [5], [21]. These convergence results employ high frequency regularization based on *second order* viscosity. In §3.2 we discuss spectral approximations based on "super-viscosity", i.e., high-frequency parabolic regularizations of order > 2. These 'super' spectral viscosities were introduced and analyzed in [30]. Extensions of the spectral super viscosity to non-periodic problems was presented in [13]. We prove the H^{-1}-stability of these spectral

---

[5] Consult the counterexamples in the introductory section of Lecture IV below.

'super-viscosity' approximations, and together with L^∞-stability, convergence follows by compensated compactness arguments [31],[16].

In §3.3 we turn to the nonperiodic case and discuss the Legendre SV method, [18]. Extensions to and applications with Chebyshev SV method can be found in [12],[2],[15]. Finally, the multidimensional problem is treated in §3.5, along the lines of [5].

We close this introduction by referring to the numerical experiments in §3.4 quoted from [18]; see also [28]. These numerical tests show that by *post-processing* the spectral (super)-viscosity approximation, the exact entropy solution is recovered within spectral accuracy. This post-processing is carried out as a highly accurate mollification and operated either in the physical space as in [10],[1],[18], or in the dual Fourier space as in [11],[19],[32]. It should be emphasized that the role of post-processing is essential in order to realize the highly accurate content of the SV solution.

For further applications in two- and three-dimensional atmospheric simulations we refer to [2],[15],[7] and the references therein.

3.2 The Fourier Spectral Viscosity (SV) method

To solve the periodic conservation law (3.145) by a spectral method, one employs an N-degree trigonometric polynomial

$$u_N(x,t) = \sum_{|k| \leq N} \widehat{u}_k(t) e^{ikx} , \qquad (3.146)$$

in order to approximate the Fourier projection of the exact entropy solution, $P_N u$.[6] Starting with $u_N(x,0) = P_N u_0(x)$, the classical spectral method lets $u_N(x,t)$ evolve according to the approximate model

$$\frac{\partial u_N}{\partial t} + \frac{\partial}{\partial x} \big[P_N \big(f(u_N) \big) \big] = 0 . \qquad (3.147)$$

As we have already noted, the convergence of u_N towards the entropy solution of (3.145), $u_N \xrightarrow[N \to \infty]{} u$, may fail, [26]. Instead, we modify (3.147) by augmenting it with high frequency viscosity regularization which amounts to

$$\frac{\partial u_N}{\partial t} + \frac{\partial}{\partial x} \big[P_N f(u_N(x,t)) \big] = \varepsilon_N (-1)^{s+1} \frac{\partial^s}{\partial x^s} \left[Q_m(x,t) * \frac{\partial^s u_N}{\partial x^s} \right], \quad s \geq 1. \qquad (3.148_s)$$

---

[6] The spectral Fourier projection of $u(x)$ is given by $\sum_{|k| \leq N} (u, e^{ikx}) e^{ikx}$; the pseudospectral Fourier projection of $u(x)$ is given by $\sum_{|k| \leq N} < u, e^{ikx} > e^{ikx}$, where $< u, e^{ikx} > := \Delta x \sum_\nu u(x_\nu) e^{-ikx_\nu}$ is collocated at the $2N+1$ equidistant gridvalues $x_\nu = 2\pi\nu\Delta x$. $P_N u$ denotes either one of these two projections.

This kind of *spectral viscosity* can be efficiently implemented in Fourier space as

$$\varepsilon_N \frac{\partial^s}{\partial x^s}\left[Q_m(x,t) * \frac{\partial^s u_N}{\partial x^s}\right] := \varepsilon \sum_{m<|k|\leq N} (ik)^{2s}\widehat{Q}_k(t)\widehat{u}_k(t)e^{ikx} . \qquad (3.149)$$

It involves the following three ingredients:

– the viscosity amplitude, $\varepsilon = \varepsilon_N$,

$$\varepsilon \equiv \varepsilon_N \sim \frac{2C_s}{N^{2s-1}}; \qquad (3.150)$$

Here, C_s is a constant which may depend on the fixed order of super-viscosity, s. (A pessimistic upper bound of this constant will be specified below — consult [5, Theorem 2.1]).

– the effective size of the *inviscid* spectrum, $m = m_N$,

$$m \equiv m_N \sim N^\theta, \qquad \theta < \frac{2s-1}{2s}; \qquad (3.151)$$

– the SV smoothing factors, $\widehat{Q}_k(t)$, which are activated only on high wavenumbers, $|k| > m_N$, satisfying

$$1 - \left(\frac{m}{|k|}\right)^{\frac{2s-1}{\theta}} \leq \widehat{Q}_k(t) \leq 1, \quad |k| > m_N. \qquad (3.152)$$

The SV method can be viewed as a compromise between the total-variation stable viscosity approximation – see (3.154) and (3.160$_s$) below – which is restricted to first order accuracy (corresponding to $\theta = 0$), and the spectrally accurate yet unstable spectral method (3.147) (corresponding to $\theta = 1$). The additional SV on the right of (3.148$_s$) is small enough to retain the *formal* spectral accuracy of the underlying spectral approximation, i.e., the following estimate holds

$$\left\|\varepsilon_N \frac{\partial^{s+p}}{\partial x^{s+p}}\left[Q_m(x,t) * \frac{\partial^s u_N}{\partial x^s}\right]\right\|_{L^2(x)} \leq \mathrm{Const} \cdot N^{-\theta(q-p-1)}\left\|\frac{\partial^q u_N}{\partial x^q}\right\|_{L^2(x)},$$

$$\forall q \geq p+1 > -\infty. \qquad (3.153)$$

At the same time this SV is shown in §3 & 4 to be large enough so that it enforces a sufficient amount of entropy dissipation, and hence — by compensated compactness arguments — [31],[16], to prevent the unstable spurious Gibbs' oscillations.

The Fourier SV method – 2nd order viscosity

The unique entropy solution of the scalar conservation law (3.145) is the one which is realized as the vanishing viscosity solution, $u = \lim_{\varepsilon \downarrow 0} u^\varepsilon$, where u^ε satisfies the standard viscosity equation

$$\frac{\partial u^\varepsilon}{\partial t} + \frac{\partial}{\partial x} f(u^\varepsilon(x,t)) = \varepsilon \frac{\partial^2}{\partial x^2} u^\varepsilon(x,t). \tag{3.154}$$

This section provides a brief review of the convergence results for the Fourier SV method (3.148_s) with $s = 1$. The convergence analysis is based on the close resemblance of the Fourier SV method (3.148_s) with $s = 1$ to the usual viscosity regularization (3.154). To quantify this similarity we rewrite (3.148_s) with $s = 1$ in the equivalent form

$$\frac{\partial u_N}{\partial t} + \frac{\partial}{\partial x} f(u_N(x,t)) = \tag{3.155}$$

$$= \varepsilon_N \frac{\partial^2 u_N}{\partial x^2} - \varepsilon_N \frac{\partial}{\partial x} \left[R_N(x,t) * \frac{\partial u_N}{\partial x} \right] + \frac{\partial}{\partial x} (I - P_N) f(u_N),$$

where

$$R_N(x,t) := \sum_{k=-N}^{N} \hat{R}_k(t) e^{ikx}, \quad \hat{R}_k(t) \equiv \begin{cases} 1 & |k| < m_N, \\ 1 - \hat{Q}_k(t) & |k| \geq m_N. \end{cases} \tag{3.156}$$

Observe that the SV approximation in (3.155) contains two additional modifications to the standard viscosity approximation in (3.154).

{i} The second term on the right of (3.155) measures the difference between the spectral viscosity, $\varepsilon_N \frac{\partial}{\partial x} \left[Q_m(x,t) * \frac{\partial u_N}{\partial x} \right]$, and the standard vanishing viscosity, $\varepsilon_N \frac{\partial^2 u_N}{\partial x^2}$. The following straightforward estimate shows this difference to be L^2-bounded $\forall \theta < \frac{1}{2}$,

$$\left\| \varepsilon_N \frac{\partial}{\partial x} \left[R_N(\cdot,t) * \frac{\partial u_N}{\partial x} \right] \right\|_{L^2} \leq \tag{3.157}$$

$$\leq \text{Const} \cdot \varepsilon_N \times \left[m_N^{1/\theta} \max_{|k| > m_N} |k|^{2-1/\theta} + m_N^2 \right] \|u_N(\cdot,t)\|_{L^2}$$

$$\leq \text{Const} \cdot N^{2\theta-1} \|u_N(\cdot,t)\|_{L^2}$$

$$\leq \text{Const} \cdot \|u_N(\cdot,t)\|_{L^2}, \qquad \forall \theta \leq \frac{1}{2}.$$

{ii} The spectral projection error contained in the third term on the right of (3.155) does not exceed

$$\|(I - P_N)f(u_N(\cdot,t))\|_{L^2} \le \text{Const}\frac{1}{N}\|\frac{\partial}{\partial x}u_N(\cdot,t)\|_{L^2}. \qquad (3.158)$$

Equipped with the last two estimates one concludes the standard entropy dissipation bound, [26], [17], [28], [5],

$$\|u_N(\cdot,t)\|_{L^2} + \sqrt{\varepsilon_N}\|\frac{\partial u_N}{\partial x}\|_{L^2_{loc}(x,t)} \le \text{Const}, \qquad \varepsilon_N \sim \frac{1}{N}. \qquad (3.159)$$

The inequality (3.159) is the usual statement of entropy stability familiar from the standard viscosity setup (3.154). For the L^∞-stability of the Fourier SV approximation consult e.g. [17],[27, §5] and [5, §4] for the one- and re-spectively, multi-dimensional problems. The convergence of the SV method then follows by compensated compactness arguments, [31],[16].

We note in passing that the the Fourier SV approximation (3.148$_s$), (3.150-(3.151) shares other familiar properties of the standard viscosity approximation (3.154), e.g., total variation boundedness, Oleinik's one-sided Lipschitz regularity (for $\theta < \frac{1}{3}$), L^1-convergence rate of order one-half, [21],[28].

Fourier SV method revisited – super viscosity

In this section we remove the restriction $\theta < \frac{1}{2}$ in (3.151), which limits the portion of the inviscid spectrum. The key is to replace the standard second-order viscosity regularization (3.154) with the "super-viscosity" regularization

$$\frac{\partial u^\varepsilon}{\partial t} + \frac{\partial}{\partial x}f(u^\varepsilon(x,t)) = \varepsilon(-1)^{s+1}\frac{\partial^{2s}}{\partial x^{2s}}u^\varepsilon(x,t). \qquad (3.160_s)$$

The convergence analysis of the spectral "super-viscosity" method (3.148$_s$) is linked to the behavior of the "super-viscosity" regularization (3.160$_s$). To this end we rewrite (3.148$_s$) in the equivalent form

$$\frac{\partial u_N}{\partial t} + \frac{\partial}{\partial x}f(u_N(x,t)) = \varepsilon_N(-1)^{s+1}\frac{\partial^{2s}u_N}{\partial x^{2s}}+$$

$$+\varepsilon_N\frac{(-\partial)^s}{\partial x^s}\left[R_N(x,t) * \frac{\partial^s u_N}{\partial x^s}\right] + \frac{\partial}{\partial x}(I - P_N)f(u_N) = \qquad (3.161)$$

$$:= \mathcal{I}_1(u_N) + \mathcal{I}_2(u_N) + \mathcal{I}_3(u_N).$$

As before, we observe that the second and third terms on the right of (3.161), $\mathcal{I}_2(u_N)$ and $\mathcal{I}_3(u_N)$, are the two additional terms which distinguish the spectral "super-viscosity" approximation (3.161) from the super-viscosity regularization (3.160$_s$). In the sequel we shall use the following upper-bounds on these two terms.

{i} The second term, $\mathcal{I}_2(u_N)$, measures the difference between the SV regularization in (3.161) and the "super-viscosity" in (3.160_s). Using the SV parameterization in (3.152), (3.151) and (3.150) (in this order), we find that this difference does not exceed

$$\|\varepsilon_N \frac{(-\partial)^s}{\partial x^s}\left[R_N(\cdot,t) * \frac{\partial^s u_N}{\partial x^s}\right]\|_{L^2} \leq \tag{3.162}$$

$$\leq \varepsilon_N \left[m_N^{2s} + m_N^{\frac{2s-1}{\theta}} \max_{|k|>m_N} |k|^{2s-\frac{2s-1}{\theta}}\right] \|u_N(\cdot,t)\|_{L^2}$$

$$\leq \text{Const} \cdot N^{2s\theta-2s+1} \|u_N(\cdot,t)\|_{L^2}$$

$$\leq \text{Const} \cdot \|u_N(\cdot,t)\|_{L^2}, \qquad \forall \theta \leq \frac{2s-1}{2s}.$$

Thus, the second term on the right of (3.161), $\mathcal{I}_2(u_N)$, is L^2-bounded :

$$\|\mathcal{I}_2(u_N)\|_{L^2(x)} \leq \text{Const}\|u_N(\cdot,t)\|_{L^2(x)}. \tag{3.163}$$

{ii} Regarding the third term, $\mathcal{I}_3(u_N)$, we shall make a frequent use of the spectral estimate which we quote from [5, §2.3], stating that[7],

$$\|\frac{\partial^p}{\partial x^p}(I - P_N)f(u_N(\cdot,t))\|_{L^2} \leq \tag{3.164}$$

$$\leq C_q \frac{1}{N^{q-p}} \|\frac{\partial^q}{\partial x^q} u_N(\cdot,t)\|_{L^2}, \qquad \forall q \geq p > -\infty, \; q > \frac{1}{2}.$$

(The restriction $q > \frac{1}{2}$ is required only for the *pseudospectral* Fourier projection, P_N, whose truncation estimate in provided in e.g., [25, Lemma 2.2]). An upper bound on the constants C_s appearing on the right of (3.164) is given by [5, Theorem 7.1]

$$C_s \sim \sum_{k=1}^{s} \|f(\cdot)\|_{C^k} \|u_N\|_{L^\infty}^{k-1}; \tag{3.165}$$

this estimate may serve as a (pessimistic) bound for the same constant used in conjunction with the viscosity amplitude, ε_N, in (3.150).

Next we turn to the behavior of the quadratic entropy of the SV solution, $U(u_N) = \frac{1}{2}u_N^2$. (A similar treatment applies to general convex entropy func-

---

[7] As usual we let $\partial_x^p w(x) := \sum_{k \neq 0}(ik)^p \hat{w}(k)e^{ikx}$. Note that if $\int w(x)dx = 0$ then $\partial_x^p w(x)$ with $p < 0$ coincides with the $|p|$-th order primitive of $w(x)$.

tions $U(u_N)$.) Multiplication of (3.161) by u_N implies

$$\frac{1}{2}\frac{\partial}{\partial t}u_N^2 + \frac{\partial}{\partial x}\int^{u_N} \xi f'(\xi)d\xi =$$

$$= u_N \mathcal{I}_1(u_N) + u_N \mathcal{I}_2(u_N) + u_N \mathcal{I}_3(u_N) = \quad (3.166)$$

$$:= II_1(u_N) + II_2(u_N) + II_3(u_N).$$

The three expressions on the right (3.166) represent the quadratic entropy dissipation + production of the SV method. Successive "differentiation by parts" enable us to rewrite the first expression as

$$II_1(u_N) \equiv \quad (3.167)$$

$$\equiv \varepsilon_N \sum_{\substack{p+q=2s-1 \\ 0 \leq p < s}} (-1)^{s+p+1}\frac{\partial}{\partial x}\left[\frac{\partial^p u_N}{\partial x^p}\frac{\partial^q u_N}{\partial x^q}\right] - \varepsilon_N \left(\frac{\partial^s u_N}{\partial x^s}\right)^2$$

$$:= II_{11}(u_N) + II_{12}(u_N).$$

Similarly, the second expression can be rewritten as

$$II_2(u_N) \equiv \quad (3.168)$$

$$\equiv \varepsilon_N \sum_{p+q=s-1} (-1)^{s+p}\frac{\partial}{\partial x}\left(\frac{\partial^p u_N}{\partial x^p}\left[\frac{\partial^q R_N(x,t)}{\partial x^q} * \frac{\partial^s u_N}{\partial x^s}\right]\right) +$$

$$+\varepsilon_N \frac{\partial^s u_N}{\partial x^s}R_N(x,t) * \frac{\partial^s u_N}{\partial x^s}$$

$$:= II_{21}(u_N) + II_{22}(u_N).$$

Finally, we have for the third expression

$$II_3(u_N) \equiv \quad (3.169)$$

$$\equiv \sum_{p=0}^{s-1}(-1)^p\frac{\partial}{\partial x}\left[\frac{\partial^p u_N}{\partial x^p}\frac{\partial^{-p}}{\partial x^{-p}}(I - P_N)f(u_N)\right] +$$

$$+(-1)^s\frac{\partial^s u_N}{\partial x^s}\frac{\partial^{-s+1}}{\partial x^{-s+1}}(I - P_N)f(u_N)$$

$$:= II_{31}(u_N) + II_{32}(u_N).$$

We arrive at the following entropy estimate which plays an essential role in the convergence analysis of the SV method.

Lemma 3.1 *Entropy dissipation estimate* There exists a constant, Const $\sim \|u_N(\cdot, 0)\|_{L^2}$, (but otherwise is independent of N), such that the following estimate holds

$$\|u_N(\cdot, t)\|_{L^2} + \sqrt{\varepsilon_N} \|\frac{\partial^s u_N}{\partial x^s}\|_{L^2_{loc}(x,t)} \leq \text{Const}, \qquad \varepsilon_N = \frac{2C_s}{N^{2s-1}}. \qquad (3.170)$$

Remark 3.1 Observe that the entropy dissipation estimate in (3.170) is considerably *weaker* in the "super-viscosity" case where $s > 1$, than in the standard viscosity regularization, $s = 1$ quoted in (3.159).

Proof. Spatial integration of (3.166) yields

$$\frac{1}{2}\frac{d}{dt}\|u_N(\cdot, t)\|_{L^2}^2 + \varepsilon_N \|\frac{\partial^s}{\partial x^s} u_N(\cdot, t)\|_{L^2}^2 = (u_N, \mathcal{I}_2(u_N))_{L^2(x)} + (u_N, \mathcal{I}_3(u_N))_{L^2(x)}. \qquad (3.171)$$

According to (3.163), the first expression on the right of the last inequality does not exceed

$$|(u_N, \mathcal{I}_2(u_N))_{L^2}| \leq \text{Const} \cdot \|u_N(\cdot, t)\|_{L^2}^2. \qquad (3.172)$$

According to (4.6c), the second expression on the right$= (-1)^s \frac{\partial^s u_N}{\partial x^s} \frac{\partial^{-s+1}}{\partial x^{-s+1}} (I - P_N) f(u_N)$, and by (3.164) it does not exceed

$$|(u_N, \mathcal{I}_3(u_N))_{L^2}| \leq \|\frac{\partial^s u_N}{\partial x^s}\|_{L^2} \cdot \frac{C_s}{N^{2s-1}} \|\frac{\partial^s u_N}{\partial x^s}\|_{L^2} \leq \qquad (3.173)$$

$$\leq \frac{1}{2}\varepsilon_N \|\frac{\partial^s}{\partial x^s} u_N(\cdot, t)\|_{L^2}^2.$$

(In fact, in the spectral case, the second expression vanishes by orthogonality). The result follows from Gronwall's inequality. $\blacksquare$

Equipped with Lemma 3.1 we now turn to the main result of this section, stating

Theorem 3.1 *Convergence* Consider the Fourier "super-viscosity" approximation (3.148_s)-(3.152), subject to L^∞-initial data, $u_N(\cdot, 0)$. Then uniformly bounded u_N converges to the unique entropy solution of the convex conservation law (3.145).

Proof. We proceed in three steps.

Step 1. (L^∞-stability). The L^∞-stability for spectral viscosity of 2nd order, $s = 1$, follows by L^p-iterations along the lines of [17] and [5], (we omit the details). The issue of an L^∞ bound for spectral viscosity of 'super' order $s > 1$ remains an open question. The intricate part of this question could be traced to the fact that already the underlying super-viscosity regularization (3.160_s), *lacks* monotonicity for $s > 1$: instead, it exhibits additional oscillations which are added to the spectral Gibbs' oscillations (Both types of oscillations are post-processed without sacrificing neither stability nor spectral accuracy).

Step 2. (H^{-1}-stability). We want to show that both — the local error on the right hand-side of (3.161), $\sum_{1 \leq j \leq 3} \mathcal{I}_j(u_N)$, and the *quadratic* entropy dissipation + production on the right of (3.166), $\sum_{1 \leq j \leq 3} \mathcal{II}_j(u_N)$, belong to a compact subset of $H_{loc}^{-1}(x, t)$.

To this end we first prepare the following. Bernstein's inequality gives us $\forall p < s \leq q$

$$\left\| \varepsilon_N \left[\frac{\partial^p u_N}{\partial x^p} \frac{\partial^q u_N}{\partial x^q} \right] \right\|_{L_{loc}^2(x,t)} \leq \text{Const} \cdot \varepsilon_N \left\| \frac{\partial^p u_N}{\partial x^p} \right\|_{L^\infty} \cdot \left\| \frac{\partial^q u_N}{\partial x^q} \right\|_{L_{loc}^2(x,t)} \leq$$

$$\ldots \text{by Bernstein inequality} \ldots \leq \text{Const} \cdot \varepsilon_N \cdot N^p \|u_N\|_{L^\infty} \times$$

$$\times N^{q-s} \left\| \frac{\partial^s u_N}{\partial x^s} \right\|_{L_{loc}^2(x,t)} \leq$$

$$\ldots \text{by Lemma 3.1} \ldots \leq \text{Const} \cdot \sqrt{\varepsilon_N} \cdot N^{p+q-s} \|u_N\|_{L^\infty} \sim$$

$$\sim \sqrt{2\mathcal{C}_s} \cdot N^{p+q-2s+\frac{1}{2}} \cdot \|u_N\|_{L^\infty}. \qquad (3.174)$$

Consider now the first two expressions, $\mathcal{I}_1(u_N)$ and $\mathcal{II}_1(u_N)$. The inequality (3.174) with $(p, q) = (0, 2s - 1)$ implies that $\mathcal{I}_1(u_N)$ tends to zero in $H_{loc}^{-1}(x, t)$, for

$$\|\mathcal{I}_1(u_N)\|_{H_{loc}^{-1}(x,t)} \leq \text{Const} \cdot \sqrt{2\mathcal{C}_s/N} \cdot \|u_N\|_{L^\infty} \to 0. \qquad (3.175)$$

We turn now to the expression $\mathcal{II}_1(u_N)$ in (3.167): its first half tends to zero in $H_{loc}^{-1}(x,t)$, for by (3.174) we have $\forall p + q = 2s - 1$,

$$\|\mathcal{II}_{11}(u_N)\|_{H_{loc}^{-1}(x,t)} \equiv \tag{3.176}$$

$$\equiv \varepsilon_N \| \cdot \sum_{\substack{p+q = 2s-1 \\ 0 \le p < s}} (-1)^{s+p} \frac{\partial}{\partial x} \left[\frac{\partial^p u_N}{\partial x^p} \frac{\partial^q u_N}{\partial x^q} \right] \|_{H_{loc}^{-1}(x,t)} \le$$

$$\le \text{Const} \cdot \sqrt{2C_s/N} \cdot \sum_{\substack{p+q = 2s-1 \\ 0 \le p < s}} \|u_N\|_{L^\infty} \le$$

$$\le \text{Const} \cdot s \sqrt{2C_s/N} \cdot \|u_N\|_{L^\infty} \to 0;$$

the second half of $\mathcal{II}_1$ in (3.167), $-\varepsilon_N \left(\frac{\partial^s u_N}{\partial x^s} \right)^2$, is bounded in $L_{loc}^1(x,t)$, consult Lemma 3.1, and hence by Murat's Lemma [16], belongs to a compact subset of $H_{loc}^{-1}(x,t)$. We conclude

$$\mathcal{II}_{12}(u_N) \underset{H_{loc}^{-1}(x,t)}{\longrightarrow} \le 0. \tag{3.177}$$

We continue with the next pair of expressions, $\mathcal{I}_2(u_N)$ and $\mathcal{II}_2(u_N)$. According to (3.163), $\mathcal{I}_2(u_N)$ — and therefore also $\mathcal{II}_2(u_N) = u_N \mathcal{I}_2(u_N)$ — are L^2-bounded, and hence belong to a compact subset of $H_{loc}^{-1}(x,t)$; in fact, by repeating our previous arguments which led to (3.163) one finds that

$$\|\mathcal{I}_2(u_N)\|_{H^{-1}(x,t)} \le \text{Const} \cdot \varepsilon_N m_N^{s-1} \| \frac{\partial^s u_N}{\partial x^s} \|_{L^2(x,t)} \le \tag{3.178}$$

$$\le \text{Const} \cdot \sqrt{\varepsilon_N} m_N^{s-1} \sim \sqrt{2C_s} \cdot N^{-\frac{2s-1}{2s}} \to 0.$$

A similar treatment shows that the first half of $\mathcal{II}_2(u_N)$ in (3.168) tends to zero in $H_{loc}^{-1}(x,t)$, for

$$\|\mathcal{II}_{21}(u_N)\|_{H_{loc}^{-1}(x,t)} = \tag{3.179}$$

$$= \varepsilon_N \| \sum_{p+q=s-1} (-1)^{s+p} \frac{\partial}{\partial x} \left(\frac{\partial^p u_N}{\partial x^p} \left[\frac{\partial^q R_N(x,t)}{\partial x^q} * \frac{\partial^s u_N}{\partial x^s} \right] \right) \|_{H_{loc}^{-1}(x,t)} \le$$

$$\le \varepsilon_N \cdot \sum_{p+q=s-1} N^p \|u_N\|_{L^\infty} \cdot m_N^q \| \frac{\partial^s u_N}{\partial x^s} \|_{L_{loc}^2(x,t)} \le$$

$$\le \text{Const} \cdot \sqrt{\varepsilon_N} \sum_{p+q=s-1} N^{p+q} \|u_N\|_{L^\infty} \le s \sqrt{2C_s/N} \cdot \|u_N\|_{L^\infty} \to 0.$$

The second half of $II_2(u_N)$ is L^1-bounded, for

$$\|II_{22}(u_N) \equiv \varepsilon_N \frac{\partial^s u_N}{\partial x^s} R_N(x,t) * \frac{\partial^s u_N}{\partial x^s}\|_{L^1} \leq \qquad (3.180)$$

$$\text{Const} \cdot \varepsilon_N \|\frac{\partial^s u_N}{\partial x^s}\|^2_{L^2_{loc}(x,t)} \leq \text{Const}.$$

Finally we treat the third pair of expressions, $I_3(u_N)$ and $II_3(u_N)$. The spectral decay estimate (3.164) with $(p,q) = (0,s)$, together with Lemma 3.1 imply that $I_3(u_N)$ tends to zero in $H^{-1}_{loc}(x,t)$; indeed

$$\|I_3(u_N) \equiv \frac{\partial}{\partial x}(I - P_N)f(u_N)\|_{H^{-1}_{loc}(x,t)} \qquad (3.181)$$

$$\leq \frac{C_s}{N^s}\|\frac{\partial^s u_N}{\partial x^s}\|_{L^2} \sim \sqrt{2C_s/N} \to 0.$$

A similar argument applies to the expression $II_3(u_N)$ given in (4.6c). Sobolev inequality – consult (3.174), followed by the spectral decay estimate (3.164) imply that the first half of $II_3(u_N)$ does not exceed

$$\|II_{31}(u_N) \equiv \qquad (3.182)$$

$$\equiv \sum_{p=0}^{s-1}(-1)^p \frac{\partial}{\partial x}\left[\frac{\partial^p u_N}{\partial x^p}\frac{\partial^{-p}}{\partial x^{-p}}(I - P_N)f(u_N)\right]\|_{H^{-1}_{loc}(x,t)} \leq$$

$$\leq \sum_{p=0}^{s-1}\|\frac{\partial^p u_N}{\partial x^p}\|_{L^\infty} \cdot \|\frac{\partial^{-p}}{\partial x^{-p}}(I - P_N)f(u_N)\|_{L^2_{loc}(x,t)} \leq$$

$$\leq \text{Const} \cdot \sum_{p=0}^{s-1}N^p\|u_N\|_{L^\infty}\frac{C_s}{N^{s+p}}\|\frac{\partial^s u_N}{\partial x^s}\|_{L^2_{loc}(x,t)} \leq$$

$$\sim \text{Const} \cdot s\sqrt{2C_s/N}\|u_N\|_{L^\infty} \to 0.$$

According to Lemma 3.1, the second half of $II_3(u_N)$ is L^1-bounded, for

$$\|II_{32}(u_N) \equiv \frac{\partial^s u_N}{\partial x^s}\frac{\partial^{-s+1}}{\partial x^{-s+1}}(I - P_N)f(u_N)\|_{L^1} \leq \qquad (3.183)$$

$$\leq \|\frac{\partial^s u_N}{\partial x^s}\|_{L^2}\frac{C_s}{N^{2s-1}}\|\frac{\partial^s u_N}{\partial x^s}\|_{L^2} \leq \text{Const},$$

and hence by Murat's Lemma [16], belongs to a compact subset of $H^{-1}_{loc}(x,t)$. We conclude that the entropy dissipation of the Fourier spectral 'super-viscosity' method, for both linear and quadratic entropies, belongs to a compact subset of $H^{-1}_{loc}(x,t)$.

Step 3. (Convergence). It follows that the SV solution u_N converges strongly
(in L^p_{loc}, $\forall p < \infty$) to a weak solution of (3.145). In fact, except for the L^1-bounded terms $II_{22}(u_N)$ and $II_{32}(u_N)$, we have shown that all the other expressions which contribute to the entropy dissipation tend either to zero or to a negative measure. Using the strong convergence of u_N it follows that $II_{22}(u_N)$ and $II_{32}(u_N)$ also tend to zero, consult [17]. Hence the convergence to the unique entropy solution. ∎

Remarks.

1. *Low pass filter* [8]. We note that the spectral "super-viscosity" in (3.148_s) allows for an increasing order of parabolicity, $s \sim N^\mu$, $\mu < 1/2$ (at least for bounded C_s's). This enables us to rewrite the spectral "super-viscosity" method in the form

$$\frac{\partial u_N}{\partial t} + \frac{\partial}{\partial x}[P_N f(u_N)] = -N \sum_{|k| \leq N} \sigma(\frac{k}{N})\hat{u}_k(t)e^{ikx}, \qquad (3.184)$$

where $\sigma(\xi)$ is a symmetric low pass filter satisfying

$$\sigma(\xi) \begin{cases} \leq |\xi|^{2s}, & |\xi| \leq 1, \\ \geq |\xi|^{2s} - \frac{1}{N}, & |\xi| > 0. \end{cases} \qquad (3.185)$$

In particular, for $s \sim N^\mu$, one is led to a low pass filter which is C^∞-tailored at the origin, consult [32].

2. *Super viscosity regularization.* The estimates outlined in Theorem 3.1 imply the convergence of the regularized 'super-viscosity' approximation u^ε in (3.160_s), to the entropy solution of the convex conservation law (3.145).

Assertion 3.1

Consider the 'super-viscosity' regularization (3.160_s),

$$\frac{\partial u^\varepsilon}{\partial t} + \frac{\partial}{\partial x}f(u^\varepsilon(x,t)) = \varepsilon(-1)^{s+1}\frac{\partial^{2s}}{\partial x^{2s}}u^\varepsilon(x,t), \qquad (3.186)$$

subject to given $L^1 \cap L^\infty$-initial data, $u(\cdot, 0)$. Assume that u^ε is uniformly bounded. Then u^ε converges to the unique entropy solution of the convex conservation law (3.145).

The question of L^∞ bound for the superviscosity case – (3.186) with $s > 1$, is open. Unlike the regular viscosity case, the solution operator associated with (3.186) with $s > 1$ is *not* monotone — here there are "spurious" oscillations, on top of the Gibbs' oscillations due to the Fourier projection. What we have shown is that the oscillations of either type do not cause instability. Moreover, these oscillations contain, in some weak sense, highly accurate information on the exact entropy solution; this could be revealed by post-processing the spectral (super)-viscosity approximation, e.g. [18].

3.3 Non-periodic boundaries

In this section we discuss the Legendre SV method, [18]. Extensions to Chebyshev SV method can be found in [12], [15]. Applications to atmospheric simulations can be found in [2].

The Legendre SV approximation

In the spectral viscosity approximation of (3.145) we seek a $I\!P_N$-polynomial of the form $u_N(x,t) = \sum_{k=0}^{N} \hat{u}_k(t) L_k(x)$, such that $\forall \varphi \in I\!P_N[-1,1]$, we have

$$(\frac{\partial}{\partial t} u_N + \frac{\partial}{\partial x} \mathcal{I}_N f(u_N), \varphi)_N = -\varepsilon_N (Q \frac{\partial}{\partial x} u_N, \frac{\partial}{\partial x} \varphi)_N + (B(u_N), \varphi)_N.$$
(3.187)

The approximation (3.187) involves the boundary operator, $B(u_N)$, and the **Spectral Viscosity operator, Q. Here, $B(u_N)$ is a forcing polynomial in $I\!P_N[-1,1]$ of the form**

$$B(u_N) = [\lambda(t)(1-x) + \mu(t)(1+x)] L'_N(x),$$
(3.188)

involving (at most) two nonzero free parameters, $\lambda(t)$ and $\mu(t)$, which should enable $u_N(x,t)$ to match *inflow* boundary data prescribed at $x = \pm 1$ whenever $\pm f'(u_N(\pm 1, t)) < 0$. And, Q denotes the spectral viscosity operator,

$$Q\varphi \equiv \sum_{k=0}^{N} \hat{Q}_k \hat{\varphi}_k L_k, \quad \forall \varphi = \sum_{k=0}^{\infty} \hat{\varphi}_k L_k,$$
(3.189)

which is associated with bounded viscosity coefficients,

$$\begin{cases} \hat{Q}_k \equiv 0 & k \leq m_N, \\ \\ 1 \geq \hat{Q}_k \geq 1 - \left(\frac{m_N}{k}\right)^4 & k > m_N. \end{cases}$$
(3.190)

The free pair of spectral viscosity parameters (ε_N, m_N) will be chosen later, such that $\varepsilon_N \downarrow 0$ and $m_N \uparrow \infty$, in order to retain the formal spectral accuracy of (3.187) with (3.145). We close this section by explaining how the SV method (3.187) can be implemented as a collocation method. Let us 'test' (3.187) against $\varphi = \varphi_i$, where φ_i is the standard characteristic polynomial of $I\!P_N[-1,1]$ satisfying $\varphi_i(\xi_j) = \delta_{ij}, 0 \leq i, j \leq N$. At the interior points we obtain

$$\frac{d}{dt} u_N(\xi_i, t) + \frac{\partial}{\partial x} \mathcal{I}_N f(u_N)(\xi_i, t) = \varepsilon_N \frac{\partial}{\partial x} Q(\frac{\partial}{\partial x} u_N)(\xi_i, t), \quad 1 \leq i \leq N-1.$$
(3.191)

These equations are augmented, at the outflow boundaries, (say at $x = +1$), with

$$\frac{d}{dt}u_N(+1,t) + \frac{\partial}{\partial x}\mathcal{I}_N f(u_N)(+1,t) = \tag{3.192}$$

$$= \varepsilon_N \frac{\partial}{\partial x}Q(\frac{\partial}{\partial x}u_N)(+1,t) - \frac{\varepsilon_N}{\omega_N}Q(\frac{\partial}{\partial x}u_N)(+1,t).$$

We note that the last term on the right of (3.192) prevents the creation of a boundary layer. Equations (3.191), (3.192) together with the prescribed inflow data (say at $x = -1$), furnish a complete equivalent statement of the pseudospectral (collocation) viscosity approximation (3.187).

The SV approximation (3.191),(3.192) enjoys *formal* spectral accuracy, i.e., its truncation error decays as fast as the global smoothness of the underlying solution permits. However, it is essential to keep in mind that this superior accuracy cannot be realized in the presence of shock discontinuities, unless the final SV solution is post-processed. The rest of this section is devoted to clarify this point.

Epilogue – on spectral post-processing

It is well-known that spectral projections like $\pi_N u$, $\mathcal{I}_N u$, etc., provide highly accurate approximations of u, provided u itself is sufficiently smooth. Indeed, these projections enjoy spectral convergence rate. This superior accuracy is destroyed if u contains discontinuities: both $\pi_N u$ and $\mathcal{I}_N u$ produce spurious $\mathcal{O}(1)$ Gibbs' oscillations which are *localized* in the neighborhoods of the discontinuities, and moreover, their *global* accuracy is deteriorated to first-order.

To accelerate the convergence rate in such cases, we follow a similar treatment in [10] for the Fourier projections of discontinuous data. We introduce a mollifier of the form

$$\Psi^{\alpha,p}(x;y) = \rho(\frac{x-y}{\alpha})K_p(x;y), \tag{3.193}$$

which consists of the following two ingredients:

- $\rho(x)$ is a $C_0^\infty(-1,1)$-localizer satisfying $\rho(0) = 1$;
- $K_p(x;y)$ is the Christoffel-Darboux kernel

$$K_p(x;y) \equiv \sum_{j=0}^{p} \frac{L_j(x)L_j(y)}{\|L_j\|^2} = \frac{(p+1)}{2}\frac{L_{p+1}(x)L_p(y) - L_{p+1}(y)L_p(x)}{x-y}. \tag{3.194}$$

We let $F^{\alpha,\beta}$ denote the smoothing filter

$$F^{\alpha,\beta}w(x) \equiv \int_{x=-1}^{1} \Psi^{\alpha,p=[N^\beta]}(x;y)w(y)dy, \tag{3.195}$$

depending on the two fixed parameters, $\alpha, \beta \in (0,1)$. Then, the following spectral error estimate was derived in [20]: $\forall s \geq 1$ there exists a constant $C_{s,\alpha}$ such that

$$|u(x) - F^{\alpha,\beta}(\pi_N u)(x)| \leq \qquad (3.196)$$

$$C_{s,\alpha} \left[N^{2\beta-(1-2\beta)s} \|u\|_{L^2[-1,1]} + N^{-(\frac{3}{4}-s)\beta} \max_{\substack{|x-y|<\alpha \\ 0 \leq j \leq s}} |D^j u(y)| \right].$$

Similar estimate holds for $\mathcal{I}_N$. These estimates show (at least for $\beta < \frac{1}{2}$) that except for a small neighborhood of the discontinuities (measured by the free parameter α), one can filter the Legendre projections, $\pi_N u$ and $\mathcal{I}_N u$, in order to recover *pointwise* values of u within spectral accuracy.

Next, let u be the desired exact solution of a given problem. The purpose of a spectral method is to compute an approximation to the *projection* of u rather than u itself. Consequently, if the underlying solution of our problem is discontinuous, then the approximation computed by a spectral method, u_N, exhibits the two difficulties of local Gibbs' oscillations, and global, low(=first)-order accuracy.

With this in mind, we now turn to discuss the present context of nonlinear conservation laws. The standard, viscous-free spectral method supports the spurious Gibbs' oscillations which render the overall approximation unstable (consult the introductory counterexamples in Lecture IV below). The task of the Spectral Viscosity is therefore two fold: to stabilize the standard spectral method (— which is otherwise unstable), and to retain the overall spectral accuracy of the underlying spectral method.

The question of stability is addressed in the following sections: we prove that Spectral Viscosity guarantees the H^{-1}-stability (and hence the convergence) of the Legendre SV approximation,

$$L^p_{loc} - \lim u_N(x,t) = u(x,t), \quad \forall p < \infty. \qquad (3.197)$$

The question of spectral accuracy requires further clarification. As noted above, the Legendre SV solution, $u_N(\cdot, t)$, should be considered as an accurate approximation of $\mathcal{I}_N u(\cdot, t)$, rather than $u(\cdot, t)$ itself. Therefore, the convergence *rate* of the SV method is limited by the first order convergence rate of $\mathcal{I}_N u(\cdot, t)$. (Of course, this limitation arises once shock-discontinuities are formed). We recall that according to (3.196), this first-order limitation can be avoided by filtering $\mathcal{I}_N u$: the *filtered* interpolant, $F^{\alpha,\beta}(\mathcal{I}_N u)$, retains a spectral convergence rate, at least in smooth regions of the discontinuous entropy solution $u(\cdot, t)$. This suggests to apply the same filtering procedure (3.195) to $u_N(\cdot, t)$, in order to accelerate the convergence rate of the SV method.

Let $\{\hat{u}_k(t)\}_{k=0}^N$ denote the computed coefficients of the Legendre SV method. The computation of the SV solution is based on adding spectral viscosity only to the "high" modes – those with wavenumbers $k > m_N$. Therefore, one *expects* the computation of the viscous-free coefficients, at least,

$$\hat{u}_k(t) \equiv \frac{(u_N, L_k)_N}{\|L_k\|_N^2}, \ k = 1, \dots, m_N, \text{ to be spectrally accurate approximation}$$

of the exact pseudospectral Legendre coefficients, $\dfrac{(u, L_k)_N}{\|L_k\|_N^2}$. Assuming that indeed this is the case, then according to (3.196) one can post-process the SV solution, $u_N(\cdot, t)$, in order to recover spectral convergence rate in smooth regions of the entropy solutions. Thus, at the final stage of the SV method, (3.191),(3.192) should be augmented with the post-processing procedure

$$F^{\alpha,\beta} u_N(x,t) = \int_{x=-1}^{1} \Psi^{\alpha,p=N^\beta}(x;y) u_N(y) dy. \tag{3.198}$$

The numerical experiments in [23] confirm that the SV method contains a spectrally accurate information about the discontinuous solution – by post-processing one recovers this information despite the presence of shock discontinuities.

We conclude by noting that the post-processing of the SV solution plays a necessary key role in realizing the spectral accuracy of the SV method *within* smooth regions of the underlying solution. The treatment of Gibbs' oscillations in the neighborhood of discontinuities requires an alternative 'one-sided' filtering procedure, which is studies in e.g., [9].

Convergence of the Legendre SV method
We want to prove the convergence of (3.187) by compensated compactness arguments. To this end we want to show that $\frac{\partial}{\partial t} U(u_N) + \frac{\partial}{\partial x} F(u_N)$ belongs to a compact subset of $H_{loc}^{-1}(x,t)$ for all convex entropy pairs $(U(u_N), F(u_N))$. Our main tool in this direction reads [18, §5]

Lemma 3.2 *A weak representation of the truncation error of the Legendre viscosity approximation (3.187) is given by*

$$\left(\frac{\partial}{\partial t} u_N + \frac{\partial}{\partial x} f(u_N), \varphi\right) = \sum_{j=1}^{6} I_j(\varphi), \qquad \varphi(x,t) \in \mathcal{D}([-1,1]), \tag{3.199}$$

where the following estimates hold:

$$\sum_{j=1}^{3} |I_j(\varphi)| \leq \mathcal{O}(\frac{1}{\sqrt{\varepsilon_N}}) \left[\|\varphi - \varphi_N\| + \frac{1}{N} \|\frac{\partial}{\partial x} \varphi_N\| \right], \tag{3.200}$$

$$|I_4(\varphi)| \leq \mathcal{O}(\varepsilon_N m_N^2 \sqrt{\ln N}) \|\frac{\partial}{\partial x} \varphi_N\|, \tag{3.201}$$

$$|I_5(\varphi) \equiv -\varepsilon_N(\frac{\partial}{\partial x}u_N, \frac{\partial}{\partial x}\varphi_N)| \leq \mathcal{O}(\sqrt{\varepsilon_N})\|\frac{\partial}{\partial x}\varphi_N\|, \qquad (3.202)$$

$$I_6(\varphi) \equiv 2(-1)^{N+1}\int_{t=0}^{T}\lambda(t)\varphi_N(-1,t)dt. \qquad (3.203)$$

Here, $\varphi_N(\cdot,t)$ is an arbitrary $\mathbb{P}_N$-polynomial at our disposal.

Appropriate choices of test functions, φ_N yield the desired convergence result.

Theorem 3.2 *Let $u_N(x,t)$ be the Legendre viscosity approximation of (3.187), (3.190), with spectral viscosity parameters (ε_N, m_N) which satisfy*

$$0 \downarrow \varepsilon_N \sim \frac{1}{N^\theta}, \quad m_N < Const\cdot N^{\frac{q}{4}} \qquad with \quad 0 < q < \theta \leq 1. \qquad (3.204)$$

Then, (a subsequence of) $u_N(x,t)$ converges strongly (in L_{loc}^p, $p < \infty$) to a weak solution of the conservation law (3.145). Moreover, if $\theta < 1$, then (the whole sequence of) $u_N(x,t)$ converges strongly to the unique entropy solution of (3.145).

3.4 Numerical results

In this section we will present numerical experiments which demonstrate the performance of the Legendre SV method for systems of conservation laws. We consider the approximate solution of the Euler equations of gas dynamics,

$$\frac{\partial}{\partial t}u(x,t) + \frac{\partial}{\partial x}f(u(x,t)) = 0, \qquad u = \begin{bmatrix} \rho \\ \rho v \\ E \end{bmatrix} \quad f(u) = \begin{bmatrix} \rho v \\ \rho v^2 + p \\ v(E + p) \end{bmatrix}, \qquad (3.205)$$

where ρ denotes the density of the gas, v its velocity, $m \equiv \rho v$ its momentum, E its energy per unit volume and $p = (\gamma - 1)\cdot(E - \frac{1}{2}\rho v^2)$ its (polytropic) pressure, $\gamma = 1.4$.

The Legendre SV approximation of this system reads

$$\frac{d}{dt}u_N(\xi_i,t) + \frac{\partial}{\partial x}\mathcal{I}_N f(u_N)(\xi_i,t) = \varepsilon_N\frac{\partial}{\partial x}Q(\frac{\partial}{\partial x}u_N)(\xi_i,t), \quad 1 \leq i \leq N - 1. \qquad (3.206)$$

Here, $u_N \equiv {}^t(\rho_N, \rho_N v_N, E_N) \in \mathbb{P}_N^3[-1,1]$ denotes the polynomial approximation of the 3-vector of (density, momentum, energy), and Q abbreviates a general 3×3 spectral viscosity matrix, $\{\hat{Q}_k^{\ell,j}\}_{k=m_N}^N$, $1 \leq \ell, j \leq 3$ which is activated only on 'high' Legendre modes, i.e., $\hat{Q}_k^{\ell,j} = 0$, $\forall k > m_N(\ell,j)$. The numerical results reported in this section were obtained using a simple scalar viscosity matrix,

$$Q(\frac{\partial}{\partial x}u_N) = {}^t(Q\frac{\partial}{\partial x}\rho_N, Q\frac{\partial}{\partial x}\rho_N v_N, Q\frac{\partial}{\partial x}E_N), \qquad (3.207)$$

with the viscosity coefficients, $\hat{Q}_k$, given by

$$\hat{Q}_k = exp\{-\frac{(k-N)^2}{(k-m_N)^2}\}, \quad k > m_N, \qquad (3.208)$$

The Legendre SV method (3.206,(3.207) amounts to a nonlinear system of $(N+1)^3$ ODEs which was integrated in time using the second order Adams-Bashforth ODE solver. We implemented the SV method for two test problems.

• *The Riemann shock tube problem* [22]. Our first example is the Riemann problem (3.205), subject to initial conditions

$$u(x,0) = \begin{cases} u_\ell = {}^t(1., \ 0, \ 2.5), & x < 0, \\ u_r = {}^t(0.125, \ 0, \ 0.25), & x > 0. \end{cases} \qquad (3.209)$$

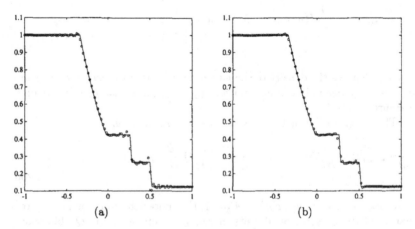

(a)   (b)

Fig. 3.16: Density ρ_N with N=128 Legendre modes. (a) before and (b) after post-processing.

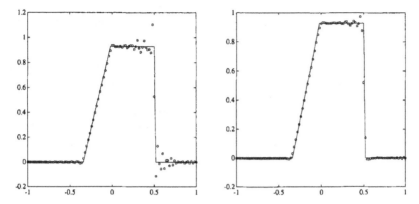

Fig. 3.17: Velocity v_N with N=128 Legendre modes. (a) before and (b) after post-processing.

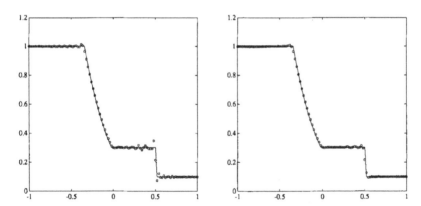

Fig. 3.18: Pressure v_N with N=128 Legendre modes. (a) before and (b) after post-processing.

Figures 3.16a, 3.17a and 3.18a display the computed density ρ_N, velocity v_N, and pressure p_N, with $N = 128$ Legendre modes. The numerical results in these figures show that the presence of Spectral Viscosity guarantees the convergence of the pseudospectral Legendre method that is otherwise unstable. However, Gibbs' oscillations which are inherited from the *projected* solution, $\mathcal{I}_N u(\cdot, t)$, are still present.

To remove these oscillations without sacrificing spectral accuracy, the SV solution on the left side of figures (3.205)-(3.209) was post-processed using the filtering procedure (3.195), $F^{\alpha,\beta}$ with $(\alpha,\beta) = (0.2, 0.85)$. Again, as in the scalar case, the post-processing leads to a dramatic improvement in the quality of the computed results, revealing the high-resolution content of the SV computation. In particular, comparing the results obtained by the post-processed SV method in figures 3.16b-3.18b, we find the representation of the rarefaction wave and the capturing of the contact discontinuity to be better than the results obtained by the finite-difference methods in [22] or the high-resolution schemes in [22]. (It is worthwhile noting that these high resolutions results of the SV computations were obtained *without* the costly characteristic decompositions which are employed in the modern high resolution finite difference approximations.)

The resolution of the shock discontinuity, however, still suffers from a smearing of spurious Gibbs' oscillations. As told by the error estimate (3.196), the oscillations in the neighborhood of the discontinuities cannot be removed by the filtering procedure (3.195). Instead, these oscillations can be avoided by using an alternative 'one-sided' filter which is currently under investigation [9].

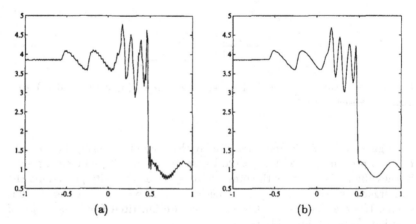

(a)   (b)

Fig. 3.19: Density ρ_N with N=220 Legendre modes. (a) before and (b) after post-processing.

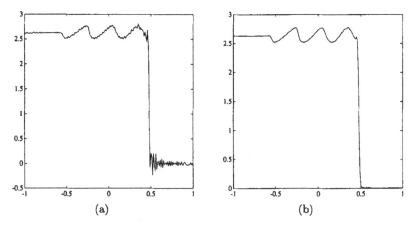

Fig. 3.20: Velocity v_N with N=220 Legendre modes. (a) before and (b) after post-processing.

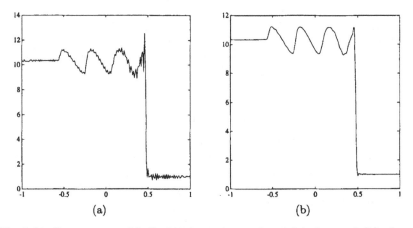

Fig. 3.21: Pressure p_N with N=220 Legendre modes. (a) before and (b) after post-processing.

• *The shock-disturbance interaction* e.g., [SO]. Our second example models the interaction of a sinusoidal disturbance and a shock wave due to initial conditions

$$(\rho(x,0), v(x,0), p(x,0)) = \begin{cases} (3.857143, \ 10.333333, \ 2.629369), \ x < -0.8, \\ (1. + 0.2\sin(5\pi x), \ 0., \ 1.), \qquad\qquad x > -0.8. \end{cases}$$
$$(3.210)$$

The exact solution of this problem, (3.205),(3.210), consists of a density wave that will emerge behind the shock discontinuity, and the fine structure of this density wave makes the current problem a suitable test case for high order methods. For example, second order MUSCL type schemes, [14], are unable to resolve the fine structure of the density wave unless the number of grid points is substantially increased.

The Legendre SV method was implemented in this case with SV parameters $(\varepsilon_N, m_N) = (\frac{1}{N}, 8\sqrt{N})$. Figures 3.19-3.21 display the numerical results of the SV approximation which was integrated in time by the second-order Adams-Bashforth method with time step $\Delta t = 2.5 \cdot 10^{-6}$.

Figures 3.19a, 3.20a, and 3.21a show the approximated density ρ_N, velocity v_N, and pressure p_N at $t = 0.36$, computed with $N = 220$ Legendre modes. These results were post-processed by the filtering procedure (3.195), $F^{\alpha,\beta}$, with $(\alpha, \beta) = (0.1, 0.89)$. Figures 3.19b, 3.20b and 3.21b present the post-processed results, which show that the velocity and pressure waves are well resolved. The density wave still contains Gibbs' oscillations in the neighborhood of the shock discontinuity, and its first extremum behind the shock is smeared by our smoothing filter. Here, a 'one-sided' filter would be recommended instead. A better resolution of the density profile near the shock was obtained by a different spectral method presented in [3]. However, the latter is a shock fitting like method which might not be easy to extend to higher dimensions.

3.5 Multidimensional Fourier SV method

We want to solve the *multidimensional* 2π-periodic initial-value problem, (3.145) by a spectral method. To this end we approximate the spectral/pseudo-spectral projection of the exact entropy solution, $P_N u(\cdot, t)$, using an N-trigonometric polynomial, $u_N(x, t) = \sum_{|\xi| \leq N} \hat{u}_\xi(t) e^{i\xi \cdot x}$, which is governed by the semi-discrete approximation

$$\frac{\partial}{\partial t} u_N(x,t) + \partial_x \cdot P_N f(u_N(x,t)) = \varepsilon_N \sum_{j,k=1}^{d} \partial_{jk}^2 Q_N^{j,k}(x,t) * u_N(x,t). \quad (3.211)$$

Together with one's favorite ODE solver, (3.211) gives a fully discrete method for the approximate solution of (3.145).

To suppress these oscillations, without sacrificing the overall spectral accuracy, we augment the standard Fourier approximation on the right-hand side of (3.211) by *spectral viscosity*, which consists of the following three ingredients:

– A vanishing viscosity amplitude, ε_N, of size

$$\varepsilon_N \sim N^{-\theta}, \quad \theta < 1. \tag{3.212}$$

– A viscosity-free spectrum of size $m_N >> 1$,

$$m_N \sim \frac{N^{\frac{\theta}{2}}}{(\log N)^{\frac{d}{2}}}, \quad \theta < 1. \tag{3.213}$$

– A family of viscosity kernels, $Q_N^{j,k}(x,t) = \sum_{|\xi|=m_N}^{N} \hat{Q}_\xi^{j,k}(t)e^{i\xi\cdot x}$, $1 \leq j, k \leq d$, activated only on high wavenumbers $|\xi| \geq m_N$, which can be conveniently implemented in the Fourier space as

$$\varepsilon_N \sum_{j,k=1}^{d} \partial_{jk}^2 Q_N^{j,k} * u_N(x,t) \equiv \tag{3.214}$$

$$\equiv -\varepsilon_N \sum_{|\xi|=m_N}^{N} < \hat{Q}_\xi \xi, \xi > \hat{u}_\xi(t)e^{i\xi\cdot x}, \quad < \hat{Q}_\xi \xi, \xi > \equiv \sum_{j,k=1}^{d} \hat{Q}_\xi^{j,k}(t)\xi_j\xi_k.$$

The viscosity kernels we deal with, $Q_N^{j,k}(x,t)$, are assumed to be spherically symmetric, that is, $\hat{Q}_\xi^{j,k} = \hat{Q}_p^{j,k}, \forall|\xi| = p$, with monotonically increasing Fourier coefficients, $\hat{Q}_p^{j,k}$, that satisfy

$$|\hat{Q}_p^{j,k} - \delta_{jk}| \leq \text{Const.} \frac{m_N^2}{p^2}, \quad \forall p \geq m_N. \tag{3.215}$$

The main convergence result, quoted from [5], are based on the following two lemmas.

Lemma 3.3 L^∞ **stability** *There exists a constant such that*

$$\|u_N(\cdot,t)\|_{L^\infty(x)} \leq \text{Const} \cdot \|u_N(\cdot,0)\|_{L^\infty(x)}, \quad \forall t \leq T. \tag{3.216}$$

Lemma 3.4 *Entropy Consistency There exists a vanishing sequence, e_N, such that*

$$\frac{\partial}{\partial t}U(u_N) + \partial_x \cdot F(u_N) \leq e_N \to 0, \quad \text{in } \mathcal{D}'. \tag{3.217}$$

Proofs of Lemma 3.3 and Lemma 3.4 can be found in [5]. Granted the L^∞-stability and the entropy consistency, we can combine DiPerna's uniqueness result for measure-valued solutions [6] with the finiteness of propagation speed (see also [24] for the case of bounded domains) to conclude the following.

Theorem 3.3 *Let u_N be the solution of the SV approximation (3.211)-(3.215), subject to bounded initial conditions satisfying*

$$\|u_N(\cdot, 0)\|_{L^\infty(x)} + \varepsilon^s \|\partial_x^s u_N(\cdot, 0)\|_{L^2(x)} \leq \text{Const.} \qquad (3.218)$$

Then u_N converges strongly to the unique entropy solution of (3.145).

References

1. S. ABARBANEL, D. GOTTLIEB & E. TADMOR, *Spectral methods for discontinuous problems*, in "Numerical Analysis for Fluid Dynamics II" (K.W. Morton and M.J. Baines, eds.), Oxford University Press, 1986, pp. 129–153.
2. Ø. ANDREASSEN, I. LIE & C.-E. WASSBERG, *The spectral viscosity method applied to simulation of waves in a stratified atmosphere*, J. Comp. Phys. 110 (1994) pp. 257-273.
3. W. CAI, D. GOTTLIEB & A. HARTEN, *Cell averaging Chebyshev method for hyperbolic problems*, Compu. and Math. with Appl., 24 (1992).
4. C. CANUTO, M.Y. HUSSAINI, A. QUARTERONI & T. ZANG, Spectral Methods with Applications to Fluid Dynamics, Springer-Verlag, 1987.
5. G.-Q. CHEN, Q. DU & E. TADMOR, *Spectral viscosity approximations to multidimensional scalar conservation laws*, Math. of Comp. 61 (1993), 629-643.
6. R. DIPERNA, *Convergence of approximate solutions to systems of conservation laws*, Arch. Rat. Mech. Anal., Vol. 82, pp. 27-70 (1983).
7. D.C. FRITTS, T. L. PALMER, Ø. ANDREASSEN, AND I. LIE, *Evolution and breakdown of Kelvin-Helmholtz billows in stratified compressible flows .1. Comparison of two- and three-dimensional flows*, J. Atmos. Sci. 53 (1996) pp. 3173-3191.
8. D. GOTTLIEB, Private communication.
9. D. GOTTLIEB, C.-W. SHU & H. VANDEVEN, *Spectral reconstruction of a discontinuous periodic function*, submitted to C. R. Acad. Sci. Paris.
10. D. GOTTLIEB & E. TADMOR, *Recovering pointwise values of discontinuous data with spectral accuracy*, in "Progress and Supercomputing in Computational Fluid Dynamics" (E. M Murman and S.S. Abarbanel eds.), Progress in Scientific Computing, Vol. 6, Birkhauser, Boston,1985, pp. 357-375.
11. H. -O. KREISS & J. OLIGER, *Stability of the Fourier method*, SINUM 16 1979, pp. 421-433.
12. H. MA, *Chebyshev-Legendre spectral viscosity method for nonlinear conservation laws*, SINUM, to appear.
13. H. MA, *Chebyshev-Legendre spectral super viscosity method for nonlinear conservation laws*, SINUM, to appear.
14. B. VAN LEER, *Toward the ultimate conservative difference schemes. V. A second order sequel to Godunov method*, J. Compt. Phys. 32, (1979), pp. 101-136.
15. I. LIE, *On the spectral viscosity method in multidomain Chebyshev discretizations*, preprint.
16. F. MURAT, *Compacité per compensation*, Ann. Scuola Norm. Sup. Disa Sci. Math. 5 (1978), pp. 489-507 and 8 (1981), pp. 69-102.
17. Y. MADAY & E. TADMOR, *Analysis of the spectral viscosity method for periodic conservation laws*, SINUM 26, 1989, pp. 854-870.

18. Y. MADAY, S.M. OULD KABER & E. TADMOR, *Legendre pseudospectral viscosity method for nonlinear conservation laws*, SINUM 30, 1993, pp. 321-342.

19. A. MAJDA, J. MCDONOUGH & S. OSHER, *The Fourier method for nonsmooth initial data*, Math. Comp. 30, 1978, pp. 1041-1081.

20. S.M. OULD KABER, *Filtrage d'ordre infini en non périodique*, in Thèse de Doctorat, Université Pierre et Marie Curie, Paris, 1991.

21. S. SCHOCHET, *The rate of convergence of spectral viscosity methods for periodic scalar conservation laws*, SINUM 27, 1990, pp. 1142-1159.

22. C.-W. SHU & S. OSHER, *Efficient Implementation of Essentially Nonoscillatory Shock-Capturing schemes, II*, J. Comp. Phys., 83 (1989), pp. 32–78.

23. C.-W. SHU & P. WONG, *A note on the accuracy of spectral method applied to nonlinear conservation laws*, J. Sci. Comput., v10 (1995), pp. 357-369.

24. A. SZEPESSY, *Measure valued solutions to scalar conservation laws with boundary conditions*, Arch. Rat. Mech. 107, 181–193 (1989).

25. E. TADMOR, *The exponential accuracy of Fourier and Chebyshev differencing methods*, SINUM 23, 1986, pp. 1-10.

26. E. TADMOR, *Convergence of spectral methods for nonlinear conservation laws*, SINUM 26, 1989, pp. 30-44.

27. E. TADMOR, *Semi-discrete approximations to nonlinear systems of conservation laws; consistency and stability imply convergence*, ICASE Report no. 88-41.

28. E. TADMOR, *Shock capturing by the spectral viscosity method*, Computer Methods in Appl. Mech. Engineer. 80 1990, pp. 197-208.

29. E. TADMOR, *Total-variation and error estimates for spectral viscosity approximations*, Math. Comp. 60, 1993, pp. 245–256.

30. E. TADMOR, *Super viscosity and spectral approximations of nonlinear conservation laws*, in "Numerical Methods for Fluid Dynamics IV", Proceedings of the 1992 Conference on Numerical Methods for Fluid Dynamics, (M. J. Baines and K. W. Morton, eds.), Clarendon Press, Oxford, 1993, pp. 69-82.

31. L. TARTAR, *Compensated compactness and applications to partial differential equations*, in *Research Notes in Mathematics 39*, Nonlinear Analysis and Mechanics, Heriott-Watt Symposium, Vol. 4 (R.J. Knopps, ed.) Pittman Press, pp. 136-211 (1975).

32. H. VANDEVEN, *A family of spectral filters for discontinuous problems*, J. Scientific Comput. 8, 1991, pp. 159–192.

4 Convergence Rate Estimates

Abstract. Let $\{v^\varepsilon(x,t)\}_{\varepsilon>0}$ be a family of approximate solutions for the convex conservation law $u_t + f(u)_x = 0$ subject to C_0^1-initial data, $u_0(x)$. The notion of *approximate solutions* is quantified in terms of Lip'-consistency: we assume that $\{v^\varepsilon(x,t)\}$ are Lip'-*consistent* in the sense its initial+truncation errors are of order $\mathcal{O}(\varepsilon)$, $\|v^\varepsilon(\cdot,0) - u_0(\cdot)\|_{Lip'(x)} + \|v_t^\varepsilon + f(v^\varepsilon)_x\|_{Lip'(x,t)} = \mathcal{O}(\varepsilon)$. Here, ε is the 'small scale' of the approximate solution, e.g., the vanishing amplitude of size ε, a gridsize of order $\varepsilon \sim \Delta x$, etc. We then prove that stability implies convergence; namely, if $\{v^\varepsilon(x,t)\}$ are Lip^+-*stable* (– in the sense that they satisfy Oleinik's E-entropy condition), then they converge to the entropy solution, and the convergence rate estimate $\|v^\varepsilon(\cdot,t) - u(\cdot,t)\|_{Lip'(x)} = \mathcal{O}(\varepsilon)$ holds. Consequently, the familiar L^p-type and new pointwise error estimates are derived. In particular, we recover classical L^1-estimates (à al Kuznetsov) of order $\mathcal{O}(\sqrt{\varepsilon})$. And we improve it to an $\mathcal{O}(\varepsilon)$ *pointwise* error estimate for all but finitely many $\mathcal{O}(\varepsilon)$-neighborhoods of shock discontinuities.

These convergence rate results are then demonstrated in the context of various approximate solutions, including Chapman-Enskog regularization, finite-difference schemes, Godunov-type methods, spectral viscosity methods, ...

4.1 Introduction

We are concerned here with the convergence *rate* of approximate solutions to the nonlinear scalar conservation law,

$$u_t + f(u)_x = 0, \tag{4.219}$$

subject to C_0^1-initial conditions,

$$u(x,0) = u_0(x). \tag{4.220}$$

In this context we first recall Strang's theorem which shows that the classical Lax-Richtmyer (LR) linear convergence theory applies for such nonlinear problem, as long as the underlying solution is sufficiently smooth e.g., [29, §5]. The generic convergence error estimate in this context reads

$$\|v^\varepsilon(\cdot,t) - u(\cdot,t)\| \le C_T\big[\|v^\varepsilon(\cdot,0) - u_0(\cdot)\| + \|v_t^\varepsilon + f(v^\varepsilon)_x\|\big], \ 0 \le t \le T. \tag{4.221}$$

Here, $\{v^\varepsilon\}$, is a family of approximate solutions which is tagged by its 'small scale', ε, e.g., a viscosity amplitude of size ε, a gridcells of size $\varepsilon \sim \Delta x$, the number of Fourier modes, $N \sim \varepsilon^{-1}$, etc. The linear Lax-Richtmyer theory tells us that if the approximate solution is *stable*, $\|v^\varepsilon(\cdot,t)\| \le Const$, then the error, $\|v^\varepsilon(\cdot,t) - u(\cdot,t)\|$ is upper bounded by the initial+truncation errors, given respectively on the right of (4.221). In particular, if the approximation is *consistent* (— in the sense that that its initial+truncation errors tend to zero as $\varepsilon \downarrow 0$), then stability implies convergence.

What norm, $\|\cdot\|$, should be used in (4.221)? The *linear* Lax-Richtmyer theory is often implemented in term of the L^2 norm; likewise, Strang's extension to nonlinear smooth problems is usually expressed in terms of higher Sobolev H^s norms. There are two main reasons for the use of the L^2 framework:

1. It is the appropriate topology to measure stability and well-posedness of hyperbolic systems;
2. The Fourier space serves as a 'mirror site' for the real space L^2-stability and error analysis. The von-Neumann stability analysis for finite-difference schemes is a classical example.

Since the solutions of the nonlinear conservation laws develop spontaneous shock-discontinuities at a finite time, however, Strang's result does not apply beyond this critical time. Indeed, the Fourier method as well as other L^2-*conservative* schemes provide simple counterexamples of consistent approximations which fail to converge (to the discontinuous entropy solution), despite their linearized L^2-stability. Here are two counterexamples in this directions (more can be found in [38,37,13].)

Counterexample 1 [38]. The Fourier approximation of the 2π-periodic equation (4.219), expressed in term of the Fourier partial sum projection S_N, reads

$$\frac{\partial}{\partial t}[v_N(x,t)] + \frac{\partial}{\partial x}[S_N f(v_N(x,t))] = 0.$$

Multiplying this by $v_N(x,t)$ and integrating over the 2π-period, we obtain that v_N–being orthogonal to $\frac{\partial}{\partial x}[(I - S_N)f(v_N(x,t))]$, satisfies

$$\frac{1}{2}\frac{d}{dt}\int_0^{2\pi} v_N^2(x,t)dx = -\int_0^{2\pi} v_N(x,t)\frac{\partial}{\partial x}[f(v_N(x,t))]dx$$

$$= -\int^{v_N(x,t)} uf'(u)du|_{x=0}^{x=2\pi} = 0.$$

Thus, the total quadratic entropy, $\eta(u) = \frac{1}{2}u^2$, is globally conserved in time

$$\frac{1}{2}\int_0^{2\pi} v_N^2(x,t)dx = \frac{1}{2}\int_0^{2\pi} v_N^2(x,0)dx, \qquad (4.222)$$

which in turn yields the existence of a weak $L^2(x)$-limit, $\overline{u}(x,t) = w\lim_{N\to\infty} v_N(x,t)$. Yet, $\overline{u}(x,t)$ *cannot* be the entropy solution of a nonlinear equation (4.219) where $f''(\cdot) \neq 0$. Otherwise, $S_N f(v_N(x,t))$ and therefore $f(v_N(x,t))$ should tend, in the weak distributional sense, to $f(\overline{u}(x,t))$; consequently, since $f(u)$ is nonlinear, $\overline{u}(x,t) = s\lim_{N\to\infty} v_N(x,t)$, which by (4.222) should satisfy $\frac{1}{2}\int_0^{2\pi}\overline{u}^2(x,t)dx = \frac{1}{2}\int_0^{2\pi}\overline{u}^2(x,0)dx$. But this is incompatible with the (quadratic) entropy inequality if $\overline{u}(x,t)$ contains shock discontinuities.

Our second example is a discrete one.

Counterexample 2. We consider the 2π-periodic conservation law

$$\frac{\partial u}{\partial t} + \frac{\partial(e^u)}{\partial x} = 0,$$

Expressed in terms of the trigonometric interpolant at the equidistant grid-points $x_\nu = \frac{2\nu\pi}{2N+1}$, the corresponding ψdospectral approximation reads

$$\frac{\partial}{\partial t}[v_N(x,t)] + \frac{\partial}{\partial x}[\psi_N e^{v_N(x,t)}] = 0.$$

Multiply this by $\psi_N e^{v_N(x,t)}$ and integrate over the 2π-period: since the trapezoidal rule is exact with integration of the $2N$-trigonometric polynomial obtained from the second brackets, we have

$$\frac{d}{dt} \sum_{\nu=0}^{2N} e^{v_N(x_\nu,t)} \Delta x = -\int_0^{2\pi} \frac{\partial}{\partial x}[\frac{1}{2}(\psi_N e^{v_N(x,t)})^2]dx = 0.$$

Thus, the total exponential entropy, $\eta(u) = e^u$, is globally conserved in time

$$\sum_{\nu=0}^{2N} \eta(v_N(x_\nu,t))\Delta x = \sum_{\nu=0}^{2N} \eta(v_N(x_\nu,0))\Delta x, \qquad \eta(u) = e^u. \qquad (4.223)$$

Hence, if $v_N(x,t)$ converges (even weakly) to a discontinuous weak solution, $\bar{u}(x,t)$, then $\psi_N e^{v_N(x,t)}$ tends (at least weakly) to $e^{\bar{u}(x,t)}$. Consequently, (4.223) would imply the global entropy conservation of $\int_0^{2\pi} e^{\bar{u}(x,t)}dx$ in time, which rules out the possibility of $\bar{u}(x,t)$ being the unique entropy solution.

In this chapter we extend the linear convergence theory into the weak regime. The extension is based on the usual two ingredients of stability and consistency. On the one hand, the counterexamples mentioned above show that one must *strengthen* the linearized L^2-stability requirement. We assume that the approximate solutions are Lip^+-stable in the sense that they satisfy a one-sided Lipschitz condition, in agreement with Oleinik's E-condition for the entropy solution. On the other hand, the lack of smoothness requires to *weaken* the consistency requirement, which is measured here in the Lip'-(semi)norm. As a guiding example, let us consider the usual viscosity approximation, v^ε, with 'truncation error' $\varepsilon v_{xx}^\varepsilon$. Localized to the neighborhood of shock discontinuities we find that $\|\varepsilon v_{xx}^\varepsilon\|_{L_{loc}^p} = \mathcal{O}(\varepsilon^{\frac{1-p}{p}})$ which rules out the L^p norms as possible measures for the a priori error estimate (4.221); instead, the *weak* Lip'-(semi)norm yields a truncation error of size $\|\varepsilon v_{xx}^\varepsilon\|_{Lip'} = \mathcal{O}(\varepsilon)$ which agrees with the fact that ε is the smallest scale present in a viscosity approximation in this case.

In §4.3 we prove for Lip^+-stable approximate solutions, that their Lip'-convergence rate to the entropy solution is of the same order as their Lip'-consistency. Thus, we show that under the assumption of Lip^+-stability, the basic Lax-Richtmyer a priori error bound (4.221) still holds when we replace the L^2 with the weaker Lip' norm.

Our Lip'-convergence rate estimates could be converted into stronger L^p convergence rate estimates. In particular, we recover the usual L^1-convergence rate of order one half, and we obtain new pointwise error estimates which depend on the *local* smoothness of the entropy solution. In fact, though the L^1-convergence rate of order $\mathcal{O}(\sqrt{\varepsilon})$ is optimal, *in practice* one obtains an L^1-rate of order $\mathcal{O}(\varepsilon)$, when there are finitely many shock discontinuities, [42],[43] (and these are the only solutions that can be computed!). In this case, we can use our Lip' theory to derive local error estimates which improve the L^1-result: using a bootstrap argument we show in [41], that the Lip^+-stable approximate solutions satisfy an $\mathcal{O}(\varepsilon)$ *pointwise* error estimate for all but finitely many $\mathcal{O}(\varepsilon)$-neighborhoods of shock discontinuities.

We now turn to the multidimensional setup. Kuznetsov [15] was the first to provide error estimates for scalar approximate solutions, $\{v^\varepsilon\}$, for both – the one- and multi-dimensional setups. Subsequently, many authors have used Kuznetsov's approach to prove convergence $+$ L^1-error estimates; we refer for the detailed treatments of [31], [22], [42],... . A more recent treatment of [7] employs the entropy dissipation estimate (1.60), which in turn, by Kuznetsov arguments, yields an L^1-convergence rate estimate of order $(\Delta x)^{\frac{1-\theta}{2}}$ (independently of the BV bound).

Kuznetsov's approach employs a *regularized* version of Kružkov's entropy pairs in (1.8), $\eta^\delta(v^\varepsilon; c) \sim |v^\varepsilon - c|$, $F^\delta(v^\varepsilon; c) \sim sgn(v^\varepsilon - c)(A(v^\varepsilon) - A(c))$. Here, one measures by how much the entropy dissipation rate of $\{v^\varepsilon\}$ fails to satisfy the entropy inequality (1.3), with Kružkov's regularized entropies. Following the general recent convergence result of [2], we consider a family of approximate solutions, $\{v^\varepsilon\}$, which satisfies

$$\partial_t |v^\varepsilon - c| + \nabla_x \cdot \{sgn(v^\varepsilon - c)(A(v^\varepsilon) - A(c))\} \leq \partial_t R_0(t,x) + \nabla_x \cdot R(t,x),$$
$$(4.224)$$

with

$$\|R_0(t,x)\|_{\mathcal{M}_{x,t}} + \|R(x,t)\|_{\mathcal{M}_{t,x}} \leq \text{Const} \cdot \varepsilon. \qquad (4.225)$$

Then, the convergence rate proof proceeds along the lines of Theorem 1.1: Using the key property of symmetry of the regularized entropy pairs, $(\eta^\delta := \varphi_\delta \eta, F^\delta := \varphi_\delta F)$, one finds $\int_x \eta^\delta(v^\varepsilon; u)dx \leq \text{Const}.\varepsilon/\delta$. In addition, there is a regularization error, $\|\eta^\delta - \eta\|_{L^1(x)}$, of size $\mathcal{O}(\delta)$, and an L^1 error estimate of order $\mathcal{O}(\sqrt{\varepsilon})$ follows (under reasonable assumptions on the L^1-initial error w.r.t. BV data), consult [2]

$$\|v^\varepsilon(\cdot,t) - u(\cdot,t)\|_{L^1_{\text{loc}(x)}} \leq \text{Const}.\sqrt{\varepsilon}.$$

Observe that this error estimate, based on (4.224)-(4.225) is the multidimensional analogue of our Lip'-consistency requirement. In general, Kuznetsov approach makes a stronger requirement of approximate entropy inequalities (i.e., in terms of *all* of Kružkov's pairs), and in return, ones obtains convergence results which apply to general, non-convex equations. The lectures by B. Cockburn provide a detailed account of Kuznetsov's L^1-convergence theory. In this chapter we therefore focus our attention on the Lip'-convergence theory mentioned above. Its multidimensional extension deals with convex Hamilton-Jacobi equations (rather than conservation laws), consult §4.3.

In §4.4 we implement these error estimates for a variety of approximate solutions. The examples we discuss include

- Regularized Chapman-Enskog approximations [33];
- Finite-difference E-schemes [24];
- Godunov-type schemes [26];
- Glimm's scheme [24];
- Spectral viscosity approximations, [40]

Other examples dealing with 2×2 systems with and without stiff relaxation coupling terms could be found in [27],[18].

4.2 Approximate solutions

We study approximate solutions of the scalar convex conservation law

$$\frac{\partial}{\partial t}u(x,t) + \frac{\partial}{\partial x}f(u(x,t)) = 0, \quad f'' \geq \alpha > 0, \tag{4.226}$$

with compactly supported initial conditions prescribed at $t = 0$,

$$u(x, t = 0) = u_0(x). \tag{4.227}$$

Let $\{v^\varepsilon(x,t)\}_{\varepsilon>0}$ be a family of approximate solutions of the conservation law (4.226), (4.227) in the following sense.

Definition 4.1 A. We say that $\{v^\varepsilon(x,t)\}_{\varepsilon>0}$ are *conservative* solutions if

$$\int_x v^\varepsilon(x,t)dx = \int_x u_0(x)dx, \quad t \geq 0. \tag{4.228}$$

B. We say that $\{v^\varepsilon(x,t)\}_{\varepsilon>0}$ are Lip'-*consistent* with the conservation law (4.226), (4.227) if the following estimates are fulfilled[8]:

(i) consistency with the initial conditions (4.227),

$$\|v^\varepsilon(x,0) - u_0(x)\|_{Lip'} \leq K_0 \cdot \varepsilon \tag{4.229}$$

(ii) consistency with the conservation law (4.226),

$$\|v_t^\varepsilon(x,t) + f(v^\varepsilon(x,t))_x\|_{Lip'(x,[0,T])} \leq K_T \cdot \varepsilon. \tag{4.230}$$

---

[8] We let$\|\varphi\|_{Lip}, \|\varphi\|_{Lip+}and\|\varphi\|_{Lip'}$ denote respectively, $esssup_{x \neq y} \left| \frac{\varphi(x)-\varphi(y)}{x-y} \right|$, $esssup_{x \neq y} \left[\frac{\varphi(x)-\varphi(y)}{x-y} \right]_+$ and $sup_\psi \frac{(\varphi-\hat{\varphi}_0, \psi)}{\|\psi\|_{Lip}}$, where $\hat{\varphi}_0 = \int_{\text{supp}\varphi} \varphi$.

We are interested in the *convergence rate* of the approximate solutions, $v^\varepsilon(x,t)$, as their small parameter $\varepsilon \downarrow 0$. This requires an appropriate stability definition for such approximate solutions. Recall that the entropy solution of the nonlinear conservation law (4.226), (4.227) satisfies the a priori estimate [4,39]

$$\|u(\cdot,t)\|_{Lip^+} \leq \frac{1}{\|u_0\|_{Lip^+}^{-1} + \alpha t}, \quad t \geq 0. \tag{4.231}$$

The case $\|u_0\|_{Lip^+} = \infty$ is included in (4.231), and it corresponds to the exact $\sim t^{-1}$ decay rate of an initial rarefaction.

Definition 4.2 We say that $\{v^\varepsilon(x,t)\}_{\varepsilon>0}$ are *Lip^+-stable* if there exists a constant $\beta \geq 0$ (independent of t and ε) such that the following estimate, analogous to (4.231), is fulfilled:

$$\|v^\varepsilon(\cdot,t)\|_{Lip^+} \leq \frac{1}{\|v^\varepsilon(\cdot,0)\|_{Lip^+}^{-1} + \beta t}, \quad t \geq 0. \tag{4.232}$$

Remarks.

1. The case of an initial rarefaction subject to the quadratic flux $f(u) = \frac{\alpha}{2}u^2$ demonstrates that the a priori decay estimate of the exact entropy solution in (4.231) is sharp. A comparison of (4.232) with (4.231) shows that a *necessary* condition for the convergence of $\{v^\varepsilon\}_{\varepsilon>0}$ is

$$0 \leq \beta \leq \alpha, \tag{4.233}$$

   for otherwise, the decay rate of $\{v^\varepsilon(\cdot,t)\}$ (and hence of its $\varepsilon \to 0$ limit) would be *faster* than that of the exact entropy solution.
2. The case $\beta > 0$ in (4.232) corresponds to a *strict Lip^+-stability* in the sense that $\|v^\varepsilon(\cdot,t)\|_{Lip^+}$ decays in time, in agreement with the decay of rarefactions indicated in (4.231).
3. In general, any a priori bound

$$\|v^\varepsilon(\cdot,t)\|_{Lip^+} \leq \text{Const}_T < \infty, \quad 0 \leq t \leq T, \tag{4.234}$$

   is a sufficient stability condition for the convergence results discussed below. In particular, we allow for $\beta = 0$ in (4.232), as long as the approximate initial conditions are Lip^+-bounded. We remark that the restriction of Lip^+-bounded initial data is indeed necessary for convergence, in view of the counterexample of Roe's scheme discussed in remark 4.2 in §4.4. Unless stated otherwise, we therefore restrict our attention to the class of Lip^+-bounded (i.e., rarefaction-free) initial conditions, where

$$L_0^+ := \max(\|u_0\|_{Lip^+}, \|v^\varepsilon(\cdot,0)\|_{Lip^+}) < \infty. \tag{4.235}$$

4. Finally, we remark that in case of strict Lip^+-stability, i.e., in case (4.232) holds with $\beta > 0$, then one can remove this restriction of Lip^+-bounded initial data and our convergence results can be extended to include general L_{loc}^∞-initial conditions, initial rarefaction are included. The discussion of this case could be found in [25], and it leads to similar error estimates discussed in this chapter, with ε being replaced by $\varepsilon \log(\varepsilon)$.

4.3 Convergence rate estimates

Convex conservation laws

We begin with the following theorem which is at the heart of matter.

Theorem 4.1 *A. Let* $\{v^\varepsilon(x,t)\}_{\varepsilon>0}$ *be a family of conservative, Lip^+-stable approximate solutions of the convex conservation law (4.226),(4.227), subject to the Lip^+- bounded initial conditions (4.235). Then the following error estimate holds*

$$\|v^\varepsilon(\cdot,T) - u(\cdot,T)\|_{Lip'} \leq C_T \left[\|v^\varepsilon(\cdot,0) - u_0(\cdot)\|_{Lip'} + \|v_t^\varepsilon + f(v^\varepsilon)_x\|_{Lip'(x,[0,T])}\right],$$
(4.236)

where

$$C_T \sim (1 + \beta L_0^+ T)^\eta, \quad \eta := \frac{\max f''}{\beta} \geq 1.$$

B. In particular, if the family $\{v^\varepsilon(x,t)\}_{\varepsilon>0}$ *is also Lip'-consistent of order* $\mathcal{O}(\varepsilon)$*, i.e., (4.229),(4.230) hold, then* $v^\varepsilon(x,t)$ *converges to the entropy solution* $u(x,t)$ *and the following convergence rate estimate holds*

$$\|v^\varepsilon(\cdot,T) - u(\cdot,T)\|_{Lip'} \leq M_T \cdot \varepsilon, \quad M_T := (K_0 + K_T)(1 + \beta L_0^+ T)^\eta.$$
(4.237)

Proof. We proceed along the lines of [39,24]. The difference, $e^\varepsilon(x,t) := v^\varepsilon(x,t) - u(x,t)$, satisfies the error equation

$$\frac{\partial}{\partial t} e^\varepsilon(x,t) + \frac{\partial}{\partial x}[\bar{a}_\varepsilon(x,t)e^\varepsilon(x,t)] = F^\varepsilon(x,t),$$
(4.238)

where $\bar{a}_\varepsilon(x,t)$ stands for the mean-value

$$\bar{a}_\varepsilon(x,t) = \int_{\xi=0}^1 a[\xi v^\varepsilon(x,t) + (1-\xi)u(x,t)]d\xi, \quad a(\cdot) \equiv f'(\cdot),$$

and $F^\varepsilon(x,t)$ is the truncation error,

$$F^\varepsilon(x,t) := v_t^\varepsilon(x,t) + f(v^\varepsilon(x,t))_x.$$

Given an arbitrary $\varphi(x) \epsilon W_0^{1,\infty}$, we let $\{\varphi^\varepsilon(x,t)\}_{0 \le t \le T}$ denote the solution of the backward transport equation

$$\varphi_t^\varepsilon(x,t) + \bar{a}_\varepsilon(x,t)\varphi_x^\varepsilon(x,t) = 0, \quad t \le T, \tag{4.239}$$

corresponding to the endvalues, $\varphi(x)$, prescribed at $t = T$,

$$\varphi^\varepsilon(x,T) = \varphi(x).$$

Here, the following a priori estimate holds [39, Theorem 2.2]

$$\|\varphi^\varepsilon(\cdot,t)\|_{Lip} \le \exp\left(\int_t^T \|\bar{a}_\varepsilon(\cdot,\tau)\|_{Lip+} d\tau\right) \cdot \|\varphi(x)\|_{Lip}, \quad 0 \le t \le T. \tag{4.240}$$

The Lip^+-stability of the entropy solution (4.231) and its approximate solutions in (4.232), provide us with the one-sided Lipschitz upper-bound required on the right-hand side of (4.240):

$$\|\bar{a}_\varepsilon(\cdot,\tau)\|_{Lip+} \le \frac{\max f''}{2}[\|v^\varepsilon(\cdot,\tau)\|_{Lip+} + \|u(\cdot,\tau)\|_{Lip+}] \le \frac{\max f''}{[L_0^+]^{-1} + \beta\tau}. \tag{4.241}$$

Equipped with (4.240), (4.241) we conclude

$$\|\varphi^\varepsilon(\cdot,t)\|_{Lip} \le \frac{(1 + \beta L_0^+ T)^\eta}{(1 + \beta L_0^+ t)^\eta}\|\varphi(x)\|_{Lip} \le \tag{4.242}$$

$$\le C_T\|\varphi(x)\|_{Lip}, \quad 0 \le t \le T, \quad C_T := (1 + \beta L_0^+ T)^\eta,$$

and employing (4.239) we also have

$$\|\varphi^\varepsilon(x,\cdot)\|_{Lip[0,T]} \le |a|_\infty \max_{0 \le t \le T}\|\varphi^\varepsilon(\cdot,t)\|_{Lip(x)} \le \tag{4.243}$$

$$\le |a|_\infty C_T\|\varphi(x)\|_{Lip}, \quad |a|_\infty := \max|f'|.$$

Of course, (4.239) is just the adjoint problem of the error equation (4.238) which gives us

$$(e^\varepsilon(\cdot,T), \varphi(\cdot)) = (e^\varepsilon(\cdot,0), \varphi^\varepsilon(\cdot,0)) + (F^\varepsilon(x,t), \varphi^\varepsilon(x,t))_{L^2(x,[0,T])}. \tag{4.244}$$

Conservation implies that $\hat{e}_0^\varepsilon \equiv \int e^\varepsilon(x,0)dx = 0$ and by (4.242) we find

$$|(e^\varepsilon(\cdot,0), \varphi^\varepsilon(\cdot,0))| \le \|e^\varepsilon(\cdot,0)\|_{Lip'}\|\varphi^\varepsilon(\cdot,0)\|_{Lip} \le$$

$$\le (1 + \beta L_0^+ T)^\eta\|e^\varepsilon(\cdot,0)\|_{Lip'} \cdot \|\varphi(x)\|_{Lip};$$

similarly, conservation implies that $\hat{F}_0^\varepsilon \equiv \int_{x,[0,T]} F^\varepsilon(x,t)dxdt = 0$ and by (4.242),(4.243) we find

$$|(F^\varepsilon(x,t), \varphi^\varepsilon(x,t))_{L^2(x,[0,T])}| \le \|F^\varepsilon(x,t)\|_{Lip'(x,[0,T])} \|\varphi^\varepsilon(x,t)\|_{Lip(x,[0,T])} \le$$

$$\le (1 + |a|_\infty) C_T \|F^\varepsilon(x,t)\|_{Lip'(x,[0,T])} \|\varphi(x)\|_{Lip}. \tag{4.245}$$

The error estimate (4.236) follows from the last two estimates together with (4.244). ■

The Lip'-convergence rate estimate (4.237) can be extended to more familiar $W^{s,p}_{loc}$-convergence rate estimates. The rest of this section is devoted to three corollaries which summarize these extensions.

We begin by noting that the conservation and Lip^+-stability of $v^\varepsilon(\cdot, t)$ imply that $v^\varepsilon(\cdot, T)$ – and consequently that the error, $v^\varepsilon(\cdot, T) - u(\cdot, T)$, have bounded variation,

$$\|v^\varepsilon(\cdot, T) - u(\cdot, T)\|_{BV} \le Const \frac{1}{[L_0^+]^{-1} + \beta T}. \tag{4.246}$$

We note in passing that the constant on the right of (4.246) depends on the *finite* size of the support of the error.
We can now interpolate between the BV-bound (4.246) and the Lip'-error estimate (4.237), to conclude the following.

Corollary 4.1 *Let $\{v^\varepsilon(x,t)\}_{\varepsilon>0}$ be a family of conservative, Lip'-consistent and Lip^+-stable approximate solutions of the conservation law (4.226), (4.227), with Lip^+-bounded initial conditions (4.235). Then the following convergence rate estimates hold $\forall p \le \infty$*

$$\|v^\varepsilon(\cdot, T) - u(\cdot, T)\|_{W^{s,p}} \le Const_T \cdot \varepsilon^{\frac{1-sp}{2p}}, \quad -1 \le s \le \frac{1}{p}. \tag{4.247}$$

The error estimate (4.247) with $(s,p) = (0,1)$ yields L^1 convergence rate of order $\mathcal{O}(\sqrt{\varepsilon})$, which is familiar from the setup of monotone difference approximations [15,22,31]. Of course, uniform convergence (which corresponds to $(s,p) = (0,\infty)$) fails in this case, due to the possible presence of shock discontinuities in the entropy solution $u(\cdot, t)$. Instead, one seeks pointwise convergence away from the singular support of $u(\cdot, t)$. To this end, we employ a $C_0^1(-1,1)$-unit mass mollifier of the form $\zeta_\delta(x) = \frac{1}{\delta}\zeta(\frac{x}{\delta})$. The error estimate (4.236) asserts that

$$|(v^\varepsilon(\cdot, T) * \zeta_\delta)(x) - (u(\cdot, T) * \zeta_\delta)(x)| \le M_T \frac{\varepsilon}{\delta^2} \|\frac{d\zeta}{dx}\|_{L^\infty}.$$

Moreover, if $\zeta(x)$ is chosen so that

$$\int x^k \zeta(x) dx = 0 \quad \text{for } k = 1, 2, \ldots, r-1, \tag{4.248}$$

then a straightforward error estimate based on Taylor's expansion yields

$$|(u(\cdot, T) * \zeta_\delta)(x) - u(x, T)| \leq \frac{\delta^r}{r!} \|\zeta\|_{L^1} \cdot |u^{(r)}|_{loc},$$

where $|u^{(r)}|_{loc}$ measures the degree of *local* smoothness of $u(\cdot, t)$,

$$|u^{(r)}|_{loc} := \|\frac{\partial^r}{\partial x^r} u(\cdot, T)\|_{L^\infty_{loc}(x + \delta \cdot \text{supp}\zeta)}.$$

The last two inequalities (with $\delta \sim \varepsilon^{\frac{1}{r+2}}$) imply

Corollary 4.2 *Let* $\{v^\varepsilon(x, t)\}_{\varepsilon > 0}$ *be a family of conservative, Lip'-consistent and Lip^+-stable approximate solutions of the conservation law (4.226), (4.227), with Lip^+-bounded initial conditions (4.235). Then, for any r-order mollifier $\zeta_\delta(x) \equiv \frac{1}{\delta}\zeta(\frac{x}{\delta})$ satisfying (4.248), the following convergence rate estimate holds*

$$|(v^\varepsilon(\cdot, T) * \zeta_\delta)(x) - u(x, T)| \leq \text{Const}(1 + \frac{|u^{(r)}|_{loc}}{r!}) \cdot \varepsilon^{\frac{r}{r+2}}. \tag{4.249}$$

Corollary 4.2 shows that by *post-processing* the approximate solutions $v^\varepsilon(\cdot, t)$, we are able to recover the pointwise values of $u(x, t)$ with an error as close to ε as the local smoothness of $u(\cdot, t)$ permits. A similar treatment enables the recovery of the derivatives of $u(x, t)$ as well, consult [39, §4].

The particular case $r = 1$ in (4.249), deserves special attention. In this case, post-processing of the approximate solution with *arbitrary* C_0^1-unit mass mollifier $\zeta(x)$, gives us

$$|(v^\varepsilon(\cdot, T) * \zeta_\delta)(x) - u(x, T)| \leq \text{Const} \cdot (1 + |u_x(\cdot, T)|_{loc}) \cdot \sqrt[3]{\varepsilon}. \tag{4.250}$$

We claim that the pointwise convergence rate of order $\mathcal{O}(\sqrt[3]{\varepsilon})$ indicated in (4.250) holds even *without* post-processing of the approximate solution. Indeed, let us consider the difference

$$v^\varepsilon(x, T) - (v^\varepsilon(\cdot, T) * \zeta_\delta)(x) = \int_y [v^\varepsilon(x, T) - v^\varepsilon(x - y, T)]\zeta_\delta(y) dy =$$

$$= \int_y \left[\frac{v^\varepsilon(x, T) - v^\varepsilon(x - y, T)}{-y}\right] \cdot -\frac{y}{\delta}\zeta(\frac{y}{\delta}) dy.$$

By choosing a positive C_0^1-unit mass mollifier $\zeta(x)$ supported on $(-1, 0)$ then, thanks to the Lip^+-stability condition (4.232), the integrand on the right does not exceed Const $\cdot \delta$, and hence

$$v^\varepsilon(x, T) - (v^\varepsilon(\cdot, T) * \zeta_\delta)(x) \leq \text{Const} \cdot \delta . \tag{4.251}$$

Similarly, a different choice of a positive C_0^1-unit mass mollifier $\zeta(x)$ supported on $(0, 1)$ leads to

$$v^\varepsilon(x, T) - (v^\varepsilon(\cdot, T) * \zeta_\delta)(x) \geq \text{Const} \cdot \delta. \tag{4.252}$$

Each of the last two inequalities (with $\delta \sim \sqrt[3]{\varepsilon}$) together with (4.250) show that the approximate solution itself converges with an $\mathcal{O}(\sqrt[3]{\varepsilon})$-rate, as asserted. We summarize what we have shown by stating the following.

Corollary 4.3 *Let $\{v^\varepsilon(x, t)\}_{\varepsilon > 0}$ be a family of conservative, Lip'-consistent and Lip^+-stable approximate solutions of the conservation law (4.226), (4.227), with Lip^+-bounded initial conditions (4.235). Then the following convergence rate estimate holds:*

$$|v^\varepsilon(x, T) - u(x, T)| \leq C_x \cdot \sqrt[3]{\varepsilon}, \quad C_x \sim |u_x(\cdot, T)|_{L^\infty(x - \sqrt[3]{\varepsilon}, x + \sqrt[3]{\varepsilon})}. \tag{4.253}$$

The above derivation of pointwise error estimates applies in more general situations. Consider, for example, a family of approximate solutions, $\{v^\varepsilon(x, t)\}_{\varepsilon > 0}$ which satisfies the stronger L^1 error estimate of order, say, $\mathcal{O}(\mu)$,

$$|(v^\varepsilon(\cdot, T) - u(\cdot, T), \varphi(\cdot))| \leq C_x \cdot \mu \|\varphi\|_{L^\infty}. \tag{4.254}$$

Then our previous arguments show how to post-process $v^\varepsilon(\cdot, T)$ in order to recover the pointwise values of the entropy solution, $u(x, T)$ with an error as close to μ as the local smoothness of $u(\cdot, T)$ permits. In particular, using (4.254) with a positive C_0^1-unit mass mollifier, $\zeta_\delta(x) = \frac{1}{\delta}\zeta(\frac{x}{\delta})$ we obtain

$$|(v^\varepsilon(\cdot, T) * \zeta_\delta)(x) - (u(\cdot, T) * \zeta_\delta)(x)| \leq C_x \cdot \frac{\mu}{\delta} \|\zeta\|_{L^\infty}. \tag{4.255}$$

Using this together with

$$|(u(\cdot, T) * \zeta_\delta)(x) - u(x, T)| \leq \delta \|\zeta\|_{L^1} \cdot \|u_x(\cdot, T)\|_{L^\infty_{loc}(x + \delta \cdot \text{supp}\zeta)}, \tag{4.256}$$

we find (with $\delta \sim \sqrt{\mu}$)

$$|(v^\varepsilon(\cdot, T) * \zeta_\delta)(x) - u(x, T)| \leq \text{Const}_T(1 + |u_x(\cdot, T)|_{loc})\sqrt{\mu}. \tag{4.257}$$

If the approximate solutions $\{v^\varepsilon(x, t)\}_{\varepsilon > 0}$ are also Lip^+-stable, then we may augment (4.257) with (4.251)-(4.252) to conclude

Corollary 4.4 *Assume that* $\{v^\varepsilon(x,t)\}$ *is a family of* Lip^+ *stable approximate solutions with global* L^1*-convergence rate of order* $\mathcal{O}(\mu)$, *(4.254). Then the following local pointwise error estimate holds*

$$|v^\varepsilon(x,T) - u(x,T)| \leq C_x \cdot \sqrt{\mu}, \quad C_x \sim |u_x(\cdot,T)|_{L^\infty(x-\sqrt{\mu},x+\sqrt{\mu})}.$$

Remarks.

1. The usual L^1-rate of order $\mu \sim \sqrt{\varepsilon}$ leads to [24]

$$|v^\varepsilon(x,T) - u(x,T)| \leq C_x \cdot \sqrt[4]{\varepsilon}, \quad C_x \sim |u_x(\cdot,T)|_{L^\infty(x-\sqrt[4]{\varepsilon},x+\sqrt[4]{\varepsilon})}. \tag{4.258}$$

2. In case $u(\cdot,t)$ has finitely many shocks, [41], one obtain an L^1-rate of order $\mu \sim \varepsilon$, [42], and hence we find a *local* error of order $\sqrt{\varepsilon}$

$$|v^\varepsilon(x,T) - u(x,T)| \leq C_x \cdot \sqrt{\varepsilon}, \quad C_x \sim |u_x(\cdot,T)|_{L^\infty(x-\sqrt{\varepsilon},x+\sqrt{\varepsilon})}. \tag{4.259}$$

3. Finally, in [41] we improved the estimate (4.259) replacing $\sqrt{\varepsilon}$ by ε. Thus, we obtain an optimal pointwise error estimate of order $\mathcal{O}(\varepsilon)$ in all but finitely many neighborhoods of shock discontinuities of width $\mathcal{O}(\varepsilon)$.

Convex Hamilton-Jacobi equations

In this section we briefly comment on the *multidimensioal* generalization of the Lip'-convergence theory outlined above, to convex Hamilton-Jacobi (HJ) equations. We consider the multidimensional Hamilton-Jacobi (HJ) equation

$$\partial_t u + H(\nabla_x u) = 0, \quad (t,x) \in R^+ \times R^d, \tag{4.260}$$

with convex Hamiltonian, $H'' > 0$. Its unique viscosity solution is identified by the one-sided concavity condition, $D_x^2 u \leq \text{Const.}$, consult [16], [20]. Given a family of approximate HJ solutions, $\{v^\varepsilon\}$, we make the analogous one-sided stability requirement of

– *Demi-concave stability.* The family $\{v^\varepsilon\}$ is demi-concave stable if

$$D_x^2 v^\varepsilon \leq \text{Const.} \tag{4.261}$$

We then have the following.

Theorem 4.2 *([19]) Assume* $\{v_1^\varepsilon\}$ *and* $\{v_2^\varepsilon\}$ *are two demi-concave stable families of approximate solutions. Then*

$$\|v_1^\varepsilon(t,\cdot) - v_2^\varepsilon(t,\cdot)\|_{L^1(x)} \leq \text{Const.}\|v_1^\varepsilon(0,\cdot) - v_2^\varepsilon(0,\cdot)\|_{L^1(x)} +$$

$$+ \text{Const.} \sum_{j=1}^{2} \|\partial_t v_j^\varepsilon + H(\nabla_x v_j^\varepsilon)\|_{L^1(t,x)} \tag{4.262}$$

If we let $v_1^\varepsilon \equiv v^1$, $v_2^\varepsilon \equiv v^2$ denote two demi-concave viscosity solutions, then (4.262) is an L^1-stability statement (compared with the usual L^∞-stability statements of viscosity solutions, [8]). If we let $\{v_1^\varepsilon\} = \{v^\varepsilon\}$ denote a given family of demi-concave approximate HJ solutions, and let v_2^ε equals the exact viscosity solution u, then (4.262) yields the L^1-error estimate

$$\|v^\varepsilon(\cdot,t) - u(\cdot,t)\|_{L^1(x)} \leq Const.\|\partial_t v^\varepsilon + H(\nabla_x v^\varepsilon)\|_{L^1(x,t)} \sim \mathcal{O}(\varepsilon). \quad (4.263)$$

This corresponds to the Lip'-error estimate of (1.46) with $(s,p) = (-1,1)$. One can then interpolate from (4.263) an L^p-error estimates of order $\mathcal{O}(\varepsilon^{\frac{1+p}{2p}})$. For a general L^∞-convergence theory for approximate solutions to HJ equations we refer to [1] and the references therein.

4.4 Examples

Regularized Chapman-Enskog expansion
Of course, the usual viscous approximation

$$\frac{\partial}{\partial t}[v^\varepsilon(x,t)] + \frac{\partial}{\partial x}[f(v^\varepsilon(x,t))] = \varepsilon\frac{\partial^2}{\partial x^2}[Q(v^\varepsilon(x,t))], \quad \varepsilon Q' \downarrow 0, \quad (4.264)$$

is the canonical example for a family of approximate solutions whose convergence rate could be analyzed in terms of our Lip' theory outlined above. Here, we concentrate on yet another, more intricate regularization of the inviscid equations of the form

$$v_t^\varepsilon + f(v^\varepsilon)_x = \left[\frac{-\varepsilon k^2}{1 + m^2\varepsilon^2 k^2}\hat{v}^\varepsilon(k)\right]^\vee,$$

or equivalently,

$$v_t^\varepsilon + f(v^\varepsilon)_x = -\frac{1}{m^2\varepsilon}(u - Q_{m\varepsilon} * u), \quad Q_\mu := \frac{1}{2\mu}e^{-|x|/\mu}. \quad (4.265)$$

Rosenau [28] has proposed this type of equation as a model for his regularized version of the Chapman-Enskog expansion for hydrodynamics. The operator on the right side looks like the usual viscosity term $\varepsilon v_{xx}^\varepsilon$ at low wave-numbers k, while for higher wave numbers it is intended to model a bounded approximation of a linearized collision operator, thereby avoiding the artificial instabilities that occur when the Chapman-Enskog expansion for such an operator is truncated after a finite number of terms [28].

We shall study the convergence rate of v^ε to the inviscid solution, along the lines of [33]. It should be pointed out that the solution of (4.265) does *not* admit all the entropy inequalities, except for the quadratic one; thus, the question of convergence in this case, is not easily answered in terms of the usual L^1-Kuznetsov theory. Instead, we use the Lip' theory outlined in §4.3. To this end, we first turn to show that the nonlinear Regularized Chapman-Enskog (RCE) equation (4.265) satisfies Oleinik's E-entropy condition.

Theorem 4.3 *Assume $f'' \geq \alpha > 0$. Then the following a priori estimate holds*

$$\|v^\varepsilon(t)\|_{Lip^+} \leq \frac{1}{\|v^\varepsilon(0)\|_{Lip^+}^{-1} + \alpha t} \quad , \quad t \geq 0. \tag{4.266}$$

Remark 4.1 The inequality (4.266) implies that the positive-variation and hence the total-variation of $v^\varepsilon(t)$ decays in time. Furthermore, this proves the zero mean-free-path convergence to the entropy solution of (4.226) for any L^∞_{loc}-initial data u_0

Proof. We add the artificial viscosity term δu_{xx} to regularize (4.265), obtaining

$$\partial_t v_\delta^\varepsilon + \partial_x f(v_\delta^\varepsilon) = -\frac{1}{m^2 \varepsilon} \{v_\delta^\varepsilon - Q_{m\varepsilon} * v_\delta^\varepsilon\} + \delta \partial_x^2 v_\delta^\varepsilon. \tag{4.267}$$

Differentiation of (4.267) yields for $w \equiv \partial_x v_\delta^\varepsilon$,

$$\partial_t w + f'(u_\varepsilon^\delta)\partial_x w + f''(u_\varepsilon^\delta)w^2 == -\frac{1}{m^2 \varepsilon}\{w - Q_{m\varepsilon} * w\} + \delta \partial_x^2 w.$$

Hence, since $f'' > \alpha > 0$, it follows that $W(t) \equiv max_x w(t)$ is governed by the differential inequality

$$\dot{W}(t) + \alpha W^2(t) \leq \frac{1}{m^2 \varepsilon}\{W(t) - Q_{m\varepsilon} * W\} \leq 0$$

and (4.266) follows by letting $\delta \downarrow 0$. ∎

Theorem 4.3 shows that solutions of the RCE equation (4.265) are Lip^+-stable. Moreover, (4.265) implies that the Lip'-size of their truncation if of order $\mathcal{O}(\varepsilon)$, for

$$\|\partial_t v^\varepsilon + \partial_x f(v^\varepsilon)\|_{Lip'} = \varepsilon\|Q_{m\varepsilon} * \partial_x v^\varepsilon\|_{L^1} \leq \varepsilon\|Q_{m\varepsilon}\|_{L^1}\|v^\varepsilon(t)\|_{BV} \leq \varepsilon\|u_\varepsilon(0)\|_{BV}.$$

Using our main result we conclude that the Lip'- convergence *rate* of the RCE solutions to the corresponding entropy solution is also of order $\mathcal{O}(\varepsilon)$.

Corollary 4.5 *Assume that $f'' \geq \alpha > 0$, and let v^ε be the unique RCE solution of (4.265) subject to C^1 initial conditions $v^\varepsilon(0) \equiv u(0)$. then v^ε converges to the unique entropy solution of (4.226) and the following error estimates hold*

$$\|v^\varepsilon(t) - u(t)\|_{W^{s,p}} \leq Const \cdot \varepsilon^{\frac{1-sp}{2p}}, \quad -1 \leq s \leq \frac{1}{p}. \tag{4.268}$$

Finite-Difference approximations
We want to solve the conservation law (4.226)-(4.227) by difference approximations. To this end we use a grid $(x_\nu := \nu \Delta x, t^n := n \Delta t)$ with a fixed mesh-ratio $\lambda \equiv \frac{\Delta t}{\Delta x} = \text{Const}$. The approximate solution at these grid points, $v_\nu^n \equiv v(x_\nu, t^n)$, is determined by a conservative difference approximation which takes the following viscosity form, e.g., [35][9]

$$v_\nu^{n+1} = v_\nu^n - \frac{\lambda}{2}[f(v_{\nu+1}^n) - f(v_{\nu-1}^n)] + \frac{1}{2}[Q_{\nu+\frac{1}{2}}^n \Delta v_{\nu+\frac{1}{2}}^n - Q_{\nu-\frac{1}{2}}^n \Delta v_{\nu-\frac{1}{2}}^n],$$
(4.269)

and is subject to Lip^+-bounded initial conditions,

$$v_\nu^0 = \frac{1}{\Delta x} \cdot \int_{x_{\nu-\frac{1}{2}}}^{x_{\nu+\frac{1}{2}}} u_0(\xi)d\xi, \quad L_0^+ = \|u_0\|_{Lip^+} < \infty.$$
(4.270)

Let $v^\varepsilon(x, t)$ be the piecewise linear interpolant of our grid solution, $v^\varepsilon(x_\nu, t^n) = v_\nu^n$, depending on the small discretization parameter $\varepsilon \equiv \Delta x \downarrow 0$. It is given by

$$v^{\Delta x}(x, t) = \sum_{j,m} v_j^m \Lambda_j^m(x, t), \quad \Lambda_j^m(x, t) := \Lambda_j(x) \Lambda^m(t),$$

where $\Lambda_j(x)$ and $\Lambda^m(t)$ denote the usual 'hat' functions,

$$\Lambda_j(x) = \frac{1}{\Delta x} \min(x - x_{j-1}, x_{j+1} - x)_+, \quad \Lambda^m(t) = \frac{1}{\Delta t} \min(t - t^{m-1}, t^{m+1} - t)_+.$$

In [24] we show that these schemes are Lip'-consistent of order $\mathcal{O}(\Delta x)$, thus arriving at

Theorem 4.4 *Assume that the difference approximation (4.269)-(4.270) is Lip^+-stable in the sense that the following one-sided Lipschitz condition is fulfilled:*

$$\max_\nu \frac{(\Delta v_{\nu+\frac{1}{2}}^n)_+}{\Delta x} \le \frac{1}{[L_0^+]^{-1} + \beta t^n}, \quad 0 \le t^n \le T.$$
(4.271)

Then the following error estimates hold:

$$\|v^{\Delta x}(\cdot, T) - u(\cdot, T)\|_{W^{s,p}} \le C_T \cdot (\Delta x)^{\frac{1-sp}{2p}}, \quad -1 \le s \le \frac{1}{p},$$
(4.272)

$$|v^{\Delta x}(x, T) - u(x, T)| \le C_x \cdot \max_{|\xi - x| \le \sqrt[3]{\Delta x}} |u_x(\xi, T)| \cdot \sqrt[3]{\Delta x}.$$
(4.273)

---

[9] We use the usual notations for forward and backward differences,
$$\Delta_\pm v_{\nu+\frac{1}{2}} := \pm(v_{\nu\pm1} - v_\nu).$$

The following first order accurate schemes (identified in a decreasing order according to their numerical viscosity coefficient, $Q_{\nu+\frac{1}{2}} \equiv Q^n_{\nu+\frac{1}{2}}$), are frequently referred to in the literature.

$$\text{Lax} - \text{Friedrichs}: \quad Q^{LxF}_{\nu+\frac{1}{2}} \equiv 1, \tag{4.274}$$

$$\text{Engquist} - \text{Osher}: \quad Q^{EO}_{\nu+\frac{1}{2}} = \frac{\lambda}{v^n_{\nu+1} - v^n_\nu} \int_{v_\nu}^{v_{\nu+1}} |f'(v)|dv, \tag{4.275}$$

$$\text{Godunov}: \quad Q^G_{\nu+\frac{1}{2}} = \lambda \max_v \left[\frac{f(v^n_{\nu+1}) + f(v^n_\nu) - 2f(v)}{v^n_{\nu+1} - v^n_\nu} \right], \tag{4.276}$$

$$\text{Roe}: \quad Q^R_{\nu+\frac{1}{2}} = \lambda \left| \frac{\Delta f^n_{\nu+\frac{1}{2}}}{\Delta v^n_{\nu+\frac{1}{2}}} \right|. \tag{4.277}$$

In [24] we prove the Lip^+ stability of these schemes, and together with their Lip' consistency (of order $\mathcal{O}(\Delta x)$) we arrive at

Corollary 4.6 *Consider the conservation law (4.226), (4.227) with Lip^+-bounded initial data (4.235). Then the Roe, Godunov, Engquist-Osher, and Lax-Friedrichs difference approximations (4.274)-(4.277) with discrete initial data (4.270) converge, and their piecewise-linear interpolants $v^{\Delta x}(x,t)$, satisfy the convergence rate estimates (4.272), (4.273).*

Remark 4.2 The Lip^+-stability (4.271) of Roe scheme with $\beta = 0$ (no decay), was proved in [3]. Note that the assumption of Lip^+-bounded initial conditions is essential for convergence to the entropy solution in this case, in view of the discrete steady-state solution, $v^0_\nu = \text{sgn}(\nu + \frac{1}{2})$, which shows that convergence of Roe scheme to the correct entropy rarefaction fails due to the fact that the initial data are *not* Lip^+-bounded.

Godunov type schemes
Godunov type schemes form a special class of transport projection methods for the approximate solution of nonlinear hyperbolic conservation laws. This class of schemes takes the following form:

$$v^{\Delta x}(\cdot, t) = \begin{cases} T_{\{t-t^{n-1}\}} v^{\Delta x}(\cdot, t^{n-1}), & t^{n-1} < t < t^n \\ P(\{I^n_j\}) v^{\Delta x}(\cdot, t^n - 0), & t = t^n = n\Delta t \end{cases} \tag{4.278}$$

where the initialization step is:

$$v^{\Delta x}(\cdot, t^0 = 0) = P(\{I^0_j\}) u_0(\cdot) . \tag{4.279}$$

These schemes are composed of the following four ingredients:

(i) The possibly variable size grid cells, $I_j^n \equiv [x_{j-\frac{1}{2}}^n, x_{j+\frac{1}{2}}^n)$, where the grid is regular in the sense that:

$$\Delta x \equiv \Delta x_{min} \leq |I_j^n| \leq \Delta x_{max} \quad ; \quad \frac{\Delta x_{max}}{\Delta x_{min}} \leq Const. \; ; \quad (4.280)$$

(ii) A conservative piecewise polynomial grid projection, $P = P(\{I_j^n\})$,

$$\int_x Pw(x)dx = \int_x w(x)dx \; ; \quad (4.281)$$

(iii) The exact entropy solution operator associated with (4.219), $T = T_t$;
(iv) The time step Δt, which is restricted by the CFL condition:

$$\lambda \max_{x,t} |f'(v^{\Delta x}(x,t))| \leq 1 \; , \quad \lambda = \frac{\Delta t}{\Delta x} \; . \quad (4.282)$$

As an example we recall here the subclass of Godunov-type schemes based on piecewise-polynomial projections, which was discussed already in the 'short guide' introduced in Lecture II.

To study the convergence rate of this class of schemes, we are required to verify the Lip'-consistency and Lip^+-stability of the scheme in question. We begin by reducing the question of Lip'-consistency to the level of a mere approximation problem, namely, measuring in Lip'-semi-norm the distance between the exact solution and its grid projection. Thus, our first theorem below enables us to avoid the delicate bookkeeping of error accumulation due to the dynamic transport part of the scheme.

Theorem 4.5 *(Lip'-consistency.) The Godunov type approximation (4.278)-(4.279) satisfies the following truncation error estimate:*

$$\|v_t^{\Delta x} + f(v^{\Delta x})_x\|_{Lip'(x,[0,T])} \leq \frac{T}{\Delta t} \max_{0 < t^n \leq T} \|(P - I)v^{\Delta x}(\cdot, t^n - 0)\|_{Lip'} \quad (4.283)$$

Remark 4.3 We emphasize that this theorem applies to both fixed and variable grid schemes.

Proof. Let N denote the number of time steps in $[0, T]$, i.e.

$$T = t^N = N\Delta t \; . \quad (4.284)$$

Then for every $\varphi \in C_0^1(\Re \times [0,T])$

$$(v_t^{\Delta x} + f(v^{\Delta x})_x, \varphi)_{x,t} = \sum_{n=1}^{N} \left[\int_{t^{n-1}}^{t^n} \int_x v_t^{\Delta x} \varphi \, dx \, dt + \int_{t^{n-1}}^{t^n} \int_x f(v^{\Delta x})_x \varphi \, dx \, dt \right] \quad .$$

Integration by parts gives that

$$(v_t^{\Delta x} + f(v^{\Delta x})_x, \varphi)_{x,t} = \sum_{n=1}^{N} (v^{\Delta x}, \varphi) \Big|_{t^n-0}^{t^n} = \sum_{n=1}^{N} ((P-I)v^{\Delta x}(\cdot, t^n - 0), \varphi(\cdot, t^n)) \quad .$$

(4.285)

But since $v^{\Delta x}$ is a weak solution in the strip $\Re \times (t^{n-1}, t^n)$, as definition (4.278) implies, then

$$\int_{t^{n-1}}^{t^n} \left((v^{\Delta x}, \varphi_t) + (f(v^{\Delta x}), \varphi_x) \right) dt = (v^{\Delta x}, \varphi) \Big|_{t^{n-1}+0}^{t^n-0} \quad . \tag{4.286}$$

Therefore, by (4.285) and (4.286),

$$(v_t^{\Delta x} + f(v^{\Delta x})_x, \varphi)_{x,t} = \sum_{n=1}^{N} \left[(v^{\Delta x}, \varphi) \Big|_{t^{n-1}}^{t^n} - (v^{\Delta x}, \varphi) \Big|_{t^{n-1}+0}^{t^n-0} \right] \quad ,$$

and since, by (4.278), $v^{\Delta x}(\cdot, t^{n-1} + 0) = v^{\Delta x}(\cdot, t^{n-1})$, we have that

$$(v_t^{\Delta x} + f(v^{\Delta x})_x, \varphi)_{x,t} = \sum_{n=1}^{N} (v^{\Delta x}, \varphi) \Big|_{t^n-0}^{t^n} = \sum_{n=1}^{N} ((P-I)v^{\Delta x}(\cdot, t^n - 0), \varphi(\cdot, t^n)) \quad .$$

Recall the conservation of P asserted in (4.281), $\int (P-I)v^{\Delta x} dx = 0$. Therefore, using the definition of the Lip'-seminorm, together with (4.284), we get

$$|(v_t^{\Delta x} + f(v^{\Delta x})_x, \varphi)_{x,t}| \le \frac{T}{\Delta t} \max_{1 \le n \le N} \|(P-I)v^{\Delta x}(\cdot, t^n - 0)\|_{Lip'} \|\varphi(\cdot, t^n)\|_{Lip}$$

Dividing by $\|\varphi(x,t)\|_{Lip}$ and taking the supremum over φ, we arrive at (4.283). ∎

Next, we turn to the question of Lip^+-stability. The standard Lip^+-seminorm, $\| \cdot \|_{Lip^+}$, is inappropriate measure for the size of *discontinuous* piecewise polynomial functions, since increasing jumps – even on the acceptable scale of the gridsize, are Lip^+-unbounded. Instead, we replace it by its discrete analogue – $\| \cdot \|_{\ell ip^+}$, requiring

$$\|v^{\Delta x}(\cdot, t^n)\|_{\ell ip^+} := \max_x \left(\frac{v^{\Delta x}(x + \Delta x, t^n) - v^{\Delta x}(x, t^n)}{\Delta x} \right)^+ \le Const.$$

(4.287)

The discrete ℓip^+ stability is weaker than Lip^+ stability, yet, as we shall show below, it will suffice for our convergence rate estimates to hold. To see this, we introduce a compactly supported non-negative unit mass mollifier,

$$\psi_\delta(x) = \frac{1}{\delta}\psi(\frac{x}{\delta}) \quad , \quad \int_x \psi_\delta(x)dx = \int_x \psi(x)dx = 1 \ . \tag{4.288}$$

The discrete ℓip^+ stability is related to the stronger Lip^+ bound on the *mollified* solution. The following lemma shows that Lip'-consistency of order $\mathcal{O}(\Delta x)$ remains invariant under a mollification with ψ_δ, $\delta = \mathcal{O}(\Delta x)$. Thus, $\mathcal{O}(\Delta x)$-mollification does not sacrifice accuracy yet we have the advantage of using the weaker discrete ℓip^+ stability.

Lemma 4.1 *Assume $v^{\Delta x}(x,t)$ has a bounded variation and is Lip'-consistent with (4.219) of order $\mathcal{O}(\Delta x)$,*

$$\|F^{\Delta x}(x,t)\|_{Lip'} = O(\Delta x) \quad , \quad F^{\Delta x}(x,t) \equiv v_t^{\Delta x} + f(v^{\Delta x})_x \ . \tag{4.289}$$

Then $v^{\Delta x,\delta} \equiv \psi_\delta * v^{\Delta x}$ is Lip'-consistent with (4.219) of order $\mathcal{O}(\Delta x)+\mathcal{O}(\delta)$.

We omit the straightforward proof (which could be found in [26]). Finally, we combine Theorem 4.5 and Lemma 4.1 to achieve our main convergence rate estimate for Godunov type schemes.

Theorem 4.6 *(Convergence rate estimates) Assume that the Godunov type approximation (4.278)-(4.279) is ℓip^+-stable, (4.287), and Lip'-consistent in the sense that*

$$\|(P - I)w\|_{Lip'} \le O(\Delta x^2)\|w\|_{BV} \ . \tag{4.290}$$

Then the following error estimates hold:

$$\|v^{\Delta x}(\cdot,t) - u(\cdot,t)\|_{W^{s,p}} = \mathcal{O}(\Delta x^{\frac{1-sp}{2p}}), -1 \le s \le \frac{1}{p}. \tag{4.291}$$

Proof. Let us denote $\tilde{v}^{\Delta x}(\cdot,t) \equiv \psi_{\Delta x} * v^{\Delta x}(\cdot,t)$, where $\psi_{\Delta x}$ is the dilated mollifier of

$$\psi(x) = \begin{cases} 1, |x| \le \frac{1}{2} \\ 0, |x| > \frac{1}{2} \end{cases} \ . \tag{4.292}$$

This choice of mollifier satisfies the Lip'-error estimate

$$\|\psi_{\Delta x} * w - w\|_{Lip'} \le \mathcal{O}(\Delta x^2)\|w\|_{BV} \ . \tag{4.293}$$

We show that $\tilde{v}^{\Delta x}$ satisfies the Lip^+-stability condition (4.232), and it is Lip'-consistent of order $\mathcal{O}(\Delta x)$.

We start with the Lip^+-stability question. The definition of the discrete ℓip^+-seminorm, (4.287), implies that $\|\tilde{v}^{\Delta x}(\cdot, t^n)\|_{Lip^+} = \|v^{\Delta x}(\cdot, t^n)\|_{\ell ip^+}$. Since $v^{\Delta x}$ is assumed to be discrete ℓip^+-stable, we conclude that at each time level t^n we have

$$\|\tilde{v}^{\Delta x}(\cdot, t^n)\|_{Lip^+} = D_n \leq C \ . \tag{4.294}$$

This, together with the fact that the intermediate exact solution operator decreases the Lip^+-seminorm, (4.231) imply Lip^+-boundedness for all $t \geq 0$:

$$\|\tilde{v}^{\Delta x}(\cdot, t)\|_{Lip^+} \leq Const. \qquad \forall t \geq 0 \ . \tag{4.295}$$

Namely, the mollified approximation $\tilde{v}^{\Delta x}$ is Lip^+-stable.

We note in passing that $v^{\Delta x}(\cdot, t)$, being compactly supported and Lip^+-bounded, has bounded variation. Turning to the question of Lip'-consistency we therefore conclude from assumption (4.290) together with the truncation error estimate (4.283), that $v^{\Delta x}$ is Lip'-consistent with (4.226) of order $\mathcal{O}(\Delta x)$, and hence by lemma 4.1, so does $\tilde{v}^{\Delta x}$,

$$\|\tilde{v}_t^{\Delta x} + f(\tilde{v}^{\Delta x})_x\| = O(\Delta x) \ .$$

Furthermore, $\tilde{v}^{\Delta x}$ is also Lip'-consistent with the initial condition (4.227), since by (4.293), (4.279) and (4.290):

$$\|\tilde{v}^{\Delta x}(\cdot, 0) - u(\cdot, 0)\|_{Lip'} \leq \|\tilde{v}^{\Delta x}(\cdot, 0) - v^{\Delta x}(\cdot, 0)\|_{Lip'} + \|v^{\Delta x}(\cdot, 0) - u_0(\cdot)\|_{Lip'}$$
$$\leq \leq O(\Delta x^2).$$

Therefore, Theorem 4.1 holds; in particular (4.236) tells us that

$$\|\tilde{v}^{\Delta x}(\cdot, T) - u(\cdot, T)\|_{Lip'} = \mathcal{O}(\Delta x) \ . \tag{4.296}$$

In addition, we have by (4.293),

$$\|\tilde{v}^{\Delta x}(\cdot, T) - v^{\Delta x}(\cdot, T)\|_{Lip'} = O(\Delta x^2) \ . \tag{4.297}$$

Combining (4.296) and (4.297) we end up with

$$\|v^{\Delta x}(\cdot, T) - u(\cdot, T)\|_{Lip'} = O(\Delta x) \ . \tag{4.298}$$

The Lip'-error estimate (4.298) may now be interpolated into the $W^{s,p}$-error estimates (4.291). ∎

Examples of the first-order Godunov and Engquist-Osher schemes as well as the second-order (upwind) MUSCL and (central) Nessyahu-Tadmor schemes are discussed in [26].

Glimm scheme

We recall the construction of Glimm approximate solution for the conservation law (4.226), see [10,32]. We let $v(x,t)$ be the entropy solution of (4.226) in the slab $t^n \leq t < t^{n+1}, n \geq 0$, subject to piecewise constant data $v(x,t^n) = \sum_\nu v_\nu^n \chi_\nu(x)$. To proceed in time, the solution is extended (in a staggered fashion) with a jump discontinuity across the lines $t^{n+1}, n \geq 0$, where $v(x, t^{n+1})$ takes the piecewise constant values

$$v(x,t^{n+1}) = \sum_\nu v_{\nu+\frac{1}{2}}^{n+1} \chi_{\nu+\frac{1}{2}}(x), \quad v_{\nu+\frac{1}{2}}^{n+1} = v(x_{\nu+\frac{1}{2}} + r^n \Delta x, t^{n+1} - 0).$$

$$(4.299)$$

Notice that in each slab, $v(x,t)$ consists of successive noninteracting Riemann solutions provided the CFL condition, $\lambda \cdot \max |a(u)| \leq \frac{1}{2}$ is met. This defines the Glimm approximate solution, $v(x,t) \equiv v^\varepsilon(x,t)$, depending on the mesh parameters $\varepsilon = \Delta x \equiv \lambda \Delta t$, and the set of random variables $\{r^n\}$, uniformly distributed in $[-\frac{1}{2}, \frac{1}{2}]$. In the deterministic version of the Glimm scheme, Liu [21] employs equidistributed rather than random sequence of numbers $\{r^n\}$. We note that in both versions, we make use of exactly *one* random or equidistributed choice per time step (independently of the spatial cells), as was first advocated by Chorin [5].

It follows that both versions of Glimm scheme share the Lip^+-stability estimate (4.232). Indeed, since the solution of a scalar Riemann problem remains in the convex hull of its initial data, we may express $v_{\nu+\frac{1}{2}}^{n+1}$ as $(1 - \theta_{\nu+\frac{1}{2}}^n)v_\nu^n + \theta_{\nu+\frac{1}{2}}^n v_{\nu+1}^n$ for some $\theta_{\nu+\frac{1}{2}}^n \in [0,1]$, and hence

$$v_{\nu+\frac{1}{2}}^{n+1} - v_{\nu-\frac{1}{2}}^{n+1} = \theta_{\nu+\frac{1}{2}}^n \Delta v_{\nu+\frac{1}{2}}^n + (1 - \theta_{\nu-\frac{1}{2}}^n)\Delta v_{\nu-\frac{1}{2}}^n.$$

We now distinguish between two cases. If either $\Delta v_{\nu-\frac{1}{2}}^n$ or $\Delta v_{\nu+\frac{1}{2}}^n$ is negative, then

$$v_{\nu+\frac{1}{2}}^{n+1} - v_{\nu-\frac{1}{2}}^{n+1} \leq \max(\Delta v_{\nu+\frac{1}{2}}^n, \Delta v_{\nu-\frac{1}{2}}^n). \tag{4.300}$$

Otherwise — when both $\Delta v_{\nu+\frac{1}{2}}^n$ and $\Delta v_{\nu-\frac{1}{2}}^n$ are positive, the two values of $v_{\nu+\frac{1}{2}}^{n+1}$ and $v_{\nu-\frac{1}{2}}^{n+1}$ are obtained as sampled values of two consecutive rarefaction waves, and a straightforward computation shows that their difference satisfies (4.300). Thus in either case, the Lip^+-stability (4.232) holds with $\beta = 0$.

Although Glimm approximate solutions are conservative "on the average," they do not satisfy the conservation requirement (4.228). We therefore need to slightly modify our previous convergence arguments in this case.

We first recall the truncation error estimate for the deterministic version of Glimm scheme [14, Theorem 3.2],

$$\left(v_t^{\Delta x} + f(v^{\Delta x})_x, \varphi(x,t)\right)_{L^2(x,[0,T])} \leq$$

$$\leq \text{Const}_T \left[\sqrt{\Delta x}|\ln \Delta x| \cdot \|\varphi\|_{L^\infty} + \Delta x \cdot \|\varphi(x,t)\|_{Lip(x,[0,T])}\right]. \tag{4.301}$$

Let $\varphi(x,t) = \varphi^{\Delta x}(x,t)$ denote the solution of the adjoint error equation (4.239). Applying (4.301) instead of (4.245) and arguing along the lines of Theorem (4.1), we conclude that Glimm scheme is Lip'-consistent (and hence has a Lip'-convergence rate) of order $\sqrt{\Delta x}|\ln \Delta x|$,

$$|(e^{\Delta x}(\cdot,T), \varphi(\cdot))| \leq \text{Const}_T \left[\sqrt{\Delta x}|\ln \Delta x| \cdot \|\varphi\|_{L^\infty} + \Delta x \cdot \|\varphi(x)\|_{Lip}\right]. \tag{4.302}$$

To obtain an L^1-convergence rate estimate we employ (4.302) with $\varphi_\delta \equiv \varphi * \frac{1}{\delta}\zeta\left(\frac{\cdot}{\delta}\right)$ yielding

$$|(e^{\Delta x}(\cdot,T), \varphi_\delta)| \leq \text{Const}_T \left[\sqrt{\Delta x}|\ln \Delta x| + \frac{\Delta x}{\delta}\right] \|\varphi(x)\|_{L^\infty}. \tag{4.303}$$

Using this estimate together with

$$(e^\varepsilon(\cdot,T), [\varphi(\cdot) - \varphi_\delta(\cdot)]) \equiv (e^\varepsilon(\cdot,T) - e_\delta^\varepsilon(\cdot,T), \varphi) \leq \text{Const}\cdot\|e^\varepsilon(\cdot,T)\|_{BV}\cdot\delta\|\varphi\|_{L^\infty},$$

imply (for $\delta \sim \sqrt{\Delta x}$), the usual L^1-convergence rate of order $O(\sqrt{\Delta x}|\ln \Delta x|)$. As noted in the closing remark of §4.3, the Lip^+-stability of Glimm's approximate solutions enables us to convert the L^1-type into pointwise convergence rate estimate.

We close this section by stating the following.

Theorem 4.7 *Consider the conservation law (4.226), (4.227) with sufficiently small Lip^+-bounded initial data (4.235). Then the (deterministic version of) Glimm approximate solution $v^{\Delta x}(x,t)$ in (4.299) converges to the entropy solution $u(x,t)$, and the following convergence rate estimates hold:*

$$\|v^{\Delta x}(\cdot,T) - u(\cdot,T)\|_{L^1} \leq \text{Const}_T \cdot \sqrt{\Delta x}|\ln \Delta x|, \tag{4.304}$$

$$|v^{\Delta x}(x,T) - u(x,T)| \leq \text{Const}_{x,T} \cdot [1 + \max_{|\xi-x|\leq\sqrt[4]{\Delta x}} |u_x(\xi,T)|] \cdot \sqrt[4]{\Delta x}|\ln \Delta x|. \tag{4.305}$$

Remarks.

1. A sharp L^1-error estimate of order $O(\sqrt{\Delta x})$ can be found in [22], improving the previous error estimates of [14].

2. Theorem 4.7 hinges on the truncation error estimate (4.301) which assumes initial data which sufficiently small variation [14]. Extensions to strong initial discontinuities for Glimm scheme and the front tracking method can be found in [6, Theorems 4.6 and 5.2].

The Spectral Viscosity method

We want to solve the 2π-periodic initial-value problem (4.219)-(4.220) by spectral methods. To this end we use an N-trigonometric polynomial, $v_N(x,t) = \sum_{k=-N}^{N} \hat{v}_k(t)e^{ikx}$, to approximate the spectral (or pseudospectral) projection of the exact entropy solution, $P_N u$. Starting with $v_N(x,0) = P_N u_0(x)$, the standard Fourier method reads,

$$\frac{\partial}{\partial t}v_N + \frac{\partial}{\partial x}P_N f(v_N) = 0. \tag{4.306}$$

Together with one's favorite ODE solver, (4.306) gives a fully discrete spectral method for the approximate solution of (4.219).

Although the spectral method (4.306) is a spectrally accurate approximation of the conservation law (4.219) in the sense that its local error does not exceed

$$\|(I - P_N)f(v_N(\cdot,t))\|_{H^{-s}} \leq \text{Const} \cdot N^{-s}\|v_N\|_{L^2}, \quad \forall s \geq 0, \tag{4.307}$$

the spectral solution, $v_N(x,t)$, need not approximate the corresponding entropy solution, $u(x,t)$. Indeed, the counterexamples in §4.1 show that the spectral approximation (4.306) lacks *entropy dissipation*, which is inconsistent with the entropy condition (4.220). Consequently, the spectral approximation (4.306) supports spurious Gibbs oscillations which prevent strong convergence to the exact solution of (1.1). To suppress these oscillations, without sacrificing the overall spectral accuracy, we consider instead the Spectral Viscosity (SV) approximation

$$\frac{\partial}{\partial t}v_N(x,t) + \frac{\partial}{\partial x}P_N f(v_N(x,t)) = \varepsilon_N \frac{\partial}{\partial x}Q_N * \frac{\partial}{\partial x}v_N(x,t). \tag{4.308}$$

The left-hand side of (4.308) is the standard spectral approximation of (4.219). On the right hand-side, it is augmented by *spectral viscosity* which consists of the following three ingredients: a vanishing viscosity amplitude of size $\varepsilon_N \downarrow 0$, a viscosity-free spectrum of size $m_N \gg 1$, and a viscosity kernel, $Q_N(x,t) = \sum_{|k|=m_N}^{N} \hat{Q}_k(t)e^{ikx}$ activated only on high wavenumbers $|k| \geq m_N$, which can be conveniently implemented in the Fourier space as

$$\varepsilon_N \frac{\partial}{\partial x}Q_N * \frac{\partial}{\partial x}v_N(x,t) \equiv -\varepsilon_N \sum_{|k|=m_N}^{N} k^2 \hat{Q}_k(t)\hat{v}_k(t)e^{ikx}.$$

We deal with real viscosity kernels $Q_N(x,t)$ with increasing Fourier coefficients, $\hat{Q}_k \equiv \hat{Q}_{|k|}$, which satisfy

$$1 - \left(\frac{m_N}{|k|}\right)^{2q} \leq \hat{Q}_k(t) \leq 1, \quad |k| \geq m_N, \quad \text{for some fixed } q \geq 1, \tag{4.309$_q$}$$

and we let the spectral viscosity parameters, (ε_N, m_N), lie in the range

$$\varepsilon_N \sim \frac{1}{N^\theta \log N}, \quad m_N \sim N^{\frac{\theta}{2q}}, \quad \theta < 1. \tag{4.310$_q$}$$

We remark that this choice of spectral viscosity parameters is small enough to retain the formal spectral accuracy of the overall approximation, since

$$\left\| \varepsilon_N \frac{\partial}{\partial x} Q_N * \frac{\partial}{\partial x} v_N(\cdot, t) \right\|_{H^{-s}} \le \mathrm{Const} \cdot N^{-\frac{\theta s}{2q}} \| v_N(\cdot, t) \|_{L^2}, \quad \forall s \ge 2. \tag{4.311}$$

At the same time, it is sufficiently large to enforce the correct amount of entropy dissipation that is missing otherwise, when either $\varepsilon_N = 0$ or $m_N = N$. Indeed, it was shown in [38],[40],[23] that the SV approximation (4.308), (4.309$_q$)-(4.310$_q$) has a bounded entropy production in the sense that

$$\varepsilon_N \left\| \frac{\partial}{\partial x} v_N(x, t) \right\|^2_{L^2_{loc}(x, t)} \le \mathrm{Const}, \tag{4.312}$$

and this together with an L^∞-bound imply – by compensated compactness arguments, that the SV approximation v_N converges to the unique entropy solution of (4.219). A detailed account on the SV method is outlined in Leture III of this volume.

Observe that in the limit case $q = \infty$, the SV method (4.308), (4.309$_q$)-(4.310$_q$), coincides with the usual viscosity approximation,

$$\frac{\partial}{\partial t} v^\varepsilon(x, t) + \frac{\partial}{\partial x} P_N f(v^\varepsilon(x, t)) = \varepsilon_N \frac{\partial^2}{\partial x^2} v^\varepsilon(x, t).$$

But of course, the spectral accuracy (4.311) is lost in this limit case.

The Lip^+-stability and Lip'-consistency (of order $\mathcal{O}(N^{-\theta})$) of the SV approximation were studies in [40]. We thus arrive at

Theorem 4.8 (Convergence rate estimates) *Consider the 2π-periodic nonlinear conservation law (4.226) with Lip^+-initial-data. Then the SV approximation (4.308), (4.309$_q$)-(4.310$_q$) with $q \ge \frac{3}{2}$ converges to the entropy solution of (4.226) and the following error estimates hold for $0 < t_0 \le \forall t \le T$:*

$$\| v_N(\cdot, t) - u(\cdot, t) \|_{W^{s,p}} \le \mathrm{Const}_T \cdot N^{-\frac{1-sp}{2p}\theta}, -1 \le s \le \frac{1}{p}; \tag{4.313}$$

$$| v_N(x, t) - u(x, t) | \le \mathrm{Const}_T \cdot N^{-\frac{\theta}{3}}, \quad 0 < t_0 \le t \le T; \tag{4.314}$$

Finally, any r-th order mollifier, (4.248), recovers the pointvalues of v_N to the order of

$$| v_N(x, t) * \psi_r - v_N(x, t) | \le C_r \cdot N^{-\frac{r}{r+2}\theta}. \tag{4.315}$$

Remarks.

1. Theorem 4.8 requires the initial data of the SV method, $v_N(x,0)$, to be Lip^+-bounded independently of N. Consequently, one might need to pre-process the prescribed initial data u_0 unless they are smooth enough to begin with. The de la Vallee Poussin pre-processing, for example, will guarantee this requirement for arbitrary Lip^+-bounded initial data u_0.

2. The error estimates (4.313),(4.314) are not uniform in time as $t_0 \downarrow 0$, unless the initial data are sufficiently smooth to guarantee the uniformity (in time) of the Lip^+ bound. For arbitrary Lip^+-initial data, u_0, an initial layer may be formed, after which the spectral viscosity becomes effective and guarantees the spectral decay of the discretization error.

3. According to (4.314) and (4.315), the pointwise convergence rate of the SV solution in smooth regions of the entropy solution is of order $\sim N^{-\frac{1}{3}}$, and by *post-processing* the SV solution this convergence rate can be made arbitrarily close to N^{-1}. In fact, numerical experiments reported in [38] show that by post-processing the SV solution using the spectrally accurate mollifier of [12], $\psi_r(x) = \psi_0(x)D_n(x), n \sim \left[\varepsilon_N^{-\frac{1}{r+2}}\right]$, we recover the pointwise values in smooth regions of the entropy solution within spectral accuracy.

4. According to (4.313) with $(s,p) = (0,1)$, the SV approximation has an L^1-convergence rate of order $\sim N^{-\frac{1}{2}}$ in agreement with [30]. This corresponds to the usual L^1-convergence rate of order $\frac{1}{2}$ for monotone difference approximations, [15],[31].

References

1. G. BARLES & P.E. SOUGANIDIS, *Convergence of approximation schemes for fully nonlinear second order equations*, Asympt. Anal. 4 (1991), 271–283.

2. F. BOUCHUT & B. PERTHAME, *Kruzkov's estimates for scalar conservation laws revisited*, Universite D'Orleans, preprint, 1996.

3. Y. BRENIER, *Roe's scheme and entropy solution for convex scalar conservation laws*, INRIA Report 423, France 1985.

4. Y. BRENIER & S. OSHER, *The discrete one-sided Lipschitz condition for convex scalar conservation laws*. 1988, SIAM J. of Num. Anal. Vol. 25, 1, pp. 8-23.

5. A. J. CHORIN, *Random choice solution of hyperbolic systems*, J. Comp. Phys., 22 (1976, pp. 517-533.

6. I. L. CHERN, *Stability theorem and truncation error analysis for the Glimm scheme and for a front tracking method for flows with strong discontinuities*, Comm. Pure Appl. Math., XLII (1989), pp. 815-844.

7. B. COCKBURN, F. COQUEL & P. LEFLOCH, *Convergence of finite volume methods for multidimensional conservation laws*, SIAM J. Numer. Anal. 32 (1995), 687–705.

8. M. G. CRANDALL & P. L. LIONS, *Viscosity solutions of Hamilton-Jacobi equations*, Trans. Amer. Math. Soc. 277 (1983), 1–42.

9. M.G. CRANDALL & A. MAJDA, *Monotone difference approximations for scalar conservation laws*, Math.Comp.,34 (1980), 1-21.

10. J. GLIMM, *Solutions in the large for nonlinear hyperbolic systems of equations*, Comm. Pure Appl. Math., 18 (1965), pp. 697-715.

11. J.B. GOODMAN & R.J. LEVEQUE, *A geometric approach to high resolution TVD schemes*, SIAM J. Numer. Anal., 25 (1988), pp. 268-284.

12. D. GOTTLIEB & E. TADMOR, *Recovering Pointwise Values of Discontinuous Data within Spectral Accuracy*, in "Progress and Supercomputing in Computational Fluid Dynamics", Progress in Scientific Computing, Vol. 6 (E. M. Murman and S. S. Abarbanel, eds.), Birkhauser, Boston, 1985, 357-375.

13. J. GOODMAN & P.D. LAX, *On dispersive difference schemes. I*, Comm. Pure Appl. Math. 41 (1988), 591-613.

14. D. HOFF & J. SMOLLER, *Error bounds for the Glimm scheme for a scalar conservation law*, Trans. Amer. Math. Soc., 289 (1988), pp. 611-642.

15. N.N. KUZNETSOV, *On stable methods for solving nonlinear first order partial differntial equations in the class of discontinuous solutions*, Topics in Num. Anal. III, Proc. Royal Irish Acad. Conf. Trinity College, Dublin (1976), pp. 183-192.

16. S.N. KRUŽKOV, *The method of finite difference for a first order non-linear equation with many independent variables*, USSR comput Math. and Math. Phys. 6 (1966), 136–151. (English Trans.)

17. S. N. KRUSHKOV, *First-order quasilinear equations in several independent variables*, Math. USSR Sb. 10 (1970), 217-243.

18. A. KURGANOV & E. TADMOR, *Stiff systems of hyperbolic conservation laws. convergence and error estimates*, SIMA, in press.

19. C.-T. LIN & E. TADMOR L^1-*Stability and error estimates for approximate Hamilton-Jacobi solutions*, preprint.

20. P. L. LIONS, Generalized Solutions of Hamilton-Jacobi Equations, Pittman, London 1982.

21. T. P. LIU, *The deterministic version of the Glimm scheme*, Comm. Math. Phys., 57 (1977), pp. 135-148.

22. B. LUCIER, *Error bounds for the methods of Glimm, Godunov, and LeVeque*, SIAM J. of Numer. Anal., 22 (1985), pp. 1074-1081.

23. Y. MADAY & E. TADMOR, *Analysis of the spectral viscosity method for periodic conservation laws*, SINUM 26, 1989, pp. 854-870.

24. H. NESSYAHU & E. TADMOR , *The convergence rate of approximate solutions for nonlinear scalar conservation laws*, SIAM J. Numer. Anal. 29 (1992), 1–15.

25. H. NESSYAHU & T. TASSA, *Convergence rates of approximate solutions to conmservation laws with initial rarefactions*, SIAM J. Numer. Anal. 31 (1994), 628–654.

26. H. NESSYAHU, E. TADMOR & T. TASSA, *The convergence rate of Godunov type schemes*, SIAM J. Numer. Anal. 31 (1994), 1–16.

27. H. NESSYAHU, *Convergence rate of approximate solutions to weakly coupled nonlinear systems*, Math. Comp. 65 (1996) pp. 575-586.

28. P. ROSENAU, *Extending hydrodynamics via the regularization of the Chapman-Enskog expansion*, Phys. Rev. A, 40(1989), 7193-6.

29. R. RICHTMYER & K.W. MORTON, Difference methods for initial-value problems, 2nd ed., Interscience, New York, 1967.

30. S. SCHOCHET, *The rate of convergence of spectral viscosity methods for periodic scalar conservation laws*, SINUM 27, 1990, pp. 1142-1159.

31. R. SANDERS, *On convergence of monotone finite difference schemes with variable spatial differencing*, Math. of Comp., 40 (1983), pp. 91-106.

32. J. SMOLLER, Shock Waves and Reaction-Diffusion Equations, Springer-Verlag, New York, 1983.

33. S. SCHOCHET & E. TADMOR, *The regularized Chapman-Enskog expansion for scalar conservation laws*, Arch. Rational Mech. Anal., 119 (1992), pp. 95-107.

34. E. TADMOR, *The large time behavior of the scalar, genuinely nonlinear Lax-Friedrichs scheme*, Math. of Comp., 43, 168 (1984), pp. 353-368.

35. E. TADMOR, *Numerical viscosity and the entropy condition for conservative difference schemes*, Math. Comp., 43 (1984), pp. 369-381.

36. E. TADMOR, *The numerical viscosity of entropy stable schemes for systems of conservation laws. I*, Math.of Comp., 49 (1987), pp. 91-103.

37. E. TADMOR, *Semi-discrete approximations to nonlinear systems of conservation laws; consistency and L^∞-stability imply convergence*,ICASE ICASE Report No. 88-41.

38. E. TADMOR, *Convergence of spectral methods for nonlinear conservation laws*, SIAM J. Numer. Anal., 26 (1989), pp. 30-44.

39. E. TADMOR, *Local error estimates for discontinuous solutions of nonlinear hyperbolic equations*, SIAM J. Numer. Anal., 28 (1991), pp. 811-906.

40. E. TADMOR, *Total variation and error estimates fo spectral viscosity approximations*, Math. Comp., 60 (1993), pp. 245-256.

41. E. TADMOR & T. TANG, *The pointwise convergence rate for piecewise smooth solutions for scalar conservatin laws*, in preparation.

42. T. TANG & Z. H. TENG, *Viscosity methods for piecewise smooth solutions to scalar conservation laws*, Math. Comp., 66 (1997), pp. 495–526.

43. T. TANG & P. -W. ZHANG, *Optimal L^1-rate of convergence for viscosity method and monotone schemes to piecewise constant solutions with shocks*, SIAM J. Numer. Anala. 34 (1997), pp 959–978.

5 Kinetic Formulations and Regularity

Abstract. We discuss the kinetic formulation of nonlinear conservation laws and related equations, a kinetic formulation which describes both the equation and the entropy criterion. This formulation is a kinetic one, involving an additional variable called velocity by analogy. We apply this formulation to derive, based upon the velocity averaging lemmas, new compactness and regularity results. In particular, we highlight the regularizing effect of nonlinear entropy solution operators, and we quantify the gained regularity in terms of the nonlinearity. Finally, we show that this kinetic formulation is in fact valid and meaningful for more general classes of equations, including equations involving nonlinear second-order terms, and the 2×2 hyperbolic system of isentropic gas dynamics, in both Eulerian or Lagrangian variables (– the so called 'p-system').

5.1 Regularizing effect in one-space dimension

We consider the convex conservation law

$$\frac{\partial}{\partial t} u(x,t) + \frac{\partial}{\partial x} A(u(x,t)) = 0, \quad A'' \geq \alpha > 0. \tag{5.316}$$

Starting with two values at the different positions, $u_\ell = u(x_\ell, t)$ and $u_r = u(x_r, t)$, we trace these values by backward characteristics. They impinge on the initial line at $x_\ell^0 = x_\ell - ta(u_\ell)$ and $x_r^0 = x_r - ta(u_r)$, respectively. Since the characteristics of entropy solutions of convex conservation laws cannot intersect, one finds that the ratio $(x_r^0 - x_\ell^0)/(x_r - x_\ell)$ remains positive for all time. After rearrangement this yields

$$\frac{a(u(x_r,t)) - a(u(x_\ell,t))}{x_r - x_\ell} \leq \frac{1}{t}.$$

Thus we conclude that the velocity of $a(u)$ satisfies the Oleĭnik's one-sided Lip condition, $a(u(\cdot,t))_x \leq 1/t$. Thanks to the convexity of A, we obtain the Lip^+ bound on u itself,

$$u_x(x,t) \leq \frac{1}{\alpha t}. \tag{5.317}$$

We recall that Lip^+ bound (5.317) served as the cornerstone for the Lip' *convergence* theory outlined in Lecture IV. Here we focus on the issue of it regularity. Granted (5.317), it follows that the solution operator associated with convex conservation laws, T_t, has a nonlinear regularizing effect, mapping

$$T_t : L_0^\infty \longrightarrow BV, \quad t > 0. \tag{5.318}$$

compact support of size $L = |\text{supp}u_0|$, one obtains $|\text{supp}u(\cdot,t)| \lesssim L + Const.t$. The Lip^+ bound (5.317) then yields an upper bound on the positive variation,

$\int u_x^+(x,t)dx \le Const.$; since the sum of the positive and negative variations is bounded,

$$\int u_x^+(x,t) + u_x^-(x,t)dx = \int u_x(x,t) \le Const.\|u_0\|_{L^\infty},$$

it follows that their difference is also bounded,

$$\|u(x,t)\|_{BV} = \int \left[u_x^+(x,t) - u_x^-(x,t) \right] dx \le Const. \tag{5.319}$$

Observe that no regularity is 'gained' in the linear case, where $A''(u) \equiv 0$. Indeed, the compactness asserted in (5.318) is a purely nonlinear regularizing phenomenon which reflects the irreversibility of nonlinear conservation laws, due to loss of entropy (information) across shock discontinuities. Here, nonlinearity is quantified in terms of convexity; in the prototype example of the inviscid Burgers' equation,

$$\frac{\partial}{\partial t}u + \frac{\partial}{\partial x}(\frac{u^2}{2}) = 0, \tag{5.320}$$

one finds a time decay, $u_x(x,t) \le 1/t$. Tartar [31] proved this regularizing effect for general nonlinear fluxes — *nonlinear* in the sense of $A''(\cdot) \ne 0, a.e.$.

The situation with *multidimensional* equations, however, is less clear. Consider the 'two-dimensional Burgers' equation', analogous to (5.320)

$$\frac{\partial}{\partial t}u + \frac{\partial}{\partial x_1}(\frac{u^2}{2}) + \frac{\partial}{\partial x_2}(\frac{u^2}{2}) = 0. \tag{5.321}$$

Since $u(x_1,x_2,t) \equiv u_0(x_1 - x_2)$ is a steady solution of (5.321) for *any* u_0, it follows that initial oscillations persist (along $x_1 - x_2 = Const$), and hence there is no regularizing effect which guarantee the compactness of the solution operator in this case. More on oscillations and discontinuities can be found in Tartar's review [32].

5.2 Velocity averaging lemmas ($m \ge 1, d \ge 1$)

We deal with solutions to transport equations

$$a(v) \cdot \nabla_x f(x,v) = \partial_v^s g(x,v). \tag{5.322}$$

The averaging lemmas, [13], [12], [11], state that in the generic non-degenerate case, averaging over the velocity space, $\bar{f}(x) := \int_v f(x,v)dv$, yields a gain of *spatial* regularity. The prototype statement reads

Lemma 5.1 *([13],[11],[22]) Let $f \in L^p(x,v)$ be a solution of the transport equation (5.322) with $g \in L^q(x,v), 1 \le q \le p \le 2$. Assume the following non-degeneracy condition holds*

$$meas_v\{v|\ |a(v)\cdot\xi/|\xi|| < \delta\} \le Const \cdot \delta^\alpha, \quad \alpha \in (0,1). \tag{5.323}$$

Then $\bar{f}(x) := \int_v f(x,v)dv$ belongs to Sobolev space $W^\theta(L^r(x))$,

$$\bar{f}(x) \in W^\theta(L^r(x)), \qquad \theta < \frac{\alpha}{\alpha(1 - \frac{p'}{q'}) + (s+1)p'}, \quad \frac{1}{r} = \frac{\theta}{q} + \frac{1-\theta}{p}.$$
(5.324)

Variants of the averaging lemmas were used by DiPerna and Lions to construct global weak (renormalized) solutions of Boltzmann, Vlasov-Maxwell and related kinetic systems, [9], [10]; in Bardos et. al., [1], averaging lemmas were used to construct solutions of the incompressible Navier-Stokes equations. We turn our attention to their use in the context of nonlinear conservation laws and related equations.

Proof. (Sketch). We shall sketch the proof in the particular case, $p = q$ which will suffice to demonstrate the general $p \neq q$ case.

Let $\Omega_\delta(\xi,v)$ denote the set where the symbol $a(v) \cdot \xi'$ is 'small',

$$\Omega_\delta(\xi,v) := \{(v,\xi)| \ |a(v) \cdot \xi'| \leq \delta\}, \quad \xi' := \frac{\xi}{|\xi|},$$
(5.325)

and decompose the average, $\bar{f}(x)$ accordingly:

$$\bar{f}(x) = \int_v f(x,v)dv =$$

$$= \int_v \mathcal{F}^{-1}|\xi|^{-1}\left[\frac{\partial_v^s \hat{g}(\xi,v)}{a(v) \cdot \xi'}\chi_{\Omega_\delta^c}(\xi,v)\right]dv + \longleftarrow \bar{f}^\delta(x)$$
(5.326)

$$+ \int_v \mathcal{F}^{-1}\left[\hat{f}(\xi,v)\chi_{\Omega_\delta}(\xi,v)\right]dv \qquad \longleftarrow \bar{f}(x) - \bar{f}^\delta(x).$$

Here, χ_Ω represents the usual *smooth* partitioning relative to Ω_δ and its complement, Ω_δ^c. On Ω^c, the symbol is 'bounded away' from zero, so we gain one derivative:

$$\|\bar{f}^\delta\|_{W^1(L^p))} \leq Const.\|g\|_{L^p(x,v)}\delta^{\frac{\alpha}{p'}-(m+1)};$$
(5.327)

On Ω – along the 'non uniformly elliptic' rays, we have no gain of regularity, but instead, our non-degeneracy assumption implies that $|\Omega|$ is a 'small' set and therefore

$$\|\bar{f} - \bar{f}^\delta\|_{L^p} \leq Const.\|f\|_{L^p(x,v)}\delta^{\frac{\alpha}{p'}}$$
(5.328)

Both (5.327) and (5.328) are straightforward for $p = 2$ and by estimating the corresponding $\mathcal{H}^1$ multipliers, the case $1 < p \leq 2$ follows by interpolation. Finally, we consider the K-functional

$$K(\bar{f},t) := \inf_{\bar{g}}\left[\|\bar{f} - \bar{g}\|_{L^p} + t\|\bar{g}\|_{W^1(L^p)}\right];$$

The behavior of this functional, $K(\bar{f}, t) \sim t^\theta$, characterize the smoothness of $\bar{f}$ in the intermediate space between L^p and $W^1(L^p)$: more precisely, $\bar{f}$ belongs to Besov space B_∞^θ with 'intermediate' smoothness of order θ.

Now set $g = \bar{f}^\delta$, then with appropriately scaled δ we find that $K(\bar{f}, t) \sim t^\theta$ with $\theta = \frac{\alpha}{(s+1)p'}$. **This means that $\bar{f}(x)$ belongs to Besov space, $\bar{f}(x)\epsilon B_\infty^\theta(L^p(x))$** and (5.324) (with $p = q = r$) follows. ∎

Remark 5.1 In the limiting case of $\alpha = 0$ in (5.323), one finds that if

$$meas_v\{v|\ |a(v) \cdot \xi'| = 0\} = 0, \qquad (5.329)$$

then averaging is a compact mapping, $\{f(x, v)\} \in L^{x,v} \hookrightarrow \{\bar{f}\} \in L^p$. The case $p = 2$ follows from Gèrard's results [12].

5.3 Regularizing effect revisited ($m = 1, d \geq 1$)

In this section we resume our discussion on the regularization effect of nonlinear conservation laws. The averaging lemma enables us to identify the proper notion of 'nonlinearity' in the multivariate case, which guarantee compactness.

The following result, adapted from [22], is in the heart of matter.

Theorem 5.1 *Consider the scalar conservation law*

$$\partial_t u + \nabla_x \cdot A(u) = 0, \quad (t, x) \in \mathbb{R}_t^+ \times \mathbb{R}_x^d. \qquad (5.330)$$

and assume that the following non-degeneracy condition holds (consult (5.323))

$$\exists \alpha \in (0, 1): \ meas_v\{v|\ |\tau + A'(v) \cdot \xi| < \delta\} \leq Const \cdot \delta^\alpha, \quad \forall \tau^2 + |\xi|^2 = 1. \qquad (5.331)$$

Let $\{u^\varepsilon\}$ be a family of approximate solutions with bounded measures of entropy production,

$$\partial_t \eta(u^\varepsilon) + \nabla_x \cdot F(u^\varepsilon) \in \mathcal{M}((0, T) \times \mathbb{R}_x^d), \qquad \forall \eta'' > 0. \qquad (5.332)$$

Then $u^\varepsilon(t, x) \in W_{loc}^{\frac{\alpha}{\alpha+4}}(L^r(t, x)), \quad r = \frac{\alpha+4}{\alpha+2}$.

Remark 5.2 Note that the bounded measure of entropy production in (5.332) need not be negative; general bounded measures will do.

Proof. To simplify notations, we use the customary 0th index for time direction,

$$x = (t \leftrightarrow x_0, x_1, \ldots, x_d), \quad A(u) = (A_0(u) \equiv 1, A_1(u), \ldots, A_d(u)).$$

The entropy condition (5.332) with Kružkov entropy pairs (1.1), reads

$$\nabla_x \cdot [sgn(u^\varepsilon - v)(A(u^\varepsilon) - A(v))] \leq 0.$$

This defines a family of non-negative measures, $m^\varepsilon(x,v)$,

$$\nabla_x \cdot [sgn(v)A(v) - sgn(u^\varepsilon - v)(A(u^\varepsilon) - A(v))] =: m^\varepsilon(x,v). \qquad (5.333)$$

Differentiate (5.333) w.r.t. v: one finds that the indicator function, $f(x,v) = \chi_{u^\varepsilon}(v)$, where

$$\chi_{u^\varepsilon}(v) := \begin{cases} +1 & 0 < v < u^\varepsilon \\ -1 & u^\varepsilon < v < 0 \\ 0 & |v| > u^\varepsilon \end{cases} \qquad (5.334)$$

satisfies the transport equation,

$$\partial_t f^\varepsilon + a(v) \cdot \nabla_x f^\varepsilon = \frac{\partial}{\partial v} m^\varepsilon(t,x,v), \qquad (5.335)$$

which corresponds to (5.322) with $s = 1, g(x,v) = m^\varepsilon(x,v) \in \mathcal{M}_{x,v}$ [10]. We now apply the averaging lemma with $(s = q = 1, p = 2)$, which tells us that $u^\varepsilon(t,x) = \int \chi_{u^\varepsilon}(v)dv \in W_{loc}^{\frac{\alpha}{\alpha+4}}(L^r(t,x))$ as asserted. ∎

It follows that if the non-degeneracy condition (5.331) holds, then the family of approximate solutions $\{u^\varepsilon\}$ is compact and strong convergence follows. In this context we refer to the convergence statement for measure-valued solutions for general multidimensional scalar conservation laws – approximate solutions measured by their nonpositive entropy production outlined in Lecture I, §1.5.

Here, Theorem 5.1 yields even more, by *quantifying* the regularity of approximate solutions with bounded entropy productions in terms of the non-degeneracy condition (5.331). In fact, more can be said if the solution operator associated with $\{u^\varepsilon\}$ is translation invariant: a bootstrap argument yields an improved regularity, [22],

$$u^\varepsilon(t > 0, \cdot) \in W^{\frac{\alpha}{\alpha+2}}(L^1(x)). \qquad (5.336)$$

In particular, if the problem is nonlinear in the sense that the non-degeneracy condition (5.329) holds,

$$meas_v\{v| \ \tau + A'(v) \cdot \xi = 0\} = 0, \qquad (5.337)$$

---

[10] Once more, it is the symmetry property (1.6) which has a key role in the derivation of the transport kinetic formulation (5.322).

then the corresponding solution operator, $T_t, t > 0$, has a *regularization* effect mapping $T_{\{t>0\}} : L_0^\infty \hookrightarrow L^1$. This could be viewed as a multidimensional generalization for Tartar's regularization result for a.e. nonlinear one-dimensional fluxes, $A''(\cdot) \neq 0$, *a.e.*.

We continue with few multidimensional examples which illustrate the relation between the non-degeneracy condition, (5.331) and regularity.

Example #1. The 'two-dimensional Burgers' equation' (5.321),

$$\frac{\partial}{\partial t}u + \frac{\partial}{\partial x_1}(\frac{u^2}{2}) + \frac{\partial}{\partial x_2}(\frac{u^2}{2}) = 0,$$

has a linearized symbol $\tau' + v\xi_1' + v\xi_2'$ which fails to satisfy the non-degeneracy/nonlinearity condition (5.331), since it vanishes $\forall v$'s along $\tau' = \xi_1' + \xi_2' = 0$. This corresponds to its persistence of oscillations along $x_1 - x_2 = const$, which excludes compactness.

Example #2. We consider

$$\frac{\partial}{\partial t}u + \frac{\partial}{\partial x_1}(\frac{u^2}{2}) + \frac{\partial}{\partial x_2}(e^u) = 0. \tag{5.338}$$

In this case the linearized symbol is given by $\tau' + v\xi_1' + e^v\xi_2'$; Here we have

$$meas\{v \mid \ |\tau' + v\xi_1' + e^v\xi_2'| \leq \delta\} \leq Const.\delta^{\frac{1}{2}}$$

(just consider the second-order touch-point at $v = 1$). Hence, the solution operator associated with (5.338) is compact (– in fact, mapping $L_0^\infty \longrightarrow W^{\frac{1}{5}}(L^1)$.)

Example #3. Consider

$$\frac{\partial}{\partial t}u + \frac{\partial}{\partial x_1}(|u^m|u) + \frac{\partial}{\partial x_2}(|u|^n u) = 0. \tag{5.339}$$

For $n \neq m$ we obtain an index of non-degeneracy/non-linearity of order $\alpha = 1/\max\{1 + m, 1 + n\}$.

Kinetic and other approximations

Theorem 5.1 provides an alternative route to analyze the convergence of *general* entropy stable multi-dimensional schemes, schemes whose convergence proof was previously accomplished by measure-valued arguments; here we refer to finite-difference, finite-volume, streamline-diffusion and spectral approximations ..., which were studied in [4,18,19,15,16,3]. Indeed, the feature in the convergence proof of all these methods is the $W_{loc}^{-1}(L^2)$-compact entropy production, (5.348). Hence, if the underlying conservation law satisfies the non-linear degeneracy condition (5.337), then the corresponding family of approximate solutions, $\{u^\varepsilon(t > 0, \cdot)\}$ becomes compact. Moreover, if the entropy production is bounded measure, then there is actually a *gain* of regularity indicated in Theorem 5.1 and respectively, in (5.336) for the translation invariant case.

Remark 5.3 Note that unlike the requirement for a *nonpositive* entropy production from measure-valued solutions (consult (1.58) in Lecture I), here we allow for an arbitrary bounded measure.

So far we have not addressed explicitly a kinetic formulation of the multidimensional conservation law (5.330). The study of regularizing effect for multidimensional conservation laws was originally carried out in [22] for the approximate solution constructed by the following BGK-like model, [28] (see also [2],[14]),

$$\frac{\partial f^\varepsilon}{\partial t} + a(v) \cdot \nabla_x f^\varepsilon = \frac{1}{\varepsilon}\left(\chi_{u^\varepsilon}(v) - f^\varepsilon\right), \quad (t, x, v) \in \mathbb{R}_t^+ \times \mathbb{R}_x^d \times \mathbb{R}_v, (5.340)$$

$$f^\varepsilon|_{t=0} = \chi_{u_0(x)}(v), \quad (x, v) \in \mathbb{R}_x^d \times \mathbb{R}_v. \tag{5.341}$$

Here, $\chi_{u^\varepsilon(t,x)}(v)$ denotes the 'pseudo-Maxwellian',

$$\chi_{u^\varepsilon}(v) := \begin{cases} +1 \ 0 < v < u^\varepsilon \\ -1 \ u^\varepsilon < v < 0 \ , \\ 0 \ |v| > u^\varepsilon \end{cases} \tag{5.342}$$

which is associated with the average of f^ε,

$$u^\varepsilon(t, x) = \bar{f}^\varepsilon := \int_\mathbb{R} f^\varepsilon(t, x, v) dv, \quad (t, x) \in R_t^+ \times R_x^d. \tag{5.343}$$

The key property of this kinetic approximation is the existence of a nonnegative measure, m^ε such that $\frac{1}{\varepsilon}\left(\chi_{u^\varepsilon}(v) - f^\varepsilon\right) = \frac{\partial m^\varepsilon}{\partial v}$ (The existence of such measures proved in [22] and is related to H-functions studied in [28] and Brenier's lemma [2].) Thus, we may rewrite (5.340) in the form

$$\frac{\partial f^\varepsilon}{\partial t} + a(v) \cdot \nabla_x f^\varepsilon = \frac{\partial m^\varepsilon}{\partial v}, \quad m^\varepsilon \in \mathcal{M}((0, T) \times \mathbb{R}_x^d \times \mathbb{R}_v^+). \tag{5.344}$$

Let (η, F) be an entropy pair associated with (5.330). Integration of (5.344) against $\eta'(v)$ implies that the corresponding macroscopic averages, $u^\varepsilon(t, x)$, satisfy

$$\partial_t \eta(u^\varepsilon) + \nabla_x \cdot F(u^\varepsilon) \le 0, \quad \forall \eta'' > 0. \tag{5.345}$$

Thus, the entropy production in this case is nonpositive and hence a bounded measure, so that Theorem 5.1 applies. Viewed as a measure-valued solution, convergence follows along DiPerna's theory [8]. If, moreover, the nondegeneracy condition (5.331) holds, then we can further quantify the W^s-regularity (of order $s = \frac{\alpha}{\alpha+2}$.)

Theorem 5.1 offers a further generalization beyond the original, 'kineticly' motivated discussion in [22]. Indeed, consideration of Theorem 5.1 reveals the intimate connection between the macroscopic assumption of bounded entropy production in (5.332), and an underlying kinetic formulation (5.335), analogous to (5.344). For a recent application of the regularizing effect for a convergence study of finite-volume schemes along these lines we refer to [24].

5.4 Degenerate parabolic equations

As an example one can treat convective equations together with (possibly degenerate) diffusive terms

$$\partial_t u^\varepsilon + \nabla_x \cdot A(u^\varepsilon) = \nabla_x \cdot (Q \nabla_x u^\varepsilon), \quad Q \geq 0. \tag{5.346}$$

Assume the problem is not linearly degenerate, in the sense that

$$meas_v\{v|\ \tau + A'(v) \cdot \xi = 0,\ \langle Q(v)\xi, \xi \rangle = 0\} = 0. \tag{5.347}$$

Let $\{u^\varepsilon\}$ be a family of approximate solutions of (5.322) with $W_{loc}^{-1}(L^2)$-compact entropy production,

$$\partial_t \eta(u^\varepsilon) + \nabla_x \cdot F(u^\varepsilon) \hookrightarrow W_{loc}^{-1}(L^2(t,x)), \quad \forall \eta'' > 0. \tag{5.348}$$

Then $\{u^\varepsilon\}$ is compact in $L_{loc}^2(t,x)$, [22].

The case $Q = 0$ corresponds to our multidimensional discussion in §5.330; the case $A = 0$ correspond possibly degenerate parabolic equations (consult [17] and the references therein, for example). According to (5.347), satisfying the ellipticity condition, $\langle Q(v)\xi, \xi \rangle > 0$ on a set of non-zero measure, guarantees regularization, compactness ...

Again, a second-order version of the averaging lemma 1.2 enables us to quantify the gained regularity which we state as

Lemma 5.2 Let $f \in L^1(x,v)$ be a solution of the diffusive equation

$$-\sum q_{ij}(v)\partial_{x_i x_j}^2 f = \frac{\partial m}{\partial v}, \quad Q := (q_{ij}) \geq 0,\ m(t,x,v) \in \mathcal{M}_+.$$

Assume the following non-degeneracy condition holds

$$meas_v\{v|\ |0 \leq \langle \xi', Q(v)\xi' \rangle < \delta\} \leq Const \cdot \delta^\alpha, \quad \alpha \in (0,1). \tag{5.349}$$

Then $\bar{f}(x) := \int_v f(x,v)dv$ belongs to Sobolev space $W^\theta(L^1(x))$,

$$\bar{f}(x) \in W^\theta(L^1(x)), \qquad \theta < \frac{8\alpha}{3\alpha + 2} \tag{5.350}$$

Example. Consider the isotropic equation

$$u_t + \Delta\psi(u) = 0, \quad \psi\uparrow .$$

Here $Q_{ij}(v) = \delta_{ij}\psi'(v)$ and the lemma 5.2 applies. The kinetic formulation of such equations was studied in [17]. In the particular case of porous media equation, $\psi(u) = u^m, m \geq 2$, (5.349) holds with $\alpha = \frac{1}{m-1} \leq 2$ and one conclude a regularizing effect of order $s < \frac{8}{2m+1}$, i.e., $u(t > 0, \cdot) : L_0^\infty \longrightarrow W^s(L^1)$.

A particular attractive advantage of the kinetic formulation in this case, is that it applies to *non-isotropic* problems as well.

5.5 The 2 × 2 isentropic equations

We consider the 2 × 2 system of isentropic equations, governing the density ρ and momentum $m = \rho u$,

$$\frac{\partial}{\partial t}\begin{pmatrix} \rho \\ \rho u \end{pmatrix} + \frac{\partial}{\partial x}\begin{pmatrix} m \\ \frac{m^2}{\rho} + p(\rho) \end{pmatrix} = 0. \tag{5.351}$$

Here $p(\rho)$ is the pressure which is assumed to satisfy the (scaled) γ law, $p(\rho) = \kappa\rho^\gamma$, $\kappa = \frac{(\gamma-1)^2}{4\gamma}$.

The question of existence for this model, depending on the γ-law, $1 < \gamma < 3$, was already studied [7],[6] by compensated compactness arguments. Here we revisit this problem with the kinetic formulation presented below which leads to existence result for $3 < \gamma < \infty$, consult [23], and is complemented with a new existence proof for $1 < \gamma < 3$, consult [21].

For the derivation of our kinetic formulation of (5.351), we start by seeking *all* weak entropy inequalities associated with the isentropic 2 × 2 system (5.351),

$$\partial_t w + \partial_x A(w) = 0, \quad w := \begin{bmatrix} \rho \\ m \end{bmatrix}, \quad A(w) := \begin{bmatrix} m \\ \frac{m^2}{\rho} + \kappa\rho^\gamma \end{bmatrix} \tag{5.352}$$

The family of entropy functions associated with (5.352) consists of those $\eta(w)$'s whose Hessians symmetrize the Jacobian, $A'(w)$; the requirement of a symmetric $\eta''(w)A'(w)$ yields the Euler-Poisson–Darboux equation, e.g, [6]

$$\eta_{\rho\rho} = \frac{(\gamma-1)^2}{4}\rho^{\gamma-3}\eta_{uu}.$$

Seeking *weak* entropy functions such that $\eta(\rho, u)|_{\rho=0} = 0$, leads to the family of weak (entropy, entropy flux) pairs, $(\eta(\rho, u), F(\rho, u))$, depending on an arbitrary φ,

$$\eta(\rho, u) = \rho \int \omega(\xi)\varphi(u + \xi\rho^\theta)d\xi,$$

$$q(\rho, u) = \rho \int \omega(\xi)\varphi(u + \xi\rho^\theta)(u + \theta\xi\rho^\theta)d\xi. \qquad (5.353)$$

Here, $\omega(\xi)$ is given by

$$\omega(\xi) := (1 - \xi^2)_+^\lambda \qquad \lambda := \frac{3 - \gamma}{2(\gamma - 1)} > 0, \ \theta := \frac{\gamma - 1}{2}.$$

We note that η is convex iff φ is. Thus by the formal change of variables, $v \longleftrightarrow u + \xi\rho^\theta$, the weight function $\omega(\xi)$ becomes the 'pseudo-Maxwellian', $\chi_{\rho,u}(v) \longleftrightarrow \rho\omega((v - u)\rho^{-\theta})$,

$$\chi_{\rho,u}(v) := (\rho^{\gamma-1} - (v - u)^2)_+^\lambda. \qquad (5.354)$$

We arrive at the kinetic formulation of (5.351) which reads

$$\partial_t \chi_{\rho,u}(v) + \partial_x [a(v, \rho, u)\chi_{\rho,u}(v)] = \partial_{vv}^2 m, \qquad m \epsilon \mathcal{M}_-. \qquad (5.355)$$

Observe that integration of (5.355) against any convex φ recovers all the weak entropy inequalities. Again, as in the scalar case, the nonpositive measure m on the right of (5.355), measures the loss of entropy which concentrates along shock discontinuities.

The transport equation (5.355) is not purely kinetic due to the dependence on the macroscopic velocity u (unless $\gamma = 3$ corresponding to $\theta = 1$),

$$a(v, \rho, u) = \theta v + (1 - \theta)u \longleftrightarrow u + \xi\theta\rho^\theta.$$

Compensated compactness arguments presented in [23] yield the following compactness result.

Theorem 5.2 *([23]) Consider the isentropic equations (5.351) with $\gamma \geq 3$ and let $(\rho_n = \rho_n(t, x), u_n = u_n(t, x))$ be a family of approximate solution with bounded entropy production and finite energy, $E_n := \rho_n u_n^2 + \rho_n^\gamma \in L^\infty(\mathbb{R}_t^+, L^1(\mathbb{R}_x))$. Then a subsequence of ρ_n (still denoted by ρ_n) converges pointwise to ρ, and (a subsequence of) u_n converges pointwise to u on the set $\{\rho(x, t) > 0\}$. In particular, $\rho_n u_n$ converges pointwise to ρu.*

Finally, we consider the 2×2 system

$$\begin{cases} \partial_t v - \partial_x w = 0, \\ \partial_t w + \partial_x p(v) = 0, & t \geq 0, x \in \mathbb{R}, \end{cases} \qquad (5.356)$$

endowed with the pressure law

$$p(v) = \kappa v^{-\gamma}, \qquad \gamma > 0, \qquad \kappa = \frac{(\gamma - 1)^2}{4\gamma}. \qquad (5.357)$$

The system (5.356)-(5.357) governs the isentropic gas dynamics written in Lagrangian coordinates. In general the equations (5.356)-(5.357) will be referred to as the p-system (see [20],[30]).

For a kinetic formulation, we first seek the (entropy,entropy flux) pairs, (η, F), associated with (5.356)-(5.357). They are determined by the relations

$$\eta_{vv} + p'(v)\,\eta_{ww} = 0, \tag{5.358}$$

where F is computed by the compatibility relations

$$F_v = \eta_w\, p'(v), \quad F_w = -\eta_v. \tag{5.359}$$

The solutions of (5.358) can be expressed in terms of the fundamental solution

$$\eta(v, w) = \int_{\mathbb{R}} \varphi(\xi)\chi_{v,w}(\xi)d\xi,$$

where the fundamental solutions, $\chi_{v,w}(\xi)$, are given by

$$\chi_{v,w}(\xi) = v\left(v^{1-\gamma} - (w - \xi)^2\right)_+^\lambda, \quad \lambda = \frac{3-\gamma}{2(\gamma - 1)}. \tag{5.360}$$

Here and below, ξ (rather than v occupied for the specific volume) denotes the kinetic variable. The corresponding kinetic fluxes are then given by

$$h_{v,w}(\xi) = \theta\frac{\xi - w}{v}\chi_{v,w}(\xi).$$

We arrive at the kinetic formulation of (5.356)-(5.357) which reads, [23]

$$\partial_t\chi_{v,w} + \partial_x[a(\xi, v, w)\chi_{v,w}(\xi)] = \partial_{\xi\xi}m, \quad m(t, x, \xi) \in \mathcal{M}_-, \tag{5.361}$$

with macroscopic velocity, $a(\xi, v, w) := \theta(\xi - w)/v$.

References

1. C. BARDOS, F. GOLSE & D. LEVERMORE, *Fluid dynamic limits of kinetic equations II: convergence proofs of the Boltzmann equations*, Comm. Pure Appl. Math. XLVI (1993), 667–754.
2. Y. BRENIER, *Résolution d'équations d'évolution quasilinéaires en dimension N d'espace à l'aide d'équations linéaires en dimension N + 1*, J. Diff. Eq. 50 (1983), 375–390.
3. G.-Q. CHEN, Q. DU & E. TADMOR, *Spectral viscosity approximation to multidimensional scalar conservation laws*, Math. of Comp. 57 (1993).
4. B. COCKBURN, F. COQUEL & P. LEFLOCH, *Convergence of finite volume methods for multidimensional conservation laws*, SIAM J. Numer. Anal. 32 (1995), 687–705.

148

5. C. Cercignani, The Boltzmann Equation and its Applications, Appl. Mathematical Sci. 67, Springer, New-York, 1988.

6. G.-Q. Chen, *The theory of compensated compactness and the system of isentropic gas dynamics*, Preprint MCS-P154-0590, Univ. of Chicago, 1990.

7. R. DiPerna, *Convergence of the viscosity method for isentropic gas dynamics*, Comm. Math. Phys. 91 (1983), 1–30.

8. R. DiPerna, *Measure-valued solutions to conservation laws*, Arch. Rat. Mech. Anal. 88 (1985), 223-270.

9. R. DiPerna & P. L. Lions, *On the Cauchy problem for Boltzmann equations: Global existence and weak stability*, Ann. Math. 130 (1989), 321–366.

10. R. DiPerna & P.L. Lions, *Global weak solutions of Vlasov-Maxwell systems*, Comm. Pure Appl. Math. 42 (1989), 729–757.

11. R. DiPerna, P.L. Lions & Y. Meyer, L^p *regularity of velocity averages*, Ann. I.H.P. Anal. Non Lin. 8(3-4) (1991), 271–287.

12. P. Gérard, *Microlocal defect measures*, Comm. PDE 16 (1991), 1761–1794.

13. F. Golse, P. L. Lions, B. Perthame & R. Sentis, *Regularity of the moments of the solution of a transport equation*, J. of Funct. Anal. 76 (1988), 110–125.

14. Y. Giga & T. Miyakawa, *A kinetic construction of global solutions of first-order quasilinear equations*, Duke Math. J. 50 (1983), 505–515.

15. C. Johnson & A. Szepessy, *Convergence of a finite element methods for a nonlinear hyperbolic conservation law*, Math. of Comp. 49 (1988), 427–444.

16. C. Johnson, A. Szepessy & P. Hansbo, *On the convergence of shock-capturing streamline diffusion finite element methods for hyperbolic conservation laws*, Math. of Comp. 54 (1990), 107–129.

17. Y. Kobayashi, *An operator theoretic method for solving* $u_t = \Delta\psi(u)$, Hiroshima Math. J. 17 (1987) 79–89.

18. D. Kröner, S. Noelle & M. Rokyta, *Convergence of higher order upwind finite volume schemes on unstructured grids for scalar conservation laws in several space dimensions*, Numer. Math. 71 (1995) 527–560.

19. D. Kröner & M. Rokyta, *Convergence of Upwind Finite Volume Schemes for Scalar Conservation Laws in two space dimensions*, SINUM 31 (1994) 324–343.

20. P.D. Lax, Hyperbolic Systems of Conservation Laws and the Mathematical Theory of Shock Waves (SIAM, Philadelphia, 1973).

21. P. L. Lions, B. Perthame & P. Souganidis, *Existence and stability of entropy solutions for the hyperbolic systems of isentropic gas dynamics in Eulerian and Lagrangian coordinates*, Comm. Pure and Appl. Math. 49 (1996), 599-638.

22. P. L. Lions, B. Perthame & E. Tadmor, *Kinetic formulation of scalar conservation laws and related equations*, J. Amer. Math. Soc. 7(1) (1994), 169–191

23. P. L. Lions, B. Perthame & E. Tadmor, *Kinetic formulation of the isentropic gas-dynamics equations and p-systems*, Comm. Math. Phys. 163(2) (1994), 415–431.

24. S. Noelle & M. Westdickenberg *Convergence of finite volume schemes. A new convergence proof for finite volume schemes using the kinetic formulation of conservation laws*, Preprint.

25. O. A. Olĕinik *Discontinuous solutions of nonlinear differential equations*, Amer. Math. Soc. Transl. (2), 26 (1963), 95–172.

26. B. PERTHAME, *Global existence of solutions to the BGK model of Boltzmann equations*, J. Diff. Eq. 81 (1989), 191-205.

27. B. PERTHAME, *Second-order Boltzmann schemes for compressible Euler equations*, SIAM J. Num. Anal. 29, (1992), 1–29.

28. B. PERTHAME & E. TADMOR, *A kinetic equation with kinetic entropy functions for scalar conservation laws*, Comm. Math. Phys.136 (1991), 501–517.

29. K. H. PRENDERGAST & K. XU, *Numerical hydrodynamics from gas-kinetic theory*, J. Comput. Phys. 109(1) (1993), 53–66.

30. J. SMOLLER, Shock Waves and Reaction-Diffusion Equations, Springer-Verlag, New York, 1983.

31. L. TARTAR, *Compensated compactness and applications to partial differential equations*, in *Research Notes in Mathematics 39*, Nonlinear Analysis and Mechanics, Heriott-Watt Symposium, Vol. 4 (R.J. Knopps, ed.) Pittman Press, (1975), 136–211.

32. L. TARTAR, *Discontinuities and oscillations*, in Directions in PDEs, Math Res. Ctr Symposium (M.G. Crandall, P.H. Rabinowitz and R.E. Turner eds.) Academic Press (1987), 211-233.

Chapter 2

An Introduction to the Discontinuous Galerkin Method for Convection-Dominated Problems

Bernardo Cockburn

School of Mathematics, University of Minnesota,
Minneapolis, Minnesota 55455, USA
E-mail: cockburn@math.umn.edu

ABSTRACT

In these notes, we study the Runge Kutta Discontinuous Galerkin method for numericaly solving nonlinear hyperbolic systems and its extension for convection-dominated problems, the so-called Local Discontinuous Galerkin method. Examples of problems to which these methods can be applied are the Euler equations of gas dynamics, the shallow water equations, the equations of magneto-hydrodynamics, the compressible Navier-Stokes equations with high Reynolds numbers, and the equations of the hydrodynamic model for semiconductor device simulation. The main features that make the methods under consideration attractive are their formal high-order accuracy, their nonlinear stability, their high parallelizability, their ability to handle complicated geometries, and their ability to capture the discontinuities or strong gradients of the exact solution without producing spurious oscillations. The purpose of these notes is to provide a short introduction to the devising and analysis of these discontinuous Galerkin methods.

Acknowledgements. The author is grateful to Alfio Quarteroni for the invitation to give a series of lectures at the CIME, June 23–28, 1997, the material of which is contained in these notes. He also thanks F. Bassi and F. Rebay, and I. Lomtev and G.E. Karniadakis for kindly providing pictures from their papers [2] and [3], and [46] and [65], respectively.

Contents

1 Preface

There are several numerical methods using a DG formulation to discretize the equations in time, space, or both. In this monograph, we consider numerical methods that use DG discretizations *in space* and combine it with an *explicit* Runge-Kutta time-marching algorithm. We thus consider the so-called Runge-Kutta discontinuous Galerkin (RKDG) introduced and developed by Cockburn and Shu [17,15,14,13,19] for *nonlinear* hyperbolic systems and the so-called local discontinuous Galerkin (LDG) for *nonlinear* convection-diffusion systems. The LDG methods are an extension of the RKDG methods to convection-diffusion problems proposed first by Bassi and Rebay [3] in the context of the compressible Navier-Stokes and recently extended to general convection-diffusion problems by Cockburn and Shu [18].

Several properties are responsible for the increasing popularity of the above mentioned methods. The use of a DG discretization *in space* gives the methods the high-order accuracy, the flexibility in handling complicated geometries, and the easy to treat boundary conditions typical of the finite element methods. Moreover, the use of *discontinuous* elements produces a block-diagonal *mass* matrix whose blocks can be easily inverted by hand. This why after discretizing in time with a high-order accurate, *explicit* Runge-Kutta method, the resulting algorithm is highly parallelizable. Finally, these methods incorporate in a very natural way the techniques of 'slope limiting' developed by van Leer [62,63] that effectively damp out the spurious oscillations that tend to be produced around the discontinuities or strong gradients of the approximate solution.

In these notes, we sudy these DG methods by following their historical development. Thus, we first study the RKDG method and then the LDG method. To study the RKDG method, we start by considering their definition for the scalar equation in one-space dimension. Then, we consider the scalar equation in several space dimensions and finally, we consider the case of multidimensional systems. The last chapter is devoted to the LDG methods.

To study the RKDG method, we take the point of view that they are formally high-order accurate 'perturbations' of the so-called 'monotone' schemes which are very stable and formally first-order accurate. Indeed, the RKDG methods were devised by trying to see if formally high-order accurate methods could be obtained that retained the remarkable stability of the monotone schemes. Of course, this approach is not new: It has been the basic idea in the devising of the so-called 'high-resolution' schemes for finite-difference and finite-volume methods for nonlinear conservation laws. Thus, the RKDG method incorporates this very successful idea into the framework of DG methods which have all the advantages of finite element methods.

2 A historical overview

2.1 The original Discontinuous Galerkin method

The original discontinuous Galerkin (DG) finite element method was introduced by Reed and Hill [54] for solving the neutron transport equation

$$\sigma u + div(\overline{a} u) = f,$$

where σ is a real number and $\overline{a}$ a constant vector. Because of the linear nature of the equation, the approximate solution given by the method of Reed and Hill can be computed element by element when the elements are suitably ordered according to the characteristic direction.

LeSaint and Raviart [41] made the first analysis of this method and proved a rate of convergence of $(\Delta x)^k$ for general triangulations and of $(\Delta x)^{k+1}$ for Cartesian grids. Later, Johnson and Pitkäränta [37] proved a rate of convergence of $(\Delta x)^{k+1/2}$ for general triangulations and Peterson [53] confirmed this rate to be optimal. Richter [55] obtained the optimal rate of convergence of $(\Delta x)^{k+1}$ for some structured two-dimensional non-Cartesian grids.

2.2 Nonlinear hyperbolic systems: The RKDG method

The success of this method for linear equations, prompted several authors to try to extend the method to nonlinear hyperbolic conservation laws

$$u_t + \sum_{i=1}^{d}(f_i(u))_{x_i} = 0,$$

equipped with suitable initial or initial–boundary conditions. However, the introduction of the nonlinearity prevents the element-by-element computation of the solution. The scheme defines a nonlinear system of equations that must be solved all at once and this renders it computationally very inefficient for hyperbolic problems.

• **The one-dimensional scalar conservation law.**

To avoid this difficulty, Chavent and Salzano [8] contructed an explicit version of the DG method in the one-dimensional scalar conservation law. To do that, they discretized in space by using the DG method with piecewise linear elements and then discretized in time by using the simple Euler forward method. Although the resulting scheme is explicit, the classical von Neumann analysis shows that it is unconditionally unstable when the ratio $\frac{\Delta t}{\Delta x}$ is held constant; it is stable if $\frac{\Delta t}{\Delta x}$ is of order $\sqrt{\Delta x}$, which is a very restrictive condition for hyperbolic problems.

To improve the stability of the scheme, Chavent and Cockburn [7] modified the scheme by introducing a suitably defined 'slope limiter' following the ideas introduced by vanLeer in [62]. They thus obtained a scheme that was proven to be total variation diminishing in the means (TVDM) and total variation bounded (TVB) under a fixed CFL number, $f' \frac{\Delta t}{\Delta x}$, that can be chosen to be less than or equal to $1/2$. Convergence of a subsequence is thus guaranteed, and the numerical results given in [7] indicate convergence to the correct entropy solutions. On the other hand, the scheme is only first order accurate in time and the 'slope limiter' has to balance the spurious oscillations in smooth regions caused by linear instability, hence adversely affecting the quality of the approximation in these regions.

These difficulties were overcome by Cockburn and Shu in [17], where the first Runge Kutta Discontinuous Galerkin (RKDG) method was introduced. This method was contructed by (i) retaining the piecewise linear DG method for the space discretization, (ii) using a special explicit TVD second order Runge-Kutta type discretization introduced by Shu and Osher in a finite difference framework [57], [58], and (iii) modifying the 'slope limiter' to maintain the formal accuracy of the scheme at extrema. The resulting explicit scheme was then proven linearly stable for CFL numbers less than $1/3$, formally uniformly second order accurate in space and time including at extrema, and TVBM. Numerical results in [17] indicate good convergence behavior: Second order in smooth regions including at extrema, sharp shock transitions (usually in one or two elements) without oscillations, and convergence to entropy solutions even for non convex fluxes.

In [15], Cockburn and Shu extended this approach to construct (formally) high-order accurate RKDG methods for the scalar conservation law. To device RKDG methods of order $k + 1$, they used (i) the DG method with polynomials of degree k for the space discretization, (ii) a TVD $(k + 1)$-th order accurate explicit time discretization, and (iii) a generalized 'slope limiter.' The generalized 'slope limiter' was carefully devised with the purpose of enforcing the TVDM property without destroying the accuracy of the scheme. The numerical results in [15], for $k = 1, 2$, indicate $(k + 1)$-th order order in smooth regions away from discontinuities as well as sharp shock transitions with no oscillations; convergence to the entropy solutions was observed in all the tests. These RKDG schemes were extended to one-dimensional systems in [14].

- **The multidimensional case.**

The extension of the RKDG method to the multidimensional case was done in [13] for the scalar conservation law. In the multidimensional case, the complicated geometry the spatial domain might have in practical applications can be easily handled by the DG space discretization. The TVD time discretizations remain the same, of course. Only the construction of the generalized 'slope limiter' represents a serious challenge. This is so, not only because of the more complicated form of the elements but also because of inherent accuracy barries imposed by the stability properties.

Indeed, since the main purpose of the 'slope limiter' is to enforce the nonlinear stability of the scheme, it is essential to realize that in the multidimensional case, the constraints imposed by the stability of a scheme on its accuracy are even greater than in the one dimensional case. Although in the one dimensional case it is possible to devise high-order accurate schemes with the TVD property, this is not true in several space dimensions since Goodman and LeVeque [28] proved that any TVD scheme is at most first order accurate. Thus, any generalized 'slope limiter' that enforces the TVD property, or the TVDM property for that matter, would unavoidably reduce the accuracy of the scheme to first-order accuracy. This is why in [13], Cockburn, Hou and Shu devised a generalized 'slope limiter' that enforced a **local** maximum principles only since they are not incompatible with high-order accuracy. No other class of schemes has a proven maximum principle for genearal nonlinearities f, and arbitrary triangulations.

The extension of the RKDG methods to general multidimensional systems was started by Cockburn and Shu in [16] and has been recently completed in [19]. Bey and Oden [5] and more recently Bassi and Rebay [2] have studied applications of the method to the Euler equations of gas dynamics.

• **The main advantages of the RKDG method.**

The resulting RKDG schemes have several important advantages. First, like finite element methods such as the SUPG-method of Hughes and Brook [29,34,30–33] (which has been analyzed by Johnson *et al* in [38–40]), the RKDG methods are better suited than finite difference methods to handle complicated geometries. Moreover, the particular finite elements of the DG space discretization allow an extremely simple treatment of the boundary conditions; no special numerical treatment of them is required in order to achieve uniform high order accuracy, as is the case for the finite difference schemes.

Second, the method can easily handle adaptivity strategies since the refining or unrefining of the grid can be done without taking into account the continuity restrictions typical of conforming finite element methods. Also, the degree of the approximating polynomial can be easily changed from one element to the other. Adaptivity is of particular importance in hyperbolic problems given the complexity of the structure of the discontinuities. In the one dimensional case the Riemann problem can be solved in closed form and discontinuity curves in the (x, t) plane are simple straight lines passing through the origin. However, in two dimensions their solutions display a very rich structure; see the works of Wagner [64], Lindquist [43], [42], Zhang and Zheng [68], and Zhang and Cheng [67]. Thus, methods which allow triangulations that can be easily adapted to resolve this structure, have an important advantage.

Third, the method is highly parallelizable. Since the elements are discontinuous, the mass matrix is block diagonal and since the order of the blocks is equal to the number of degrees of freedom inside the corresponding elements, the blocks can be inverted by hand once and for all. Thus, at each Runge-Kutta inner step, to update the degrees of freedom inside a given element, only the degrees of freedom of the elements sharing a face are involved; communication between processors is thus kept to a minimum. Extensive studies of adaptivity and parallelizability issues of the RKDG method were started by Biswas, Devine, and Flaherty [6] and then continued by deCougny *et al.* [20], Devine *et al.* [22,21] and by Özturan *et al.* [52].

2.3 Convection-diffusion systems: The LDG method

The first extensions of the RKDG method to nonlinear, convection-diffusion systems of the form

$$\partial_t \mathbf{u} + \nabla \cdot \mathbf{F}(\mathbf{u}, D \mathbf{u}) = 0, \text{ in } (0, T) \times \Omega,$$

were proposed by Chen *et al.* [10], [9] in the framework of hydrodynamic models for semiconductor device simulation. In these extensions, approximations of second and third-order derivatives of the discontinuous approximate solution were obtained by using simple projections into suitable finite elements spaces. This projection requires the inversion of global mass matrices, which in [10] and [9] are 'lumped' in order to maintain the high parallelizability of the method. Since in [10] and [9] polynomials of degree one are used, the 'mass lumping' is justified; however, if polynomials of higher degree were used, the 'mass lumping' needed to enforce the full parallelizability of the method could cause a degradation of the formal order of accuracy.

Fortunately, this is not an issue with the methods proposed by Bassi and Rebay [3] (see also Bassi *et al* [2]) for the compressible Navier-Stokes equations. In these methods, the original idea of the RKDG method is applied to *both u and D u* which are now considered as *independent* unknowns. Like the RKDG methods, the resulting methods are highly parallelizable methods of high-order accuracy which are very efficient for time-dependent, convection-dominated flows. The LDG methods considered by Cockburn and Shu [18] are a generalization of these methods.

The basic idea to construct the LDG methods is to *suitably rewrite* the original system as a larger, degenerate, first-order system and then discretize it by the RKDG method. By a careful choice of this rewriting, nonlinear stability can be achieved even without slope limiters, just as the RKDG method in the purely hyperbolic case; see Jiang and Shu [36].

The LDG methods [18] are very different from the so-called Discontinuous Galerkin (DG) method for parabolic problems introduced by Jamet [35] and studied by Eriksson, Johnson, and Thomée [27], Eriksson and Johnson [23–26], and more recently by Makridakis and Babuška [50]. In the DG method, the approximate solution is discontinuous only in time, not in space; in fact, the space discretization is the standard Galerkin discretization with *continuous* finite elements. This is in strong contrast with the space discretizations of the LDG methods which use *discontinuous* finite elements. To emphasize this difference, those methods are called **Local** Discontinuous Galerkin methods. The large amount of degrees of freedom and the restrictive conditions of the size of the time step for explicit time-discretizations, render the LDG methods inefficient for diffusion-dominated problems; in this situation, the use of methods with continuous-in-space approximate solutions is recommended. However, as for the successful RKDG methods for purely hyperbolic problems, the extremely local domain of dependency of the LDG methods allows a very efficient parallelization that by far compensates for the extra amount of degrees of freedom in the case of convection-dominated flows.

Karniadakis *et al.* have implemented and tested these methods for the compressible Navier Stokes equations in two and three space dimensions with impressive results; see [44], [45], [46], [47], and [65].

2.4 The content of these notes

In these notes, we study the RKDG and LDG methods. Our exposition will be based on the papers by Cockburn and Shu [17], [15], [14], [13], and [19] in which the RKDG method was developed and on the paper by Cockburn and Shu [18] which is devoted to the LDG methods. Numerical results from the papers by Bassi and Rebay [2], on the Euler equations of gas dynamics, and [3], on the compressible Navier-Stokes equations, are also included.

The emphasis in these notes is on *how the above mentioned schemes were devised*. As a consequence, the sections that follow reflect that development. Thus, section 2, in which the RKDG schemes for the one-dimensional scalar conservation law are constructed, constitutes the core of the notes because it contains all the important ideas for the devicing of the RKDG methods; section 3 contains the extension to multidimensional systems; and section 4, the extension to convection-diffusion problems.

We would like to emphasize that the guiding principle in the devicing of the RKDG methods for scalar conservation laws is to consider them as *perturbations of the so-called monotone schemes*. As it is well-known, monotone schemes for scalar conservation laws are stable and converge to the entropy solution but are only first-order accurate. Following a widespread approach in the field of numerical schemes for nonlinear conservation laws, the RKDG are constructed in such a way that they are high-order accurate schemes that 'become' a monotone scheme when a piecewise-constant approximation is used. Thus, to obtain high-order accurate RKDG schemes, we 'perturb' the piecewise-constant approximation and allow it to be piecewise a polynomial of arbitrary degree. Then, the conditions under which the stability properties of the monotone schemes are still valid are sought and enforced by means of the generalized 'slope limiter.' The fact that it is possible to do so without destroying the accuracy of the RKDG method is the crucial point that makes this method both robust and accurate.

The issues of parallelization and adaptivity developed by Biswas, Devine, and Flaherty [6], deCougny *et al.* [20], Devine *et al.* [22,21] and by Özturan *et al.* [52] are certainly very important. Another issue of importance is how to render the method computationaly more efficient, like the quadrature rule-free versions of the RKDG method recently studied by Atkins and Shu [1].

However, these topics fall beyond the scope of these notes whose main intention is to provide a simple introduction to the topic of discontinuous Galerkin methods for convection-dominated problems.

3 The scalar conservation law in one space dimension

3.1 Introduction

In this section, we introduce and study the RKDG method for the following simple model problem:

$$u_t + f(u)_x = 0, \qquad \text{in } (0,1) \times (0,T), \qquad (3.1)$$
$$u(x,0) = u_0(x), \qquad \forall\, x \in (0,1), \qquad (3.2)$$

and periodic boundary conditions. This section has material drawn from [17] and [15].

3.2 The discontinuous Galerkin-space discretization

3.3 The weak formulation

To discretize in space, we proceed as follows. For each partition of the interval $(0,1)$, $\{x_{j+1/2}\}_{j=0}^N$, we set $I_j = (x_{j-1/2}, x_{j+1/2})$, $\Delta_j = x_{j+1/2} - x_{j-1/2}$ for $j = 1, \ldots, N$, and denote the quantity $\max_{1 \le j \le N} \Delta_j$ by Δx .

We seek an approximation u_h to u such that for each time $t \in [0,T]$, $u_h(t)$ belongs to the finite dimensional space

$$V_h = V_h^k \equiv \{ v \in L^1(0,1) : v|_{I_j} \in P^k(I_j),\ j = 1, \ldots, N \}, \qquad (3.3)$$

where $P^k(I)$ denotes the space of polynomials in I of degree at most k. In order to determine the approximate solution u_h, we use a weak formulation that we obtain as follows. First, we multiply the equations (3.1) and (3.2) by arbitrary, smooth functions v and integrate over I_j, and get, after a simple formal integration by parts,

$$\int_{I_j} \partial_t u(x,t)\, v(x)\, dx - \int_{I_j} f(u(x,t))\, \partial_x v(x)\, dx \qquad (3.4)$$
$$+ f(u(x_{j+1/2}, t))\, v(x_{j+1/2}^-) - f(u(x_{j-1/2}, t))\, v(x_{j-1/2}^+) = 0,$$

$$\int_{I_j} u(x,0)\, v(x)\, dx = \int_{I_j} u_0(x)\, v(x)\, dx. \qquad (3.5)$$

Next, we replace the smooth functions v by test functions v_h belonging to the finite element space V_h, and the exact solution u by the approximate solution

u_h. Since the function u_h is discontinuous at the points $x_{j+1/2}$, we must also replace the nonlinear 'flux' $f(u(x_{j+1/2}, t))$ by a *numerical* 'flux' that depends on the two values of u_h at the point $(x_{j+1/2}, t)$, that is, by the function

$$h(u)_{j+1/2}(t) = h(u(x_{j+1/2}^-, t), u(x_{j+1/2}^+, t)), \tag{3.6}$$

that will be suitably chosen later. Note that *we always use the same numerical flux regardless of the form of the finite element space*. Thus, the approximate solution given by the DG-space discretization is defined as the solution of the following weak formulation:

$$\forall\, j = 1, \ldots, N, \qquad \forall\, v_h \in P^k(I_j):$$

$$\int_{I_j} \partial_t u_h(x, t)\, v_h(x)\, dx - \int_{I_j} f(u_h(x, t))\, \partial_x v_h(x)\, dx \tag{3.7}$$
$$+ h(u_h)_{j+1/2}(t)\, v_h(x_{j+1/2}^-) - h(u_h)_{j-1/2}(t)\, v_h(x_{j-1/2}^+) = 0,$$

$$\int_{I_j} u_h(x, 0)\, v_h(x)\, dx = \int_{I_j} u_0(x)\, v_h(x)\, dx. \tag{3.8}$$

3.4 Incorporating the monotone numerical fluxes

To complete the definition of the approximate solution u_h, it only remains to choose the numerical flux h. To do that, we invoke our main point of view, namely, that *we want to construct schemes that are perturbations of the so-called monotone schemes* because monotone schemes, although only first-order accurate, are very stable and converge to the entropy solution. More precisely, we want that in the case $k = 0$, that is, when the approximate solution u_h is a piecewise-constant function, our DG-space discretization gives rise to a monotone scheme.

Since in this case, for $x \in I_j$ we can write

$$u_h(x, t) = u_j^0,$$

we can rewrite our weak formulation (3.7), (3.8) as follows:

$$\forall\ j = 1, \ldots, N :$$

$$\partial_t\, u_j^0(t) + \{h(u_j^0(t), u_{j+1}^0(t)) - h(u_{j-1}^0(t), u_j^0(t))\}/\Delta_j = 0,$$

$$u_j^0(0) = \frac{1}{\Delta_j} \int_{I_j} u_0(x)\, dx,$$

and it is well-known that this defines a monotone scheme if $h(a, b)$ is a Lipschitz, consistent, monotone flux, that is, if it is,

(i) locally Lipschitz and consistent with the flux $f(u)$, i.e., $h(u, u) = f(u)$,
(ii) a nondecreasing function of its first argument, and
(iii) a nonincreasing function of its second argument.

The best-known examples of numerical fluxes satisfying the above properties are the following:

(i) The Godunov flux:

$$h^G(a, b) = \begin{cases} \min_{a \leq u \leq b} f(u)\,, & \text{if } a \leq b, \\ \max_{a \geq u \geq b} f(u)\,, & \text{if } a > b; \end{cases}$$

(ii) The Engquist-Osher flux:

$$h^{EO}(a, b) = \int_0^b \min(f'(s), 0)\, ds + \int_0^a \max(f'(s), 0)\, ds + f(0);$$

(iii) The Lax-Friedrichs flux:

$$h^{LF}(a, b) = \frac{1}{2}\, [f(a) + f(b) - C\,(b - a)],$$

$$C = \max_{\inf u^0(x) \leq s \leq \sup u^0(x)} |f'(s)|;$$

(iv) The local Lax–Friedrichs flux:

$$h^{LLF}(a, b) = \frac{1}{2}\, [f(a) + f(b) - C(b - a)],$$

$$C = \max_{\min(a,b) \leq s \leq \max(a,b)} |f'(s)|;$$

(v) The Roe flux with 'entropy fix':

$$h^R(a, b) = \begin{cases} f(a), & \text{if } f'(u) \geq 0 \quad \text{for} \quad u \in [\min(a, b),\ \max(a, b)], \\ f(b), & \text{if } f'(u) \leq 0 \quad \text{for} \quad u \in [\min(a, b), \max(a, b)], \\ h^{LLF}(a, b), & \text{otherwise.} \end{cases}$$

For the flux h, we can use the Godunov flux h^G since it is well-known that this is the numerical flux that produces the smallest amount of artificial viscosity. The local Lax-Friedrichs flux produces more artificial viscosity than the Godunov flux, but their performances are remarkably similar. Of course, if f is too complicated, we can always use the Lax-Friedrichs flux. However, numerical experience suggests that as the degree k of the approximate solution increases, the choice of the numerical flux does not have a significant impact on the quality of the approximations.

3.5 Diagonalizing the mass matrix

If we choose the Legendre polynomials P_ℓ as local basis functions, we can exploit their L^2-orthogonality, namely,

$$\int_{-1}^{1} P_\ell(s) P_{\ell'}(s)\, ds = \left(\frac{2}{2\ell + 1} \right) \delta_{\ell \ell'},$$

and obtain a *diagonal* mass matrix. Indeed, if for $x \in I_j$, we express our approximate solution u_h as follows:

$$u_h(x, t) = \sum_{\ell=0}^{k} u_j^\ell\, \varphi_\ell(x),$$

where

$$\varphi_\ell(x) = P_\ell(2\,(x - x_j)/\Delta_j),$$

the weak formulation (3.7), (3.8) takes the following simple form:

$$\forall\ j = 1, \ldots, N \text{ and } \ell = 0, \ldots, k :$$

$$\left(\frac{1}{2\ell + 1} \right) \partial_t u_j^\ell(t) - \frac{1}{\Delta_j} \int_{I_j} f(u_h(x, t))\, \partial_x \varphi_\ell(x)\, dx$$

$$+ \frac{1}{\Delta_j} \left\{ h(u_h(x_{j+1/2}))(t) - (-1)^\ell\, h(u_h(x_{j-1/2}))(t) \right\} = 0,$$

$$u_j^\ell(0) = \frac{2\ell + 1}{\Delta_j} \int_{I_j} u_0(x)\, \varphi_\ell(x)\, dx,$$

where we have use the following properties of the Legendre polynomials:

$$P_\ell(1) = 1, \qquad P_\ell(-1) = (-1)^\ell.$$

This shows that after discretizing in space the problem (3.1), (3.2) by the DG method, we obtain a system of ODEs for the degrees of freedom that we can rewrite as follows:

$$\frac{d}{dt} u_h = L_h(u_h), \qquad \text{in } (0, T), \tag{3.9}$$

$$u_h(t = 0) = u_{0h}. \tag{3.10}$$

The element $L_h(u_h)$ of V_h is, of course, the approximation to $-f(u)_x$ provided by the DG-space discretization.

Note that if we choose a different local basis, the local mass matrix could be a full matrix but it will always be a matrix of order $(k + 1)$. By inverting it by means of a symbolic manipulator, we can always write the equations for the degrees of freedom of u_h as an ODE system of the form above.

3.6 Convergence analysis of the linear case

In the linear case $f(u) = c\,u$, the $L^\infty(0, T; L^2(0, 1))$-accuracy of the method (3.7), (3.8) can be established by using the $L^\infty(0, T; L^2(0, 1))$-stability of the method and the approximation properties of the finite element space V_h.

Note that in this case, all the fluxes displayed in the examples above coincide and are equal to

$$h(a, b) = c\,\frac{a + b}{2} - \frac{|c|}{2}(b - a). \tag{3.11}$$

The following results are thus for this numerical flux.

We state the L²-stability result in terms of the jumps of u_h across $x_{j+1/2}$ which we denote by

$$[u_h]_{j+1/2} \equiv u_h(x_{j+1/2}^+) - u_h(x_{j+1/2}^-).$$

Proposition 3.1 *(L²-stability) We have,*

$$\tfrac{1}{2} \| u_h(T) \|_{L^2(0,1)}^2 + \Theta_T(u_h) \le \tfrac{1}{2} \| u_0 \|_{L^2(0,1)}^2,$$

where

$$\Theta_T(u_h) = \tfrac{|c|}{2} \int_0^T \sum_{1 \le j \le N} [u_h(t)]_{j+1/2}^2 \, dt.$$

Note how the jumps of u_h are controled by the L^2-norm of the initial condition. This control reflects the subtle built-in dissipation mechanism of the DG-methods and is what allows the DG-methods to be more accurate than the standard Galerkin methods. Indeed, the standard Galerkin method has an order of accuracy equal to k whereas the DG-methods have an order of accuray equal to $k + 1/2$ for the same smoothness of the initial condition.

Theorem 3.1 *Suppose that the initial condition u_0 belongs to $H^{k+1}(0,1)$. Let e be the approximation error $u - u_h$. Then we have,*

$$\| e(T) \|_{L^2(0,1)} \leq C \, | u_0 |_{H^{k+1}(0,1)} (\varDelta x)^{k+1/2},$$

where C depends solely on k, $|c|$, and T.

It is also possible to prove the following result if we assume that the initial condition is more regular. Indeed, we have the following result.

Theorem 3.2 *Suppose that the initial condition u_0 belongs to $H^{k+2}(0,1)$. Let e be the approximation error $u - u_h$. Then we have,*

$$\| e(T) \|_{L^2(0,1)} \leq C \, | u_0 |_{H^{k+2}(0,1)} (\varDelta x)^{k+1},$$

where C depends solely on k, $|c|$, and T.

The Theorem 3.1 is a simplified version of a more general result proven in 1986 by Johnson and Pitkäranta [37] and the Theorem 3.2 is a simplified version of a more general result proven in 1974 by LeSaint and Raviart [41]. To provide a simple introduction to the techniques used in these more general results, we give *new* proofs of these theorems in an appendix to this section.

The above theorems show that the DG-space discretization results in a $(k+1)$th-order accurate scheme, at least in the linear case. This gives a strong indication that the same order of accuracy should hold in the nonlinear case when the exact solution is smooth enough, of course.

Now that we know that the DG-space discretization produces a high-order accurate scheme for smooth exact solutions, we consider the question of how does it behave when the flux is a nonlinear function.

3.7   Convergence analysis in the nonlinear case

To study the convergence properties of the DG-method, we first study the convergence properties of the solution w of the following problem:

$$w_t + f(w)_x = (\nu(w) \, w_x)_x, \qquad \text{in } (0,1) \times (0,T), \qquad (3.12)$$

$$w(x,0) = u_0(x), \qquad \forall \, x \in (0,1), \qquad (3.13)$$

and periodic boundary conditions. We then mimic the procedure to study the convergence of the DG-method for the piecewise-constant case. The general DG-method will be considered later after having introduced the Runge-Kutta time-discretization.

The continuous case as a model. In order to compare u and w, it is *enough* to have (i) an entropy inequality and (ii) uniform boundedness of $\| w_x \|_{L^1(0,1)}$. Next, we show how to obtain these properties in a formal way.

We start with the entropy inequality. To obtain such an inequality, the basic idea is to multiply the equation (3.12) by $U'(w-c)$, where $U(\cdot)$ denotes the absolute value function and c denotes an arbitrary real number. Since

$$U'(w-c)\,w_t = U(w-c)_t,$$
$$U'(w-c)\,f(w)_x = \left(U'(w-c)\,(f(w)-f(c)) \right) \equiv F(w,c)_x,$$
$$U'(w-c)\,(\nu(w)\,w_x)_x = \left(\int_c^w U'(\rho - c)\,\nu(\rho)\,d\rho \right)_{xx} - U''(w-c)\,\nu(w)\,(w_x)^2$$
$$\equiv \Phi(w,c)_{xx} - U''(w-c)\,\nu(w)\,(w_x)^2,$$

we obtain

$$U(w-c)_t + F(w,c)_x - \Phi(w,c)_{xx} \le 0, \qquad \text{in } (0,1) \times (0,T),$$

which is nothing but the entropy inequality we wanted.

To obtain the uniform boundedness of $\| w_x \|_{L^1(0,1)}$, the idea is to multiply the equation (3.12) by $-(U'(w_x))_x$ and integrate on x from 0 to 1. Since

$$\int_0^1 -(U'(w_x))_x\,w_t = \int_0^1 U'(w_x)\,(w_x)_t = \frac{d}{dt} \| w_x \|_{L^1(0,1)},$$
$$\int_0^1 -(U'(w_x))_x\,f(w)_x = -\int_0^1 U''(w_x)\,w_{xx}\,f'(w)\,w_x = 0,$$
$$\int_0^1 -(U'(w_x))_x\,(\nu(w)\,w_x)_x = -\int_0^1 U''(w_x)\,w_{xx}\,(\nu'(w)\,(w_x)^2 + \nu(w)\,w_{xx})$$
$$= -\int_0^1 U''(w_x)\,\nu(w)\,(w_{xx})^2 \le 0,$$

we immediately get that

$$\frac{d}{dt} \| w_x \|_{L^1(0,1)} \le 0,$$

and so,

$$\| w_x \|_{L^1(0,1)} \le \| (u_0)_x \|_{L^1(0,1)}, \qquad \forall\, t \in (0,T).$$

When the function u_0 has discontinuities, the same result holds with the total variation of u_0, $| u_0 |_{TV(0,1)}$, replacing the quantity $\| (u_0)_x \|_{L^1(0,1)}$; these two quantities coincide when $u_0 \in W^{1,1}(0,1)$.

With the two above ingredients, the following error estimate, obtained in 1976 by Kuznetsov, can be proved:

Theorem 3.3 *We have*

$$\| u(T) - w(T) \|_{L^1(0,1)} \leq | u_0 |_{TV(0,1)} \sqrt{8\,T\,\nu},$$

where $\nu = \sup_{s \in [\inf u_0, \sup u_0]} \nu(s)$.

The piecewise-constant case. Let consider the simple case of the DG-method that uses a piecewise-constant approximate solution:

$$\forall\, j = 1, \ldots, N :$$

$$\partial_t\, u_j + \{ h(u_j, u_{j+1}) - h(u_{j-1}, u_j) \}/\Delta_j = 0,$$
$$u_j(0) = \frac{1}{\Delta_j} \int_{I_j} u_0(x)\, dx,$$

where we have dropped the superindex '0.' We pick the numerical flux h to be the Engquist-Osher flux.

According to the model provided by the continuous case, we must obtain (i) an entropy inequality and (ii) the uniform boundedness of the total variation of u_h.

To obtain the entropy inequality, we multiply our equation by $U'(u_j - c)$:

$$\partial_t\, U(u_j - c) + U'(u_j - c)\{ h(u_j, u_{j+1}) - h(u_{j-1}, u_j) \}/\Delta_j = 0.$$

The second term in the above equation needs to be carefully treated. First, we rewrite the Engquist-Osher flux in the following form:

$$h^{EO}(a, b) = f^+(a) + f^-(b),$$

and, accordingly, rewrite the second term of the equality above as follows:

$$ST_j = U'(u_j - c)\{ f^+(u_j) - f^+(u_{j-1}) \} + U'(u_j - c)\{ f^-(u_{j+1}) - f^-(u_j) \}.$$

Using the simple identity

$$U'(a - c)(g(a) - g(b)) = G(a, c) - G(b, c) + \int_a^b (g(b) - g(\rho))\, U''(\rho - x)\, d\rho$$

where $G(a, c) = \int_c^a U'(\rho - c) \, g(\rho) \, d\rho$, we get

$$ST_j = F^+(u_j, c) - F^+(u_{j-1}, c) + \int_{u_j}^{u_{j-1}} \left(f^+(u_{j-1}) - f^+(\rho) \right) U''(\rho - x) \, d\rho$$

$$+ F^-(u_{j+1}, c) - F^-(u_j, c) - \int_{u_j}^{u_{j+1}} \left(f^-(u_{j+1}) - f^-(\rho) \right) U''(\rho - x) \, d\rho$$

$$= F(u_j, u_{j+1}; c) - F(u_{j-1}, u_j; c) + \Theta_{diss,j}$$

where

$$F(a, b; c) = F^+(a, c) + F^-(b, c),$$

$$\Theta_{diss,j} = + \int_{u_j}^{u_{j-1}} \left(f^+(u_{j-1}) - f^+(\rho) \right) U''(\rho - x) \, d\rho$$

$$- \int_{u_j}^{u_{j+1}} \left(f^-(u_{j+1}) - f^-(\rho) \right) U''(\rho - x) \, d\rho.$$

We thus get

$$\partial_t U(u_j - c) + \left\{ F(u_j, u_{j+1}; c) - F(u_{j-1}, u_j; c) \right\} / \Delta_j + \Theta_{diss,j} / \Delta_j = 0.$$

Since, f^+ and $-f^-$ are nondecreasing functions, we easily see that

$$\Theta_{diss,j} \geq 0,$$

and we obtain our entropy inequality:

$$\partial_t U(u_j - c) + \left\{ F(u_j, u_{j+1}; c) - F(u_{j-1}, u_j; c) \right\} / \Delta_j \leq 0.$$

Next, we obtain the uniform boundedness on the total variation. To do that, we follow our model and multiply our equation by a discrete version of $-(U'(w_x))_x$, namely,

$$v_j^0 = -\frac{1}{\Delta_j} \left\{ U'\left(\frac{u_{j+1} - u_j}{\Delta_{j+1/2}} \right) - U'\left(\frac{u_j - u_{j-1}}{\Delta_{j-1/2}} \right) \right\},$$

where $\Delta_{j+1/2} = (\Delta_j + \Delta_{j+1})/2$, multiply it by Δ_j and sum over j from 1 to N. We easily obtain

$$\frac{d}{dt} | u_h |_{TV(0,1)} + \sum_{1 \leq j \leq N} v_j^0 \left\{ h(u_j, u_{j+1}) - h(u_{j-1}, u_j) \right\} = 0,$$

where

$$|u_h|_{TV(0,1)} \equiv \sum_{1 \le j \le N} |u_{j+1} - u_j|.$$

According to our continuous model, the second term in the above equality should be positive. Let us see that this is indeed the case:

$$v_j^0 \{h(u_j, u_{j+1}) - h(u_{j-1}, u_j)\} = v_j^0 \{f^+(u_j) - f^+(u_{j-1})\} \\ + v_j^0 \{f^-(u_{j+1}) - f^-(u_j)\} \ge 0,$$

by the definition of v_j^0, f^+, and f^-. This implies that

$$|u_h(t)|_{TV(0,1)} \le |u_h(0)|_{TV(0,1)} \le |u_0|_{TV(0,1)}.$$

With the two above ingredients, the following error estimate, obtained in 1976 by Kuznetsov, can be proved:

Theorem 3.4 *We have*

$$\|u(T) - u_h(T)\|_{L^1(0,1)} \le \|u_0 - u_h(0)\|_{L^1(0,1)} + C|u_0|_{TV(0,1)}\sqrt{T \Delta x}.$$

3.8 The TVD-Runge-Kutta time discretization

To discretize our ODE system in time, we use the TVD Runge Kutta time discretization introduced in [60]; see also [57] and [58].

3.9 The discretization

Thus, if $\{t^n\}_{n=0}^N$ is a partition of $[0, T]$ and $\Delta t^n = t^{n+1} - t^n, n = 0, ..., N-1$, our time-marching algorithm reads as follows:

- Set $u_h^0 = u_{0h}$;
- For $n = 0, ..., N-1$ compute u_h^{n+1} from u_h^n as follows:
  1. set $u_h^{(0)} = u_h^n$;
  2. for $i = 1, ..., k+1$ compute the intermediate functions:

$$u_h^{(i)} = \left\{ \sum_{l=0}^{i-1} \alpha_{il} u_h^{(l)} + \beta_{il} \Delta t^n L_h(u_h^{(l)}) \right\};$$

  3. set $u_h^{n+1} = u_h^{(k+1)}$.

Note that this method is very easy to code since *only a single subroutine defining $L_h(u_h)$ is needed*. Some Runge-Kutta time discretization parameters are displayed on the table below.

Table 1

Parameters of some practical Runge-Kutta time discretizations			
order	α_{il}	β_{il}	$\max\{\beta_{il}/\alpha_{il}\}$
2	1 $\frac{1}{2}$ $\frac{1}{2}$	1 0 $\frac{1}{2}$	1
3	1 $\frac{3}{4}$ $\frac{1}{4}$ $\frac{1}{3}$ 0 $\frac{2}{3}$	1 0 $\frac{1}{4}$ 0 0 $\frac{2}{3}$	1

3.10 The stability property

Note that all the values of the parameters α_{il} displayed in the table below are nonnegative; this is not an accident. Indeed, this is a condition on the parameters α_{il} that ensures the stability property

$$|u_h^{n+1}| \leq |u_h^n|,$$

provided that the 'local' stability property

$$|w| \leq |v|, \tag{3.14}$$

where w is obtained from v by the following 'Euler forward' step,

$$w = v + \delta\, L_h(v), \tag{3.15}$$

holds for values of $|\delta|$ smaller than a given number δ_0.

For example, the second-order Runke-Kutta method displayed in the table above can be rewritten as follows:

$$u_h^{(1)} = u_h^n + \Delta t\, L_h(u_h^n),$$
$$w_h = u_h^{(1)} + \Delta t\, L_h(u_h^{(1)}),$$
$$u_h^{n+1} = \frac{1}{2}(u_h^n + w_h).$$

Now, assuming that the stability property (3.14), (3.15) is satisfied for

$$\delta_0 = |\Delta t \ \max\{\beta_{il}/\alpha_{il}\}| = \Delta t,$$

we have

$$|u_h^{(1)}| \le |u_h^n|, \qquad |w_h| \le |u_h^{(1)}|,$$

and so,

$$|u_h^{n+1}| \le \frac{1}{2}(|u_h^n| + |w_h|) \le |u_h^n|.$$

Note that we can obtain this result because the coefficients α_{il} are positive! Runge-Kutta methods of this type of order up to order 5 can be found in [58].

The above example shows how to prove the following more general result.

Theorem 3.5 *Assume that the stability property for the single 'Euler forward' step (3.14), (3.15) is satisfied for*

$$\delta_0 = \max_{0 \le n \le N} |\Delta t^n \ \max\{\beta_{il}/\alpha_{il}\}|.$$

Assume also that all the coeficients α_{il} are nonnegative and satisfy the following condition:

$$\sum_{l=0}^{i-1} \alpha_{il} = 1, \qquad i = 1, \ldots, k+1.$$

Then

$$|u_h^n| \le |u_h^0|, \qquad \forall n \ge 0.$$

This stability property of the TVD-Runge-Kutta methods is crucial since it allows us to obtain the stability of the method from the stability of a single 'Euler forward' step.

Proof of Theorem 3.5. We start by rewriting our time discretization as follows:

– Set $u_h^0 = u_{0h}$;
– For $n = 0, \ldots, N-1$ compute u_h^{n+1} from u_h^n as follows:
  1. set $u_h^{(0)} = u_h^n$;
  2. for $i = 1, \ldots, k+1$ compute the intermediate functions:

$$u_h^{(i)} = \sum_{l=0}^{i-1} \alpha_{il} w_h^{(il)},$$

  where

$$w_h^{(il)} = u_h^{(l)} + \frac{\beta_{il}}{\alpha_{il}} \Delta t^n L_h(u_h^{(l)});$$

  3. set $u_h^{n+1} = u_h^{(k+1)}$.

We then have

$$|u_h^{(i)}| \le \sum_{l=0}^{i-1} \alpha_{il} |w_h^{(il)}|, \quad \text{since } \alpha_{il} \ge 0,$$

$$\le \sum_{l=0}^{i-1} \alpha_{il} |u_h^{(l)}|, \quad \text{by the stability property (3.14), (3.15),}$$

$$\le \max_{0 \le l \le i-1} |u_h^{(l)}|, \quad \text{since } \sum_{l=0}^{i-1} \alpha_{il} = 1.$$

It is clear now that that Theorem 3.5 follows from the above inequality by a simple induction argument.   □

3.11   Remarks about the stability in the linear case

For the linear case $f(u) = cu$, Chavent and Cockburn [7] proved that for the case $k = 1$, i.e., for piecewise-linear approximate solutions, the single 'Euler forward' step is *unconditionally* $L^\infty(0, T; L^2(0, 1))$-unstable for any fixed ratio $\Delta t / \Delta x$. On the other hand, in [17] it was shown that if a Runge-Kutta method of second order is used, the scheme is $L^\infty(0, T; L^2(0, 1))$-stable provided that

$$c \frac{\Delta t}{\Delta x} \le \frac{1}{3}.$$

This means that we cannot deduce the stability of the complete Runge-Kutta method from the stability of the single 'Euler forward' step. As a consequence, we cannot apply Theorem 3.5 and we must consider the complete method at once.

Our numerical experiments show that when polynomial of degree k are used, a Runge-Kutta of order $(k+1)$ must be used. In this case, the $L^\infty(0, T; L^2(0, 1))$-stability condition is the following:

$$c \frac{\Delta t}{\Delta x} \le \frac{1}{2k+1}.$$

There is no rigorous proof of this fact yet.

At a first glance, this stability condition, also called the Courant-Friedrichs-Levy (CFL) condition, seems to compare unfavorably with that of the well-known finite difference schemes. However, we must remember that in the DG-methods there are $(k+1)$ degrees of freedom in each element of size Δx whereas for finite difference schemes there is a single degree of freedom of each cell of size Δx. Also, if a finite difference scheme is of order $(k+1)$ its so-called stencil must be of at least $(2k+1)$ points, whereas the DG-scheme has a stencil of $(k+1)$ elements only.

3.12 Convergence analysis in the nonlinear case

Now, we explore what is the impact of the explicit Runge-Kutta time-discretization on the convergence properties of the methods under consideration. We start by considering the piecewise-constant case.

The piecewise-constant case. Let us begin by considering the simplest case, namely,

$$\forall \; j = 1, \ldots, N :$$

$$(u_j^{n+1} - u_j^n)/\Delta t + \{ h(u_j^n, u_{j+1}^n) - h(u_{j-1}^n, u_j^n) \}/\Delta_j = 0,$$

$$u_j(0) = \frac{1}{\Delta_j} \int_{I_j} u_0(x)\, dx,$$

where we pick the numerical flux h to be the Engquist-Osher flux.

According to the model provided by the continuous case, we must obtain (i) an entropy inequality and (ii) the uniform boundedness of the total variation of u_h.

To obtain the entropy inequality, we proceed as in the semidiscrete case and obtain the following result; see [12] for details.

Theorem 3.6 *We have*

$$\{ U(u_j^{n+1} - c) - U(u_j^n - c) \}/\Delta t + \{ F(u_j^n, u_{j+1}^n; c) - F(u_{j-1}^n, u_j^n; c) \}/\Delta_j$$
$$+ \Theta_{diss,j}^n/\Delta t = 0,$$

where

$$\Theta_{diss,j}^n = \int_{u_j^{n+1}}^{u_j^n} (p_j(u_j^n) - p_j(\rho)) \, U''(\rho - x)\, d\rho$$

$$+ \frac{\Delta t}{\Delta_j} \int_{u_j^{n+1}}^{u_{j-1}^n} (f^+(u_{j-1}^n) - f^+(\rho)) \, U''(\rho - x)\, d\rho$$

$$- \frac{\Delta t}{\Delta_j} \int_{u_j^{n+1}}^{u_{j+1}^n} (f^-(u_{j+1}^n) - f^-(\rho)) \, U''(\rho - x)\, d\rho,$$

and

$$p_j(w) = w - \frac{\Delta t}{\Delta_j}(f^+(w) - f^-(w)).$$

Moreover, if the following CFL condition is satisfied

$$\max_{1 \le j \le N} \frac{\Delta t}{\Delta_j} |f'| \le 1,$$

then $\Theta_{diss,j}^n \ge 0$, *and the following entropy inequality holds:*

$$\{U(u_j^{n+1} - c) - U(u_j^n - c)\}/\Delta t + \{F(u_j^n, u_{j+1}; c) - F(u_{j-1}, u_j; c)\}/\Delta_j \le 0.$$

Note that $\Theta_{diss,j}^n \ge 0$ because f^+, $-f^-$, are nondecreasing and because p_j is also nondecreasing under the above CFL condition.

Next, we obtain the uniform boundedness on the total variation. Proceeding as before, we easily obtain the following result.

Theorem 3.7 *We have*

$$|u_h^{n+1}|_{TV(0,1)} - |u_h^n|_{TV(0,1)} + \Theta_{TV}^n = 0,$$

where

$$
\Theta_{TV}^n = \sum_{1 \le j \le N} \left(U'_{j+1/2}^n - U'_{j+1/2}^{n+1} \right) (p_{j+1/2}(u_{j+1}^n) - p_{j+1/2}(u_j^n))
$$
$$
+ \sum_{1 \le j \le N} \frac{\Delta t}{\Delta_j} \left(U'_{j-1/2}^n - U'_{j+1/2}^{n+1} \right) (f^+(u_j^n) - f^+(u_{j-1}^n))
$$
$$
- \sum_{1 \le j \le N} \frac{\Delta t}{\Delta_j} \left(U'_{j+1/2}^n - U'_{j-1/2}^{n+1} \right) (f^-(u_{j+1}^n) - f^-(u_j^n))
$$

where

$$U'_{i+1/2}^m = U' \left(\frac{u_{i+1}^m - u_i^m}{\Delta_{i+1/2}} \right),$$

and

$$p_{j+1/2}(w) = s - \frac{\Delta t}{\Delta_{j+1}} f^+(w) + \frac{\Delta t}{\Delta_j} f^-(w).$$

Moreover, if the following CFL condition is satisfied

$$\max_{1 \le j \le N} \frac{\Delta t}{\Delta_j} |f'| \le 1,$$

then $\Theta_{TV}^n \ge 0$, *and we have*

$$|u_h^n|_{TV(0,1)} \le |u_0|_{TV(0,1)}.$$

With the two above ingredients, the following error estimate, obtained in 1976 by Kuznetsov, can be proved:

Theorem 3.8 *We have*

$$\| u(T) - u_h(T) \|_{L^1(0,1)} \leq \| u_0 - u_h(0) \|_{L^1(0,1)} + C \, | u_0 |_{TV(0,1)} \sqrt{T \, \Delta x}.$$

The general case. The study of the general case is much more difficult than the study of the monotone schemes. In these notes, we restrict ourselves to the study of the stability of the RKDG schemes. Hence, we restrict ourselves to the task of studying under what conditions the total variation of the *local means* is uniformly bounded.

If we denote by $\bar{u}_j$ the mean of u_h on the interval I_j, by setting $v_h = 1$ in the equation (3.7), we obtain,

$$\forall \, j = 1, \ldots, N :$$

$$(\bar{u}_j)_t + \{ h(u^-_{j+1/2}, u^+_{j+1/2}) - h(u^-_{j-1/2}, u^+_{j-1/2}) \} / \Delta_j = 0,$$

where $u^-_{j+1/2}$ denotes the limit from the left and $u^+_{j+1/2}$ the limit from the right. We pick the numerical flux h to be the Engquist-Osher flux.

This shows that if we set w_h equal to the Euler forward step $u_h + \delta \, L_h(u_h)$, we obtain

$$\forall \, j = 1, \ldots, N :$$

$$(\bar{w}_j - \bar{u}_j) / \delta + \{ h(u^-_{j+1/2}, u^+_{j+1/2}) - h(u^-_{j-1/2}, u^+_{j-1/2}) \} / \Delta_j = 0.$$

Proceeding exactly as in the piecewise-constant case, we obtain the following result for the total variation of the avergages,

$$| \bar{u}_h |_{TV(0,1)} \equiv \sum_{1 \leq j \leq N} | \bar{u}_{j+1} - \bar{u}_j |.$$

Theorem 3.9 *We have*

$$| \bar{w}_h |_{TV(0,1)} - | \bar{u}_h |_{TV(0,1)} + \Theta_{TVM} = 0,$$

where

$$\Theta_{TVM} = \sum_{1 \le j \le N} \left(U'_{j+1/2} - U'_{j+1/2} \right) \left(p_{j+1/2}(u_h|_{I_{j+1}}) - p_{j+1/2}(u_h|_{I_j}) \right)$$

$$+ \sum_{1 \le j \le N} \frac{\delta}{\Delta_j} \left(U'_{j-1/2} - U'_{j+1/2} \right) (f^+(u^-_{j+1/2}) - f^+(u^-_{j-1/2}))$$

$$- \sum_{1 \le j \le N} \frac{\delta}{\Delta_j} \left(U'_{j+1/2} - U'_{j-1/2} \right) (f^-(u^+_{j+1/2}) - f^-(u^+_{j-1/2}))$$

where

$$U'_{i+1/2} = U'\left(\frac{u_{i+1} - u_i}{\Delta_{i+1/2}} \right),$$

and

$$p_{j+1/2}(u_h|_{I_m}) = \overline{u}_m - \frac{\delta}{\Delta_{j+1}} f^+(u^-_{m+1/2}) + \frac{\delta}{\Delta_j} f^-(u^+_{m-1/2}).$$

From the above result, we see that the total variation of the means of the Euler forward step is nonincreasing if the following three conditions are satisfied:

$$sgn(\overline{u}_{j+1} - \overline{u}_j) = sgn(p_{j+1/2}(u_h|_{I_{j+1}}) - p_{j+1/2}(u_h|_{I_j})), \qquad (3.16)$$

$$sgn(\overline{u}_j - \overline{u}_{j-1}) = sgn(u^{n,-}_{j+1/2} - u^{n,-}_{j-1/2}), \qquad (3.17)$$

$$sgn(\overline{u}_{j+1} - \overline{u}_j) = sgn(u^{n,+}_{j+1/2} - u^{n,+}_{j-1/2}). \qquad (3.18)$$

Note that if the properties (3.16) and (3.17) are satisfied, then the property (3.18) can always be satisfied for a small enough values of $|\delta|$.

Of course, the numerical method under consideration does not provide an approximate solution automatically satisfying the above conditions. It is thus necessary to *enforce* them by means of a suitably defined generalized slope limiter,' $\Lambda \Pi_h$.

3.13 The generalized slope limiter

High-order accuracy versus the TVDM property: Heuristics The ideal generalized slope limiter $\Lambda \Pi_h$

- Maintains the conservation of *mass* element by element,
- Satifies the properties (3.16), (3.17), and (3.18),
- Does not degrade the accuracy of the method.

The first requirement simply states that the slope limiting must not change the total mass contained in each interval, that is, if $u_h = \Lambda \Pi_h(v_h)$,

$$\overline{u}_j = \overline{v}_j, \qquad j = 1, \dots, N.$$

This is, of course a very sensible requirement because after all we are dealing with consevation laws. It is also a requirement very easy to satisfy.

The second requirement, states that if $u_h = \Lambda\Pi_h(v_h)$ and $w_h = u_h + \delta L_h(u_h)$ then

$$|\overline{w}_h|_{TV(0,1)} \leq |\overline{u}_h|_{TV(0,1)},$$

for small enough values of $|\delta|$.

The third requirement deserves a more delicate discussion. Note that if u_h is a very good approximation of a smooth solution u in a neigborhood of the point x_0, it behaves (asymptotically as Δx goes to zero) as a straight line if $u_x(x_0) \neq 0$. If x_0 is an isolated extrema of u, then it behaves like a parabola provided $u_{xx}(x_0) \neq 0$. Now, if u_h is a straightline, it trivially satisfies conditions (3.16) and (3.17). However, if u_h is a parabola, conditions (3.16) and (3.17) are not always satisfied. This shows that it is impossible to construct the above ideal generalized 'solpe limiter,' or, in other words, that in order to enforce the TVDM property, we must loose high-order accuracy at the local extrema. This is a very well-known phenomenon for TVD finite difference schemes!

Fortunatelly, it is still possible to construct generalized slope limiters that do preserve high-order accuracy even at local extrema. The resulting scheme will then not be TVDM but total variation bounded in the means (TVBM) as we will show.

In what follows we first consider generalized slope limiters that render the RKDG schemes TVDM. Then we suitably modify them in order to obtain TVBM schemes.

Constructing TVDM generalized slope limiters Next, we look for simple, sufficient conditions on the function u_h that imply the conditions (3.16), (3.17), and (3.18). These conditions will be stated in terms of the *minmod* function m defined as follows:

$$m(a_1, \ldots, a_\nu) = \begin{cases} s \min_{1 \leq n \leq \nu} |a_n|, & \text{if } s = sign(a_1) = \cdots = sign(a_\nu), \\ 0, & \text{otherwise.} \end{cases}$$

Theorem 3.10 *Suppose the the following CFL condition is satisfied:*

$$|\delta|\left(\frac{|f^+|_{Lip}}{\Delta_{j+1}} + \frac{|f^-|_{Lip}}{\Delta_j}\right) \leq 1/2, \qquad j = 1, \ldots, N. \tag{3.19}$$

Then, conditions (3.16), (3.17), and (3.18) are satisfied if, for all $j = 1, \ldots, N$, we have that

$$u^-_{j+1/2} - \overline{u}_j = m\,(u^-_{j+1/2} - \overline{u}_j,\, \overline{u}_j - \overline{u}_{j-1},\, \overline{u}_{j+1} - \overline{u}_j) \qquad (3.20)$$

$$\overline{u}_j - u^+_{j-1/2} = m\,(\overline{u}_j - u^+_{j-1/2},\, \overline{u}_j - \overline{u}_{j-1},\, \overline{u}_{j+1} - \overline{u}_j). \qquad (3.21)$$

Proof. Let us start by showing that the property (3.17) is satisfied. We have:

$$u^-_{j+1/2} - u^-_{j-1/2} = (u^-_{j+1/2} - \overline{u}_j) + (\overline{u}_j - \overline{u}_{j-1}) + (\overline{u}_{j-1} - u^-_{j-1/2})$$
$$= \Theta\,(\overline{u}_j - \overline{u}_{j-1}),$$

where

$$\Theta = 1 + \frac{u^-_{j+1/2} - \overline{u}_j}{\overline{u}_j - \overline{u}_{j-1}} - \frac{u^-_{j-1/2} - \overline{u}_{j-1}}{\overline{u}_j - \overline{u}_{j-1}} \in [0,2],$$

by conditions (3.20) and (3.21). This implies that the property (3.17) is satisfied. Properties (3.18) and (3.16) are proven in a similar way. This completes the proof. □

Examples of TVDM generalized slope limiters

a. The MUSCL limiter. In the case of piecewise linear approximate solutions, that is,

$$v_h|_{I_j} = \overline{v}_j + (x - x_j)\,v_{x,j}, \qquad j = 1, \ldots, N,$$

the following generalized slope limiter does satisfy the conditions (3.20) and (3.21):

$$u_h|_{I_j} = \overline{v}_j + (x - x_j)\,m\left(v_{x,j},\, \frac{\overline{v}_{j+1} - \overline{v}_j}{\Delta_j},\, \frac{\overline{v}_j - \overline{v}_{j-1}}{\Delta_j}\right).$$

This is the well-known slope limiter of the MUSCL schemes of vanLeer [62,63].

b. The less restrictive limiter $\Lambda\Pi^1_h$. The following less restrictive slope limiter also satisfies the conditions (3.20) and (3.21):

$$u_h|_{I_j} = \overline{v}_j + (x - x_j)\,m\left(v_{x,j},\, \frac{\overline{v}_{j+1} - \overline{v}_j}{\Delta_j/2},\, \frac{\overline{v}_j - \overline{v}_{j-1}}{\Delta_j/2}\right).$$

Moreover, it can be rewritten as follows:

$$u^-_{j+1/2} = \overline{v}_j + m\,(\,v^-_{j+1/2} - \overline{v}_j,\, \overline{v}_j - \overline{v}_{j-1},\, \overline{v}_{j+1} - \overline{v}_j) \qquad (3.22)$$

$$u^+_{j-1/2} = \overline{v}_j - m\,(\,\overline{v}_j - v^+_{j-1/2},\, \overline{v}_j - \overline{v}_{j-1},\, \overline{v}_{j+1} - \overline{v}_j). \qquad (3.23)$$

We denote this limiter by $\Lambda\Pi^1_h$.

Note that we have that

$$\| \overline{v}_h - \Lambda\Pi^1_h(v_h) \|_{L^1(0,1)} \le \frac{\Delta x}{2}\,|\overline{v}_h|_{TV(0,1)}.$$

See Theorem 3.13 below.

c. The limiter $\Lambda\Pi^k_h$. In the case in which the approximate solution is piecewise a polynomial of degree k, that is, when

$$v_h(x,t) = \sum_{\ell=0}^{k} v^\ell_j\, \varphi_\ell(x),$$

where

$$\varphi_\ell(x) = P_\ell(2\,(x - x_j)/\Delta_j),$$

and P_ℓ are the Legendre polynomials, we can define a generalized slope limiter in a very simple way. To do that, we need the define what could be called the P^1-part of v_h:

$$v^1_h(x,t) = \sum_{\ell=0}^{1} v^\ell_j\, \varphi_\ell(x),$$

We define $u_h = \Lambda\Pi_h(v_h)$ as follows:

- For $j = 1, ..., N$ compute $u_h|_{I_j}$ as follows:
  1. Compute $u^-_{j+1/2}$ and $u^+_{j-1/2}$ by using (3.22) and (3.23),
  2. If $u^-_{j+1/2} = v^-_{j+1/2}$ and $u^+_{j-1/2} = v^+_{j-1/2}$ set $u_h|_{I_j} = v_h|_{I_j}$,
  3. If not, take $u_h|_{I_j}$ equal to $\Lambda\Pi^1_h(v^1_h)$.

d. The limiter $\Lambda\Pi^k_{h,\alpha}$. When instead of (3.22) and (3.23), we use

$$u^-_{j+1/2} = \overline{v}_j + m\,(\,v^-_{j+1/2} - \overline{v}_j,\, \overline{v}_j - \overline{v}_{j-1},\, \overline{v}_{j+1} - \overline{v}_j, C\,(\Delta x)^\alpha) \quad (3.24)$$

$$u^+_{j-1/2} = \overline{v}_j - m\,(\,\overline{v}_j - v^+_{j-1/2},\, \overline{v}_j - \overline{v}_{j-1},\, \overline{v}_{j+1} - \overline{v}_j, C\,(\Delta x)^\alpha), \quad (3.25)$$

for some fixed constant C and $\alpha \in (0,1)$, we obtain a generalized slope limiter we denote by $\Lambda\Pi^k_{h,\alpha}$.

This generalized slope limiter is never used in practice, but we consider it here because it is used for theoretical purposes; see Theorem 3.13 below.

The complete RKDG method Now that we have our generalized slope limiters, we can display the complete RKDG method. It is contained in the following algorith:

- Set $u_h^0 = \Lambda\Pi_h \, P_{V_h}(u_0)$;
- For $n = 0, ..., N - 1$ compute u_h^{n+1} as follows:
  1. set $u_h^{(0)} = u_h^n$;
  2. for $i = 1, ..., k + 1$ compute the intermediate functions:

$$u_h^{(i)} = \Lambda\Pi_h \left\{ \sum_{l=0}^{i-1} \alpha_{il} \, u_h^{(l)} + \beta_{il} \Delta t^n L_h(u_h^{(l)}) \right\};$$

  3. set $u_h^{n+1} = u_h^{(k+1)}$.

This algorithm describes the complete RKDG method. Note how the generalized slope limiter has to be applied at each intermediate computation of the Runge-Kutta method. This way of appying the generalized slope limiter in the time-marching algorithm ensures that the scheme is TVDM, as we next show.

The TVDM property of the RKDG method To do that, we start by noting that if we set

$$u_h = \Lambda\Pi_h(v_h), \qquad w_h = u_h + \delta \, L_h(u_h),$$

then we have that

$$|\overline{u}_h|_{TV(0,1)} \leq |\overline{v}_h|_{TV(0,1)}, \tag{3.26}$$

$$|\overline{w}_h|_{TV(0,1)} \leq |\overline{u}_h|_{TV(0,1)}, \qquad \forall |\delta| \leq \delta_0, \tag{3.27}$$

where

$$\delta_0^{-1} = 2 \max_j \left(\frac{|f^+|_{Lip}}{\Delta_{j+1}} + \frac{|f^-|_{Lip}}{\Delta_j} \right) \qquad j = 1, \ldots, N,$$

by Theorem 3.10. By using the above two properties of the generalized slope limiter,' it is possible to show that the RKDG method is TVDM.

Theorem 3.11 *Assume that the generalized slope limiter $\Lambda\Pi_h$ satisfies the properties (3.26) and (3.27). Assume also that all the coeficients α_{il} are non-negative and satisfy the following condition:*

$$\sum_{l=0}^{i-1} \alpha_{il} = 1, \qquad i = 1, \ldots, k + 1.$$

Then

$$|\overline{u}_h^n|_{TV(0,1)} \leq |u_0|_{TV(0,1)}, \qquad \forall n \geq 0.$$

Proof of Theorem 3.11. The proof of this result is very similar to the proof of Theorem 3.5. Thus, we start by rewriting our time discretization as follows:

- Set $u_h^0 = u_{0h}$;
- For $n = 0, ..., N-1$ compute u_h^{n+1} from u_h^n as follows:
  1. set $u_h^{(0)} = u_h^n$;
  2. for $i = 1, ..., k+1$ compute the intermediate functions:

$$u_h^{(i)} = \Lambda \Pi_h \left\{ \sum_{l=0}^{i-1} \alpha_{il}\, w_h^{(il)} \right\},$$

  where

$$w_h^{(il)} = u_h^{(l)} + \frac{\beta_{il}}{\alpha_{il}}\, \Delta t^n\, L_h(u_h^{(l)});$$

  3. set $u_h^{n+1} = u_h^{(k+1)}$.

Then have,

$$|\overline{u}_h^{(i)}|_{TV(0,1)} \leq |\sum_{l=0}^{i-1} \alpha_{il}\, \overline{w}_h^{(il)}|_{TV(0,1)}, \quad \text{by (3.26),}$$

$$\leq \sum_{l=0}^{i-1} \alpha_{il}\, |\overline{w}_h^{(il)}|_{TV(0,1)}, \quad \text{since } \alpha_{il} \geq 0,$$

$$\leq |\sum_{l=0}^{i-1} \alpha_{il}\, \overline{u}_h^{(l)}|_{TV(0,1)}, \quad \text{by (3.27),}$$

$$\leq \max_{0 \leq l \leq i-1} |\overline{u}_h^{(l)}|_{TV(0,1)}, \quad \text{since } \sum_{l=0}^{i-1} \alpha_{il} = 1.$$

It is clear now that that the inequality

$$|\overline{u}_h^n|_{TV(0,1)} \leq |\overline{u}_h^0|_{TV(0,1)}, \qquad \forall n \geq 0.$$

follows from the above inequality by a simple induction argument. To obtain the result of the theorem, it is enough to note that we have

$$|\overline{u}_h^0|_{TV(0,1)} \leq |u_0|_{TV(0,1)},$$

by the definition of the initial condition u_h^0. This completes the proof. $\quad\square$

TVBM generalized slope limiters As was pointed out before, it is possible to modify the generalized slope limiters displayed in the examples above in such a way that the degradation of the accuracy at local extrema is avoided. To achieve this, we follow Shu [59] and modify the definition of the generalized slope limiters by simply replacing the *minmod* function m by the TVB corrected *minmod* function $\bar{m}$ defined as follows:

$$\bar{m}(a_1, ..., a_m) = \begin{cases} a_1, & if \ |a_1| \le M(\Delta x)^2, \\ m(a_1, ..., a_m), & otherwise, \end{cases} \qquad (3.28)$$

where M is a given constant. We call the generalized slope limiters thus constructed, TVBM slope limiters.

The constant M is, of course, an upper bound of the absolute value of the second-order derivative of the solution at local extrema. In the case of the nonlinear conservation laws under consideration, it is easy to see that, if the initial data is piecewise C^2, we can take

$$M = \sup\{\,|(u_0)_{xx}(y)|, y : (u_0)_x(y) = 0\}.$$

See [15] for other choices of M.

Thus, if the constant M is is taken as above, there is no degeneracy of accuracy at the extrema and the resulting RKDG scheme retains its optimal accuracy. Moreover, we have the following stability result.

Theorem 3.12 *Assume that the generalized slope limiter $\Lambda\Pi_h$ is a TVBM slope limiter. Assume also that all the coeficients α_{il} are nonnegative and satisfy the following condition:*

$$\sum_{l=0}^{i-1} \alpha_{il} = 1, \qquad i = 1, \ldots, k+1.$$

Then

$$|\bar{u}_h^n|_{TV(0,1)} \le |\bar{u}_0|_{TV(0,1)} + C\,M, \qquad \forall\,n \ge 0,$$

where C depends on k only.

Convergence in the nonlinear case By using the stability above stability results, we can use the Ascoli-Arzelá theorem to prove the following convergence result.

Theorem 3.13 *Assume that the generalized slope limiter $\Lambda\Pi_h$ is a TVDM or a TVBM slope limiter. Assume also that all the coeficients α_{il} are nonnegative and satisfy the following condition:*

$$\sum_{l=0}^{i-1} \alpha_{il} = 1, \qquad i = 1, \ldots, k+1.$$

Then there is a subsequence $\{\bar{u}_{h'}\}_{h'>0}$ of the sequence $\{\bar{u}_h\}_{h>0}$ generate by the RKDG scheme that converges in $L^\infty(0,T;L^1(0,1))$ to a weak solution of the problem (3.1), (3.2).

Moreover, if the TVBM version of the slope limiter $\Lambda\Pi^k_{h,\alpha}$ is used, the weak solution is the entropy solution and the whole sequence converges.

Finally, if the generalized slope limiter $\Lambda\Pi_h$ is such that

$$\|\bar{v}_h - \Lambda\Pi_h(v_h)\|_{L^1(0,1)} \leq C\,\Delta x\,|\bar{v}_h|_{TV(0,1)},$$

then the above results hold not only to the sequence of the means $\{\bar{u}_h\}_{h>0}$ but to the sequence of the functions $\{u_h\}_{h>0}$.

3.14 Computational results

In this subsection, we display the performance of the RKDG schemes in a simple but typical test problem. We use piecewise linear $(k=1)$ and piecewise quadratic $(k=2)$ elements; the $\Lambda\Pi^k_h$ generalized slope limter is used. Our purpose is to show that (i) when the constant M is properly chosen, the RKDG method using polynomials of degree k is is order $k+1$ in the uniform norm away from the discontinuities, that (ii) it is computationally more efficient to use high-degree polynomial approximations, and that (iii) shocks are captured in a few elements without production of spurious oscillations

We solve the Burger's equation with a periodic boundary condition:

$$u_t + (\frac{u^2}{2})_x = 0,$$

$$u(x,0) = u_0(x) = \frac{1}{4} + \frac{1}{2}\,\sin(\pi(2\,x-1)).$$

The exact solution is smooth at $T=.05$ and has a well developed shock at $T=0.4$. Notice that there is a sonic point. In Tables 1,2, and 3, the history of convergence of the RKDG method using piecewise linear elements is dsplayed and in Tables 4,5, and 6, the history of convergence of the RKDG method using piecewise quadratic elements. It can be seen that when the TVDM generalized slope limiter is used, i.e., when we take $M=0$, there is degradation of the accuracy of the scheme, whereas when the TVBM generalized slope limiter is used with a properly chosen constant M, i.e., when $M=20 \geq 2\pi^2$, the scheme is uniformly high order in regions of smoothness that include critical and sonic points.

Next, we compare the efficiency of the RKDG schemes for $k=1$ and $k=2$ for the case $M=20$ and $T=0.05$. We define the inverse of the efficiency of the method as the product of the error times the number of operations. Since the RKDG method that uses quadratic elements has $0.3/0.2$ times more time steps, $3/2$ times more inner iterations per time step, and $3/2$ time more unknowns in space, its number of operations is $27/8$ times bigger than the one of the RKDH method using linear elements. Hence, the ratio of the efficiency of the RKDG method with quadratic elements to that of the RKDG method with linear elements is

$$r = \frac{8}{27} \frac{error(RKDG(k=1))}{error(RKDG(k=2))}.$$

The results are displayed in Table 7. We can see that the efficiency of the RKDG scheme with quadratic polynomials is several times that of the RKDG scheme with linear polynomials even for very small values of Δx. We can also see that the ratio r of efficiencies is proportional to $(\Delta x)^{-1}$, which is expected for smooth solutions. This indicates that it is indeed more efficient to work with RKDG methods using polynomials of higher degree.

That this is also true when the solution displays discontinuities can be seen figures 3.22, and 3.23. In the figure 3.22, it can be seen that the shock is captured in essentially two elements. A zoom of these figures is shown in figure 3.23, where the approximation right in front of the shock is shown. It is clear that the approximation using quadratic elements is superior to the approximation using linear elements.

3.15   Concluding remarks

In this subsection, which is the core of these notes, we have devised the general RKDG method for nonlinear scalar conservation laws with periodic boundary conditions.

We have seen that the RKDG are constructed in three steps. First, the Discontinuous Galerkin method is used to discretize in space the conservation law. Then, an explicit TVB-Runge-Kutta time discretizationis used to discretize the resulting ODE system. Finally, a generalized slope limiter is introduced that enforces nonlinear stability without degrading the accuracy of the scheme.

We have seen that the numerical results show that the RKDG methods using polynomials of degree k, $k = 1, 2$ are uniformly $(k+1)$-th order accurate away from discontinuities and that the use of high degree polynomials render the RKDG method more efficient, even close to discontinuities.

All these results can be extended to the initial boundary value problem, see [15]. In what follows, we extend the RKDG methods to multidimensional systems.

Table 1

P^1, $M = 0$, CFL= 0.3, $T = 0.05$.

Δx	$L^1(0,1) - error$		$L^\infty(0,1) - error$	
	$10^5 \cdot error$	order	$10^5 \cdot error$	order
1/10	1286.23	-	3491.79	-
1/20	334.93	1.85	1129.21	1.63
1/40	85.32	1.97	449.29	1.33
1/80	21.64	1.98	137.30	1.71
1/160	5.49	1.98	45.10	1.61
1/320	1.37	2.00	14.79	1.61
1/640	0.34	2.01	4.85	1.60
1/1280	0.08	2.02	1.60	1.61

Table 2

P^1, $M = 20$, CFL= 0.3, $T = 0.05$.

Δx	$L^1(0,1) - error$		$L^\infty(0,1) - error$	
	$10^5 \cdot error$	order	$10^5 \cdot error$	order
1/10	1073.58	-	2406.38	-
1/20	277.38	1.95	628.12	1.94
1/40	71.92	1.95	161.65	1.96
1/80	18.77	1.94	42.30	1.93
1/160	4.79	1.97	10.71	1.98
1/320	1.21	1.99	2.82	1.93
1/640	0.30	2.00	0.78	1.86
1/1280	0.08	2.00	0.21	1.90

Table 3
Errors in smooth region $\Omega = \{x : |x - shock| \geq 0.1\}$.
P^1, $M = 20$, CFL= 0.3, $T = 0.4$.

	$L^1(\Omega) - error$		$L^\infty(\Omega) - error$	
Δx	$10^5 \cdot error$	order	$10^5 \cdot error$	order
1/10	1477.16	-	17027.32	-
1/20	155.67	3.25	1088.55	3.97
1/40	38.35	2.02	247.35	2.14
1/80	9.70	1.98	65.30	1.92
1/160	2.44	1.99	17.35	1.91
1/320	0.61	1.99	4.48	1.95
1/640	0.15	2.00	1.14	1.98
1/1280	0.04	2.00	0.29	1.99

Table 4
P^2, $M = 0$, CFL= 0.2, $T = 0.05$.

	$L^1(0,1) - error$		$L^\infty(0,1) - error$	
Δx	$10^5 \cdot error$	order	$10^5 \cdot error$	order
1/10	2066.13	-	16910.05	-
1/20	251.79	3.03	3014.64	2.49
1/40	42.52	2.57	1032.53	1.55
1/80	7.56	2.49	336.62	1.61

Table 5

P^2, $M = 20$, CFL= 0.2, $T = 0.05$.

	$L^1(0,1) - error$		$L^\infty(0,1) - error$	
Δx	$10^5 \cdot error$	order	$10^5 \cdot error$	order
1/10	37.31	-	101.44	-
1/20	4.58	3.02	13.50	2.91
1/40	0.55	3.05	1.52	3.15
1/80	0.07	3.08	0.19	3.01

Table 6

Errors in smooth region $\Omega = \{x : |x - shock| \geq 0.1\}$.
P^2, $M = 20$, CFL= 0.2, $T = 0.4$.

	$L^1(\Omega) - error$		$L^\infty(\Omega) - error$	
Δx	$10^5 \cdot error$	order	$10^5 \cdot error$	order
1/10	786.36	-	16413.79	-
1/20	5.52	7.16	86.01	7.58
1/40	0.36	3.94	15.49	2.47
1/80	0.06	2.48	0.54	4.84

Table 7

Comparison of the efficiencies of RKDG schemes for $k = 2$ and $k = 1$
$M = 20$, $T = 0.05$.

	L^1-norm		L^∞-norm	
Δx	$eff.ratio$	order	$eff.ratio$	order
1/10	8.52	-	7.03	-
1/20	17.94	-1.07	46.53	-2.73
1/40	38.74	-1.11	106.35	-1.19
1/80	79.45	-1.04	222.63	-1.07

3.16 Appendix: Proof of the L^2-error estimates in the linear case

Proof of the L^2-stability In this subsection, we prove the the stability result of Proposition 3.1. To do that, we first show how to obtain the corresponding stability result for the exact solution and then mimic the argument to obtain Proposition 3.1.

The continuous case as a model. We start by rewriting the equations (3.4) in *compact form*. If in the equations (3.4) we replace $v(x)$ by $v(x,t)$, sum on j from 1 to N, and integrate in time from 0 to T, we obtain

$$\mathbb{B}(u,v) = 0, \qquad \forall\, v : v(t) \text{ is smooth} \quad \forall\, t \in (0,T), \qquad (3.29)$$

where

$$\mathbb{B}(u,v) = \int_0^T \int_0^1 \left\{ \partial_t u(x,t)\, v(x,t) - c\, u(x,t)\, \partial_x v(x,t) \right\} dx\, dt. \quad (3.30)$$

Taking $v = u$, we easily see that we see that

$$\mathbb{B}(u,u) = \frac{1}{2} \| u(T) \|_{L^2(0,1)}^2 - \frac{1}{2} \| u_0 \|_{L^2(0,1)}^2,$$

and since

$$\mathbb{B}(u,u) = 0,$$

by (3.29), we immediately obtain the following L^2-stability result:

$$\frac{1}{2} \| u(T) \|_{L^2(0,1)}^2 = \frac{1}{2} \| u_0 \|_{L^2(0,1)}^2.$$

This is the argument we have to mimic in order to prove Proposition 3.1.

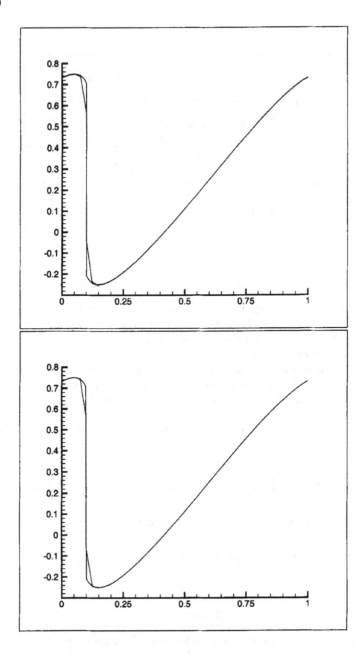

Fig. 3.22: Comparison of the exact and the approximate solution obtained with $M = 20$, $\Delta x = 1/40$ at $T = .4$: Piecewise linear elements (top) and piecewise quadratic elements (bottom)

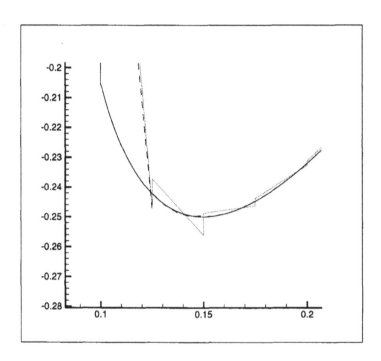

Fig. 3.23: Detail of previous figure. Behavior of the approximate solutions four elements in front of the shock: Exact solution (solid line), piecewise linear solution (dotted line), and piecewise quadratic solution (dashed line).

The discrete case. Thus, we start by finding the discrete version of the form $\mathbb{B}(\cdot,\cdot)$. If we replace $v(x)$ by $v_h(x,t)$ in the equation (3.7), sum on j from 1 to N, and integrate in time from 0 to T, we obtain

$$\mathbb{B}_h(u_h,v_h) = 0, \qquad \forall\, v_h : v_h(t) \in V_h^k \quad \forall\, t \in (0,T). \qquad (3.31)$$

where

$$\mathbb{B}_h(u_h,v_h) = \int_0^T \int_0^1 \partial_t u_h(x,t)\, v_h(x,t)\, dx\, dt \qquad (3.32)$$

$$- \int_0^T \sum_{1 \le j \le N} \int_{I_j} c\, u_h(x,t)\, \partial_x v_h(x,t)\, dx\, dt$$

$$- \int_0^T \sum_{1 \le j \le N} h(u_h)_{j+1/2}(t)\, [\,v_h(t)\,]_{j+1/2}\, dt.$$

Following the model provided by the continuous case, we next obtain an expression for $\mathbb{B}_h(w_h, w_h)$. It is contained in the following result which will proved later.

Lemma 3.1 *We have*

$$\mathbb{B}_h(w_h,w_h) = \frac{1}{2}\| w_h(T) \|_{L^2(0,1)}^2 + \Theta_T(w_h) - \frac{1}{2}\| w_h(0) \|_{L^2(0,1)}^2,$$

where

$$\Theta_T(w_h) = \frac{|c|}{2} \int_0^T \sum_{1 \le j \le N} [\,w_h(t)\,]_{j+1/2}^2\, dt.$$

Taking $w_h = u_h$ in the above result and noting that by (3.31),

$$\mathbb{B}_h(u_h,u_h) = 0,$$

we get the equality

$$\frac{1}{2}\| u_h(T) \|_{L^2(0,1)}^2 + \Theta_T(u_h) = \frac{1}{2}\| u_h(0) \|_{L^2(0,1)}^2,$$

from which Proposition 3.1 easily follows, since

$$\frac{1}{2}\| u_h(T) \|_{L^2(0,1)}^2 \le \frac{1}{2}\| u_0 \|_{L^2(0,1)}^2,$$

by (3.8). It only remains to prove Lemma 3.1.

Proof of Lemma 3.1. After setting $u_h = v_h = w_h$ in the definition of $\mathbb{B}_h$, (3.32), we get

$$\mathbb{B}_h(w_h, w_h) = \frac{1}{2} \| w_h(T) \|^2_{L^2(0,1)} + \int_0^T \Theta_{diss}(t) \, dt - \frac{1}{2} \| w_h(0) \|^2_{L^2(0,1)},$$

where

$$\Theta_{diss}(t) = - \sum_{1 \le j \le N} \left\{ h(w_h)_{j+1/2}(t) \, [w_h(t)]_{j+1/2} + \int_{I_j} c \, w_h(x,t) \, \partial_x w_h(x,t) \, dx \right\}$$

We only have to show that $\int_0^T \Theta_{diss}(t) \, dt = \Theta_T(w_h)$. To do that, we proceed as follows. Dropping the dependence on the variable t and setting

$$\overline{w}_h(x_{j+1/2}) = \frac{1}{2} (w_h(x^-_{j+1/2}) + w_h(x^+_{j+1/2})),$$

we have, by the definition of the flux h, (3.11),

$$- \sum_{1 \le j \le N} \int_{I_j} h(w_h)_{j+1/2} \, [w_h]_{j+1/2} = - \sum_{1 \le j \le N} \left\{ c \overline{w}_h \, [w_h] - \frac{|c|}{2} [w_h]^2 \right\}_{j+1/2},$$

and

$$- \sum_{1 \le j \le N} \int_{I_j} c \, w_h(x) \, \partial_x w_h(x) \, dx = \frac{c}{2} \sum_{1 \le j \le N} [w_h^2]_{j+1/2}$$

$$= c \sum_{1 \le j \le N} \{ \overline{w}_h \, [w_h] \}_{j+1/2}$$

Hence

$$\Theta_{diss}(t) = \frac{|c|}{2} \sum_{1 \le j \le N} [u_h(t)]^2_{j+1/2},$$

and the result follows. This completes the proof of Lemma 3.1. □
This completes the proof of Proposition 3.1.

Proof of the Theorem 3.1 In this subsection, we prove the error estimate of Theorem 3.1 which holds for the linear case $f(u) = c \, u$. To do that, we first show how to estimate the error between the solutions $w_\nu = (u_\nu, q_\nu)^t$, $\nu = 1, 2$, of

$$\partial_t u_\nu + \partial_x f(u_\nu) = 0 \quad \text{in } (0, T) \times (0, 1),$$
$$u_\nu(t = 0) = u_{0,\nu}, \quad \text{on } (0, 1).$$

Then, we mimic the argument in order to prove Theorem 3.1.

The continuous case as a model. By the definition of the form $\mathbb{B}(\cdot,\cdot)$, (3.30), we have, for $\nu = 1, 2$,

$$\mathbb{B}(w_\nu, v) = 0, \qquad \forall\ v : v(t) \text{ is smooth} \quad \forall\ t \in (0, T).$$

Since the form $\mathbb{B}(\cdot,\cdot)$ is bilinear, from the above equation we obtain the so-called *error equation*:

$$\mathbb{B}(e, v) = 0, \qquad \forall\ v : v(t) \text{ is smooth} \quad \forall\ t \in (0, T). \tag{3.33}$$

where $e = w_1 - w_2$. Now, since

$$\mathbb{B}(e, e) = \frac{1}{2} \| e(T) \|^2_{L^2(0,1)} - \frac{1}{2} \| e(0) \|^2_{L^2(0,1)},$$

and

$$\mathbb{B}(e, e) = 0,$$

by the error equation (3.33), we immediately obtain the error estimate we sought:

$$\frac{1}{2} \| e(T) \|^2_{L^2(0,1)} = \frac{1}{2} \| u_{0,1} - u_{0,2} \|^2_{L^2(0,1)}.$$

To prove Theorem 3.1, we only need to obtain a discrete version of this argument.

The discrete case. Since,

$$\mathbb{B}_h(u_h, v_h) = 0, \qquad \forall\ v_h : v(t) \in V_h \quad \forall\ t \in (0, T),$$
$$\mathbb{B}_h(u, v_h) = 0, \qquad \forall\ v_h : v_h(t) \in V_h \quad \forall\ t \in (0, T),$$

by (3.7) and by equations (3.4), respectively, we easily obtain our *error equation*:

$$\mathbb{B}_h(e, v_h) = 0, \qquad \forall\ v_h : v_h(t) \in V_h \quad \forall\ t \in (0, T), \tag{3.34}$$

where $e = w - w_h$.

Now, according to the continuous case argument, we should consider next the quantity $\mathbb{B}_h(e, e)$; however, since $e(t)$ is not in the finite element space V_h, it is more convenient to consider $\mathbb{B}_h(\mathbb{P}_h(e), \mathbb{P}_h(e))$, where $\mathbb{P}_h(e(t))$ is the L^2-projection of the error $e(t)$ into the finite element space V_h^k.

The L^2-projection of the function $p \in L^2(0, 1)$ into V_h, $\mathbb{P}_h(p)$, is defined as the only element of the finite element space V_h such that

$$\int_0^1 \big(\mathbb{P}_h(p)(x) - p(x) \big)\, v_h(x)\, dx = 0, \qquad \forall\ v_h \in V_h. \tag{3.35}$$

Note that in fact $u_h(t = 0) = \mathbb{P}_h(u_0)$, by (3.8).

Thus, by Lemma 3.1, we have

$$\mathbb{B}_h(\mathbb{P}_h(e), \mathbb{P}_h(e)) = \frac{1}{2} \| \mathbb{P}_h(e(T)) \|_{L^2(0,1)}^2 + \Theta_T(\mathbb{P}_h(e)) - \frac{1}{2} \| \mathbb{P}_h(e(0)) \|_{L^2(0,1)}^2,$$

and since

$$\mathbb{P}_h(e(0)) = \mathbb{P}_h(u_0 - u_h(0)) = \mathbb{P}_h(u_0) - u_h(0) = 0,$$

and

$$\mathbb{B}_h(\mathbb{P}_h(e), \mathbb{P}_h(e)) = \mathbb{B}_h(\mathbb{P}_h(e) - e, \mathbb{P}_h(e)) = \mathbb{B}_h(\mathbb{P}_h(u) - u, \mathbb{P}_h(e)),$$

by the *error equation* (3.34), we get

$$\frac{1}{2} \| \mathbb{P}_h(e(T)) \|_{L^2(0,1)}^2 + \Theta_T(\mathbb{P}_h(e)) = \mathbb{B}_h(\mathbb{P}_h(u) - u, \mathbb{P}_h(e)). \qquad (3.36)$$

It only remains to estimate the right-hand side

$$\mathbb{B}(\mathbb{P}_h(u) - u, \mathbb{P}_h(e)),$$

which, according to our continuous model, should be small.

Estimating the right-hand side. To show that this is so, we must suitably treat the term $\mathbb{B}(\mathbb{P}_h(w) - w, \mathbb{P}_h(e))$. We start with the following remarkable result.

Lemma 3.2 *We have*

$$\mathbb{B}_h(\mathbb{P}_h(u) - u, \mathbb{P}_h(e)) = -\int_0^T \sum_{1 \leq j \leq N} h(\mathbb{P}_h(u) - u)_{j+1/2}(t) \, [\mathbb{P}_h(e)(t)]_{j+1/2} \, dt.$$

Proof Setting $p = \mathbb{P}_h(u) - u$ and $v_h = \mathbb{P}_h(e)$ and recalling the definition of $\mathbb{B}_h(\cdot, \cdot)$, (3.32), we have

$$\begin{aligned}
\mathbb{B}_h(p, v_h) = &\int_0^T \int_0^1 \partial_t p(x, t) \, v_h(x, t) \, dx \, dt \\
&- \int_0^T \sum_{1 \leq j \leq N} \int_{I_j} c \, p(x, t) \, \partial_x v_h(x, t) \, dx \, dt \\
&- \int_0^T \sum_{1 \leq j \leq N} h(p)_{j+1/2}(t) \, [v_h(t)]_{j+1/2} \, dt \\
= &- \int_0^T \sum_{1 \leq j \leq N} h(p)_{j+1/2}(t) \, [v_h(t)]_{j+1/2} \, dt,
\end{aligned}$$

by the definition of the L^2-projection (3.35). This completes the proof. $\quad\square$

Now, we can see that a simple application of Young's inequality and a standard approximation result should give us the estimate we were looking for. The approximation result we need is the following.

Lemma 3.3 *If $w \in H^{k+1}(I_j \cup I_{j+1})$, then*

$$| h(\mathbb{P}_h(w) - w)(x_{j+1/2}) | \leq c_k (\Delta x)^{k+1/2} \frac{|c|}{2} | w |_{H^{k+1}(I_j \cup I_{j+1})},$$

where the constant c_k depends solely on k.

Proof. Dropping the argument $x_{j+1/2}$ we have, by the definition (3.11) of the flux h,

$$| h(\mathbb{P}(w) - w) | = \frac{c}{2}(\mathbb{P}_h(w)^+ + \mathbb{P}_h(w)^-) - \frac{|c|}{2}(\mathbb{P}_h(w)^+ - \mathbb{P}_h(w)^-) - cw$$

$$= \frac{c - |c|}{2}(\mathbb{P}_h(w)^+ - w) + \frac{c + |c|}{2}(\mathbb{P}_h(w)^- - w)$$

$$\leq |c| \max\{| \mathbb{P}_h(w)^+ - w |, | \mathbb{P}_h(w)^- - w |\}$$

and the result follows from the properties of $\mathbb{P}_h$ after a simple application of the Bramble-Hilbert lemma; see [11]. This completes the proof. $\square$

An immediate consequence of this result is the estimate we wanted.

Lemma 3.4 *We have*

$$B_h(\mathbb{P}_h(u) - u, \mathbb{P}_h(e)) \leq c_k^2 (\Delta x)^{2k+1} \frac{|c|}{2} T | u_0 |_{H^{k+1}(0,1)}^2 + \frac{1}{2} \Theta_T(\mathbb{P}_h(e)),$$

where the constant c_k depends solely on k.

Proof. After using Young's inequality in the right-hand side of Lemma 3.2, we get

$$B_h(\mathbb{P}_h(u) - u, \mathbb{P}_h(e)) \leq \int_0^T \sum_{1 \leq j \leq N} \frac{1}{|c|} | h(\mathbb{P}_h(u) - u)_{j+1/2}(t) |^2$$

$$+ \int_0^T \sum_{1 \leq j \leq N} \frac{|c|}{4} [\mathbb{P}_h(e)(t)]_{j+1/2}^2 dt.$$

By Lemma 3.3 and the definition of the form Θ_T, we get

$$B_h(\mathbb{P}_h(u) - u, \mathbb{P}_h(e)) \leq c_k^2 (\Delta x)^{2k+1} \frac{|c|}{4} \int_0^T \sum_{1 \leq j \leq N} | u |_{H^{k+1}(I_j \cup I_{j+1})}^2 + \frac{1}{2} \Theta_T(\mathbb{P}_h(e))$$

$$\leq c_k^2 (\Delta x)^{2k+1} \frac{|c|}{2} T | u_0 |_{H^{k+1}(0,1)}^2 + \frac{1}{2} \Theta_T(\mathbb{P}_h(e)).$$

This completes the proof. $\square$

Conclusion. Finally, inserting in the equation (3.36) the estimate of its right hand side obtained in Lemma 3.4, we get

$$\| \mathbb{P}_h(e(T)) \|_{L^2(0,1)}^2 + \Theta_T(\mathbb{P}_h(e)) \leq c_k (\Delta x)^{2k+1} | c | T | u_0 |_{H^{k+1}(0,1)}^2,$$

Theorem 3.1 now follows from the above estimate and from the following inequality:

$$\| e(T) \|_{L^2(0,1)} \leq \| u(T) - \mathbb{P}_h(u(T)) \|_{L^2(0,1)} + \| \mathbb{P}_h(e(T)) \|_{L^2(0,1)}$$
$$\leq c_k' (\Delta x)^{k+1} | u_0 |_{H^{k+1}(0,1)} + \| \mathbb{P}_h(e(T)) \|_{L^2(0,1)}.$$

Proof of the Theorem 3.2 To prove Theorem 3.2, we only have to suitably modify the proof of Theorem 3.1. The modification consists in *replacing* the L^2-projection of the error, $\mathbb{P}_h(e)$, by another projection that we denote by $\mathbb{R}_h(e)$.

Given a function $p \in L^\infty(0,1)$ that is continuous on each element I_j, we define $\mathbb{R}_h(p)$ as the only element of the finite element space V_h such that

$$\forall j = 1, \ldots, N : \qquad \mathbb{R}_h(p)(x_{j,\ell}) - p(x_{j,\ell}) = 0, \qquad \ell = 0, \ldots, k, \quad (3.37)$$

where the points $x_{j,\ell}$ are the Gauss-Radau quadrature points of the interval I_j. We take

$$x_{j,k} = x_{j+1/2}, \quad \text{if } c > 0, \quad \text{and} \quad x_{j,0} = x_{j-1/2}, \quad \text{if } c < 0. \quad (3.38)$$

The special nature of the Gauss-Radau quadrature points is captured in the following property:

$$\forall \varphi \in P^\ell(I_j), \quad \ell \leq k, \quad \forall p \in P^{2k-\ell}(I_j) :$$
$$\int_{I_j} (\mathbb{R}_h(p)(x) - p(x)) \varphi(x) \, dx = 0. \qquad (3.39)$$

Compare this equality with (3.35).

The quantity $\mathbb{B}_h(\mathbb{R}_h(e), \mathbb{R}_h(e))$. To prove our error estimate, we start by considering the quantity $\mathbb{B}_h(\mathbb{R}_h(e), \mathbb{R}_h(e))$. By Lemma 3.1, we have

$$\mathbb{B}_h(\mathbb{R}_h(e), \mathbb{R}_h(e)) = \frac{1}{2} \| \mathbb{R}_h(e(T)) \|^2_{L^2(0,1)} + \Theta_T(\mathbb{R}_h(e)) - \frac{1}{2} \| \mathbb{R}_h(e(0)) \|^2_{L^2(0,1)},$$

and since

$$\mathbb{B}_h(\mathbb{R}_h(e), \mathbb{R}_h(e)) = \mathbb{B}_h(\mathbb{R}_h(e) - e, \mathbb{R}_h(e)) = \mathbb{B}_h(\mathbb{R}_h(u) - u, \mathbb{R}_h(e)),$$

by the *error equation* (3.34), we get

$$\frac{1}{2} \| \mathbb{R}_h(e(T)) \|^2_{L^2(0,1)} + \Theta_T(\mathbb{R}_h(e)) = \frac{1}{2} \| \mathbb{R}_h(e(0)) \|^2_{L^2(0,1)} + \mathbb{B}_h(\mathbb{R}_h(u) - u, \mathbb{R}_h(e)).$$

Next, we estimate the term $\mathbb{B}(\mathbb{R}_h(u) - u, \mathbb{R}_h(e))$.

Estimating $\mathbb{B}(\mathbb{R}_h(u) - u, \mathbb{R}_h(e))$. The following result corresponds to Lemma 3.2.

Lemma 3.5 *We have*

$$\mathbb{B}_h(\mathbb{R}_h(u) - u, v_h) = \int_0^T \int_0^1 \left(\mathbb{R}_h(\partial_t u)(x,t) - \partial_t u(x,t) \right) v_h(x,t)\, dx\, dt$$

$$- \int_0^T \sum_{1 \le j \le N} \int_{I_j} c\left(\mathbb{R}_h(u)(x,t) - u(x,t) \right) \partial_x v_h(x,t)\, dx\, dt.$$

Proof Setting $p = \mathbb{R}_h(u) - u$ and $v_h = \mathbb{R}_h(e)$ and recalling the definition of $\mathbb{B}_h(\cdot,\cdot)$, (3.32), we have

$$\mathbb{B}_h(p, v_h) = \int_0^T \int_0^1 \partial_t p(x,t)\, v_h(x,t)\, dx\, dt$$

$$- \int_0^T \sum_{1 \le j \le N} \int_{I_j} c\, p(x,t)\, \partial_x v_h(x,t)\, dx\, dt$$

$$- \int_0^T \sum_{1 \le j \le N} h(p)_{j+1/2}(t) \left[v_h(t) \right]_{j+1/2}\, dt.$$

But, from the definition (3.11) of the flux h, we have

$$h(\mathbb{R}(u) - u) = \frac{c}{2}(\mathbb{R}_h(u)^+ + \mathbb{R}_h(u)^-) - \frac{|c|}{2}(\mathbb{R}_h(u)^+ - \mathbb{R}_h(u)^-) - c\, u$$

$$= \frac{c - |c|}{2}(\mathbb{R}_h(u)^+ - u) + \frac{c + |c|}{2}(\mathbb{R}_h(u)^- - u)$$

$$= 0,$$

by (3.38) and the result follows. $\square$

Next, we need some approximation results.

Lemma 3.6 *If* $w \in H^{k+2}(I_j)$, *and* $v_h \in P^k(I_j)$, *then*

$$\left| \int_{I_j} (\mathbb{R}_h(w) - w)(x)\, v_h(x)\, dx \right| \le c_k\, (\Delta x)^{k+1}\, |w|_{H^{k+1}(I_j)}\, \|v_h\|_{L^2(I_j)},$$

$$\left| \int_{I_j} (\mathbb{R}_h(w) - w)(x)\, \partial_x v_h(x)\, dx \right| \le c_k\, (\Delta x)^{k+1}\, |w|_{H^{k+2}(I_j)}\, \|v_h\|_{L^2(I_j)},$$

where the constant c_k *depends solely on* k.

Proof. The first inequality follows from the property (3.39) with $\ell = k$ and from standard approximation results. The second follows in a similar way from the property 3.39 with $\ell = k - 1$ and a standard scaling argument. This completes the proof. $\square$

An immediate consequence of this result is the estimate we wanted.

Lemma 3.7 *We have*

$$\mathbb{B}_h(\mathbb{R}_h(u) - u, \mathbb{R}_h(e)) \leq c_k \, (\Delta x)^{k+1} \, |\, u_0 \,|_{H^{k+2}(0,1)} \int_0^T \| \, \mathbb{R}_h(e(t)) \, \|_{L^2(0,1)} \, dt,$$

where the constant c_k depends solely on k and $|c|$.

Conclusion. Finally, inserting in the equation (3.36) the estimate of its right hand side obtained in Lemma 3.7, we get

$$\| \, \mathbb{R}_h(e(T)) \, \|^2_{L^2(0,1)} + \Theta_T(\mathbb{R}_h(e)) \leq \| \, \mathbb{R}_h(e(0)) \, \|^2_{L^2(0,1)}$$
$$+ c_k \, (\Delta x)^{k+1} \, |\, u_0 \,|_{H^{k+2}(0,1)} \int_0^T \| \, \mathbb{R}_h(e(t)) \, \|_{L^2(0,1)} \, dt.$$

After applying a simple variation of the Gronwall lemma, we obtain

$$\| \, \mathbb{R}_h(e(T)) \, \|_{L^2(0,1)} \leq \| \, \mathbb{R}_h(e(0))(x) \, \|_{L^2(0,1)} + c_k \, (\Delta x)^{k+1} \, T \, |\, u_0 \,|_{H^{k+2}(0,1)}$$
$$\leq c'_k (\Delta x)^{k+1} \, |\, u_0 \,|_{H^{k+2}(0,1)}.$$

Theorem 3.2 now follows from the above estimate and from the following inequality:

$$\| \, e(T) \, \|_{L^2(0,1)} \leq \| \, u(T) - \mathbb{R}_h(u(T)) \, \|_{L^2(0,1)} + \| \, \mathbb{R}_h(e(T)) \, \|_{L^2(0,1)}$$
$$\leq c'_k (\Delta x)^{k+1} \, |\, u_0 \,|_{H^{k+1}(0,1)} + \| \, \mathbb{R}_h(e(T)) \, \|_{L^2(0,1)}.$$

4 The RKDG method for multidimensional systems

4.1 Introduction

In this section, we extend the RKDG methods to multidimensional systems:

$$u_t + \nabla f(u) = 0, \qquad \text{in } \Omega \times (0, T), \tag{4.1}$$

$$u(x, 0) = u_0(x), \qquad \forall\, x \in \Omega, \tag{4.2}$$

and periodic boundary conditions. For simplicity, we assume that Ω is the unit cube.

This section is essentially devoted to the description of the algorithms and their implementation details. The practitioner should be able to find here all the necessary information to completely code the RKDG methods.

This section also contains two sets of numerical results for the Euler equations of gas dynamics in two space dimensions. The first set is devoted to transient computations and domains that have corners; the effect of using triangles or rectangles and the effect of using polynomials of degree one or two are explored. The main conclusions from these computations are that (i) the RKDG method works as well with triangles as it does with rectangles and that (ii) the use of high-order polynomials does not deteriorate the approximation of strong shocks and is advantageous in the approximation of contact discontinuities.

The second set concerns steady state computations with smooth solutions. For these computations, no generalized slope limiter is needed. The effect of (i) the quality of the approximation of curved boundaries and of (ii) the degree of the polynomials on the quality of the approximate solution is explored. The main conclusions from these computations are that (i) a high-order approximation of the curve boundaries introduces a dramatic improvement on the quality of the solution and that (ii) the use of high-degree polynomials is advantageous when smooth solutions are shought.

This section contains material from the papers [14], [13], and [19]. It also contains numerical results from the paper by Bassi and Rebay [2] in two dimensions and from the paper by Warburton, Lomtev, Kirby and Karniadakis [65] in three dimensions.

4.2 The general RKDG method

The RKDG method for multidimensional systems has the same structure it has for one-dimensional scalar conservation laws, that is,

- Set $u_h^0 = \Lambda \Pi_h\, P_{V_h}(u_0)$;

– For $n = 0, ..., N - 1$ compute u_h^{n+1} as follows:

1. set $u_h^{(0)} = u_h^n$;
2. for $i = 1, ..., k + 1$ compute the intermediate functions:

$$u_h^{(i)} = \Lambda \Pi_h \left\{ \sum_{l=0}^{i-1} \alpha_{il} u_h^{(l)} + \beta_{il} \Delta t^n L_h(u_h^{(l)}) \right\};$$

3. set $u_h^{n+1} = u_h^{(k+1)}$.

In what follows, we describe the operator L_h that results form the DG-space discretization, and the generalized slope limiter $\Lambda \Pi_h$.

The Discontinuous Galerkin space discretization To show how to discretize in space by the DG method, it is enough to consider the case in which u is a scalar quantity since to deal with the general case in which u, we apply the same procedure component by component.

Once a triangulation $\mathbf{T}_h$ of Ω has been obtained, we determine $L_h(\cdot)$ as follows. First, we multiply (4.1) by v_h in the finite elemen space V_h, integrate over the element K of the triangulation $\mathbf{T}_h$ and replace the exact solution u by its approximation $u_h \in V_h$:

$$\frac{d}{dt} \int_K u_h(t, x) v_h(x) \, dx + \int_K div \, f(u_h(t, x)) v_h(x) \, dx = 0, \ \forall v_h \in V_h.$$

Integrating by parts formally we obtain

$$\frac{d}{dt} \int_K u_h(t, x) v_h(x) \, dx + \sum_{e \in \partial K} \int_e f(u_h(t, x)) \cdot n_{e,K} v_h(x) \, d\Gamma$$
$$- \int_K f(u_h(t, x)) \cdot \text{grad } v_h(x) \, dx = 0, \quad \forall v_h \in V_h,$$

where $n_{e,K}$ is the outward unit normal to the edge e. Notice that $f(u_h(t, x)) \cdot n_{e,K}$ does not have a precise meaning, for u_h is discontinuous at $x \in e \in \partial K$. Thus, as in the one dimensional case, we replace $f(u_h(t, x)) \cdot n_{e,K}$ by the function $h_{e,K}(u_h(t, x^{int(K)}), u_h(t, x^{ext(K)}))$. The function $h_{e,K}(\cdot, \cdot)$ is any consistent two–point monotone Lipschitz flux, consistent with $f(u) \cdot n_{e,K}$.

In this way we obtain

$$\frac{d}{dt} \int_K u_h(t, x) v_h(x) \, dx + \sum_{e \in \partial K} \int_e h_{e,K}(t, x) v_h(x) \, d\Gamma$$
$$- \int_K f(u_h(t, x)) \cdot \text{grad } v_h(x) \, dx = 0, \quad \forall v_h \in V_h.$$

Finally, we replace the integrals by quadrature rules that we shall choose as follows:

$$\int_e h_{e,K}(t,x)\, v_h(x)\, d\Gamma \approx \sum_{l=1}^{L} \omega_l\, h_{e,K}(t,x_{el})\, v(x_{el})|e|, \qquad (4.3)$$

$$\int_K f(u_h(t,x))\cdot \text{grad }v_h(x)\, dx \approx$$

$$\sum_{j=1}^{M} \omega_j\, f(u_h(t,x_{Kj}))\cdot \text{grad }v_h(x_{Kj})|K|. \qquad (4.4)$$

Thus, we finally obtain the weak formulation:

$$\frac{d}{dt}\int_k u_h(t,x)v_h(x)dx + \sum_{e\in\partial K}\sum_{l=1}^{L}\omega_l\, h_{e,K}(t,x_{el})\, v(x_{el})|e|$$

$$-\sum_{j=1}^{M}\omega_j\, f(u_h(t,x_{Kj}))\cdot \text{grad}v_h(x_{Kj})|K| = 0, \quad \forall v_h \in V_h, \quad \forall K \in \mathbb{T}_h.$$

These equations can be rewritten in ODE form as $\frac{d}{dt}u_h = L_h(u_h,\gamma_h)$. This defines the operator $L_h(u_h)$, which is a discrete approximation of $-div\, f(u)$. The following result gives an indication of the quality of this approximation.

Proposition 4.1 *Let* $f(u) \in W^{k+2,\infty}(\Omega)$, *and set* $\gamma = trace(u)$. *Let the quadrature rule over the edges be exact for polynomials of degree* $(2k+1)$, *and let the one over the element be exact for polynomials of degree* $(2k)$. *Assume that the family of triangulations* $\mathbb{F} = \{\mathbb{T}_h\}_{h>0}$ *is regular, i.e., that there is a constant* σ *such that:*

$$\frac{h_K}{\rho_K} \geq \sigma, \quad \forall K \in \mathbb{T}_h, \quad \forall \mathbb{T}_h \in \mathbb{F}, \qquad (4.5)$$

where h_K *is the diameter of* K, *and* ρ_K *is the diameter of the biggest ball included in* K. *Then, if* $V(K) \supset P^k(K), \quad \forall\, K \in \mathbb{T}_h$:

$$\|L_h(u,\gamma) + div\, f(u)\|_{L^\infty(\Omega)} \leq C\, h^{k+1}|f(u)|_{W^{k+2,\infty}(\Omega)}.$$

For a proof, see [13].

The form of the generalized slope limiter $\Lambda\Pi_h$. The construction of generalized slope limiters $\Lambda\Pi_h$ for several space dimensions is not a trivial matter and will not be discussed in these notes; we refer the interested reader to the paper by Cockburn, Hou, and Shu [13].

In these notes, we restrict ourselves to displaying very simple, practical, and effective generalized slope limiters $\Lambda\Pi_h$ which are closely related to the generalized slope limiters $\Lambda\Pi_h^k$ of the previous section.

To compute $\Lambda\Pi_h u_h$, we rely on the *assumption* that spurious oscillations are present in u_h only if they are present in its P^1 part u_h^1, which is its L^2-projection into the space of piecewise linear functions V_h^1. Thus, if they are not present in u_h^1, i.e., if

$$u_h^1 = \Lambda\Pi_h u_h^1,$$

then we assume that they are not present in u_h and hence do not do any limiting:

$$\Lambda\Pi_h u_h = u_h.$$

On the other hand, if spurious oscillations are present in the P^1 part of the solution u_h^1, i.e., if

$$u_h^1 \neq \Lambda\Pi_h u_h^1,$$

then we chop off the higher order part of the numerical solution, and limit the remaining P^1 part:

$$\Lambda\Pi_h u_h = \Lambda\Pi_h u_h^1.$$

In this way, in order to define $\Lambda\Pi_h$ for arbitrary space V_h, we only need to actually define it for piecewise linear functions V_h^1. The exact way to do that, both for the triangular elements and for the rectangular elements, will be discussed in the next section.

4.3 Algorithm and implementation details

In this section we give the algorithm and implementation details, including numerical fluxes, quadrature rules, degrees of freedom, fluxes, and limiters of the RKDG method for both piecewise-linear and piecewise-quadratic approximations in both triangular and rectangular elements.

Fluxes The numerical flux we use is the simple Lax-Friedrichs flux:

$$h_{e,K}(a,b) = \frac{1}{2}\left[\mathbf{f}(a)\cdot n_{e,K} + \mathbf{f}(b)\cdot n_{e,K} - \alpha_{e,K}(b-a)\right].$$

The numerical viscosity constant $\alpha_{e,K}$ should be an estimate of the biggest eigenvalue of the Jacobian $\frac{\partial}{\partial u}\mathbf{f}(u_h(x,t))\cdot n_{e,K}$ for (x,t) in a neighborhood of the edge e.

For the triangular elements, we use the local Lax-Friedrichs recipe:

– Take $\alpha_{e,K}$ to be the larger one of the largest eigenvalue (in absolute value) of $\frac{\partial}{\partial u}\mathbf{f}(\bar{u}_K)\cdot n_{e,K}$ and that of $\frac{\partial}{\partial u}\mathbf{f}(\bar{u}_{K'})\cdot n_{e,K}$, where $\bar{u}_K$ and $\bar{u}_{K'}$ are the means of the numerical solution in the elements K and K' sharing the edge e.

For the rectangular elements, we use the local Lax-Friedrichs recipe :

– Take $\alpha_{e,K}$ to be the largest of the largest eigenvalue (in absolute value) of $\frac{\partial}{\partial u}\mathbf{f}(\bar{u}_{K''})\cdot n_{e,K}$, where $\bar{u}_{K''}$ is the mean of the numerical solution in the element K'', which runs over all elements on the same line (horizontally or vertically, depending on the direction of $n_{e,K}$) with K and K' sharing the edge e.

Quadrature rules According to the analysis done in [13], the quadrature rules for the edges of the elements, (4.3), must be exact for polynomials of degree $2k+1$, and the quadrature rules for the interior of the elements, (4.4), must be exact for polynomials of degree $2k$, if P^k methods are used. Here we discuss the quadrature points used for P^1 and P^2 in the triangular and rectangular element cases.

The rectangular elements For the edge integral, we use the following two point Gaussian rule

$$\int_{-1}^{1} g(x)dx \approx g\left(-\frac{1}{\sqrt{3}}\right) + g\left(\frac{1}{\sqrt{3}}\right), \qquad (4.1)$$

for the P^1 case, and the following three point Gaussian rule

$$\int_{-1}^{1} g(x)dx \approx \frac{5}{9}\left[g\left(-\frac{3}{5}\right) + g\left(\frac{3}{5}\right)\right] + \frac{8}{9}g(0), \qquad (4.2)$$

for the P^2 case, suitably scaled to the relevant intervals.

For the interior of the elements, we could use a tensor product of (4.1), with four quadrature points, for the P^1 case. But to save cost, we "recycle" the values of the fluxes at the element boundaries, and only add one new quadrature point in the middle of the element. Thus, to approximate the integral $\int_{-1}^{1}\int_{-1}^{1} g(x,y)dxdy$, we use the following quadrature rule:

$$\approx \frac{1}{4}\left[g\left(-1,\frac{1}{\sqrt{3}}\right) + g\left(-1,-\frac{1}{\sqrt{3}}\right) + g\left(-\frac{1}{\sqrt{3}},-1\right) + g\left(\frac{1}{\sqrt{3}},-1\right)\right.$$
$$\left. +g\left(1,-\frac{1}{\sqrt{3}}\right) + g\left(1,\frac{1}{\sqrt{3}}\right) + g\left(\frac{1}{\sqrt{3}},1\right) + g\left(-\frac{1}{\sqrt{3}},1\right)\right] + 2\,g(0,0).$$

For the P^2 case, we use a tensor product of (4.2), with 9 quadrature points.

The triangular elements For the edge integral, we use the same two point or three point Gaussian quadratures as in the rectangular case, (4.1) and (4.2), for the P^1 and P^2 cases, respectively.

For the interior integrals (4.4), we use the three mid-point rule

$$\int_{K} g(x,y)dxdy \approx \frac{|K|}{3}\sum_{i=1}^{3} g(m_i),$$

where m_i are the mid-points of the edges, for the P^1 case. For the P^2 case, we use a seven-point quadrature rule which is exact for polynomials of degree 5 over triangles.

Basis and degrees of freedom We emphasize that the choice of basis and degrees of freedom does not affect the algorithm, as it is completely determined by the choice of function space $V(h)$, the numerical fluxes, the quadrature rules, the slope limiting, and the time discretization. However, a suitable choice of basis and degrees of freedom may simplify the implementation and calculation.

The rectangular elements For the P^1 case, we use the following expression for the approximate solution $u_h(x, y, t)$ inside the rectangular element $[x_{i-\frac{1}{2}}, x_{i+\frac{1}{2}}] \times [y_{j-\frac{1}{2}}, y_{j+\frac{1}{2}}]$:

$$u_h(x, y, t) = \bar{u}(t) + u_x(t)\varphi_i(x) + u_y(t)\psi_j(y) \tag{4.3}$$

where

$$\varphi_i(x) = \frac{x - x_i}{\Delta x_i/2}, \qquad \psi_j(y) = \frac{y - y_j}{\Delta y_j/2}, \tag{4.4}$$

and

$$\Delta x_i = x_{i+\frac{1}{2}} - x_{i-\frac{1}{2}}, \qquad \Delta y_j = y_{j+\frac{1}{2}} - y_{j-\frac{1}{2}}.$$

The degrees of freedoms, to be evolved in time, are then

$$\bar{u}(t), \ u_x(t), \ u_y(t).$$

Here we have omitted the subscripts ij these degrees of freedom should have, to indicate that they belong to the element ij which is $[x_{i-\frac{1}{2}}, x_{i+\frac{1}{2}}] \times [y_{j-\frac{1}{2}}, y_{j+\frac{1}{2}}]$.

Notice that the basis functions

$$1, \ \varphi_i(x), \ \psi_j(y),$$

are orthogonal, hence the local mass matrix is diagonal:

$$M = \Delta x_i \Delta y_j \, diag\left(1, \frac{1}{3}, \frac{1}{3}\right).$$

For the P^2 case, the expression for the approximate solution $u_h(x, y, t)$ inside the rectangular element $[x_{i-\frac{1}{2}}, x_{i+\frac{1}{2}}] \times [y_{j-\frac{1}{2}}, y_{j+\frac{1}{2}}]$ is:

$$u_h(x, y, t) = \bar{u}(t) + u_x(t)\varphi_i(x) + u_y(t)\psi_j(y) + u_{xy}(t)\varphi_i(x)\psi_j(y)$$
$$+ u_{xx}(t)\left(\varphi_i^2(x) - \frac{1}{3}\right) + u_{yy}(t)\left(\psi_j^2(y) - \frac{1}{3}\right), \tag{4.5}$$

where $\varphi_i(x)$ and $\psi_j(y)$ are defined by (4.4). The degrees of freedoms, to be evolved in time, are

$$\bar{u}(t), \ u_x(t), \ u_y(t), \ u_{xy}(t), \ u_{xx}(t), \ u_{yy}(t).$$

Again the basis functions

$$1, \quad \varphi_i(x), \quad \psi_j(y), \quad \varphi_i(x)\psi_j(y), \quad \varphi_i^2(x) - \frac{1}{3}, \quad \psi_j^2(y) - \frac{1}{3},$$

are orthogonal, hence the local mass matrix is diagonal:

$$M = \Delta x_i \Delta y_j \, diag \left(1, \frac{1}{3}, \frac{1}{3}, \frac{1}{9}, \frac{4}{45}, \frac{4}{45}\right).$$

The triangular elements For the P^1 case, we use the following expression for the approximate solution $u_h(x, y, t)$ inside the triangle K:

$$u_h(x, y, t) = \sum_{i=1}^{3} u_i(t)\varphi_i(x, y)$$

where the degrees of freedom $u_i(t)$ are values of the numerical solution at the midpoints of edges, and the basis function $\varphi_i(x, y)$ is the linear function which takes the value 1 at the mid-point m_i of the i-th edge, and the value 0 at the mid-points of the two other edges. The mass matrix is diagonal

$$M = |K| diag \left(\frac{1}{3}, \frac{1}{3}, \frac{1}{3}\right).$$

For the P^2 case, we use the following expression for the approximate solution $u_h(x, y, t)$ inside the triangle K:

$$u_h(x, y, t) = \sum_{i=1}^{6} u_i(t)\xi_i(x, y)$$

where the degrees of freedom, $u_i(t)$, are values of the numerical solution at the three midpoints of edges and the three vertices. The basis function $\xi_i(x, y)$, is the quadratic function which takes the value 1 at the point i of the six points mentioned above (the three midpoints of edges and the three vertices), and the value 0 at the remaining five points. The mass matrix this time is not diagonal.

Limiting We construct slope limiting operators $\Lambda\Pi_h$ on piecewise linear functions u_h in such a way that the following properties are satisfied:

1. Accuracy: if u_h is linear then $\Lambda\Pi_h u_h = u_h$.
2. Conservation of mass: for every element K of the triangulation $\mathbb{T}_h$, we have:

$$\int_K \Lambda\Pi_h u_h = \int_K u_h.$$

3. Slope limiting: on each element K of $\mathbb{T}_h$, the gradient of $\Lambda\Pi_h u_h$ is not bigger than that of u_h.

The actual form of the slope limiting operators is closely related to that of the slope limiting operators studied in [15] and [13].

The rectangular elements The limiting is performed on u_x and u_y in (4.3), using the differences of the means. For a scalar equation, u_x would be limited (replaced) by

$$\bar{m}\left(u_x, \bar{u}_{i+1,j} - \bar{u}_{ij}, \bar{u}_{ij} - \bar{u}_{i-1,j}\right) \tag{4.6}$$

where the function $\bar{m}$ is the TVB corrected *minmod* function defined in the previous section.

The TVB correction is needed to avoid unnecessary limiting near smooth extrema, where the quantity u_x or u_y is on the order of $O(\Delta x^2)$ or $O(\Delta y^2)$. For an estimate of the TVB constant M in terms of the second derivatives of the function, see [15]. Usually, the numerical results are not sensitive to the choice of M in a large range. In all the calculations in this paper we take M to be 50.

Similarly, u_y is limited (replaced) by

$$\bar{m}(u_y, \bar{u}_{i,j+1} - \bar{u}_{ij}, \bar{u}_{ij} - \bar{u}_{i,j-1}).$$

with a change of Δx to Δy in (4.6).

For systems, we perform the limiting in the local characteristic variables. To limit the vector u_x in the element ij, we proceed as follows:

- Find the matrix R and its inverse R^{-1}, which diagonalize the Jacobian evaluated at the mean in the element ij in the x-direction:

$$R^{-1}\frac{\partial f_1(\bar{u}_{ij})}{\partial u}R = \Lambda,$$

where Λ is a diagonal matrix containing the eigenvalues of the Jacobian. Notice that the columns of R are the right eigenvectors of $\frac{\partial f_1(\bar{u}_{ij})}{\partial u}$ and the rows of R^{-1} are the left eigenvectors.
- Transform all quantities needed for limiting, i.e., the three vectors $u_{x\,ij}$, $\bar{u}_{i+1,j} - \bar{u}_{ij}$ and $\bar{u}_{ij} - \bar{u}_{i-1,j}$, to the characteristic fields. This is achieved by left multiplying these three vectors by R^{-1}.
- Apply the scalar limiter (4.6) to each of the components of the transformed vectors.
- The result is transformed back to the original space by left multiplying R on the left.

The triangular elements To construct the slope limiting operators for triangular elements, we proceed as follows. We start by making a simple observation. Consider the triangles in Figure 4.1, where m_1 is the mid-point of the edge on the boundary of K_0 and b_i denotes the barycenter of the triangle K_i for $i = 0, 1, 2, 3$.

Since we have that

$$m_1 - b_0 = \alpha_1 (b_1 - b_0) + \alpha_2 (b_2 - b_0),$$

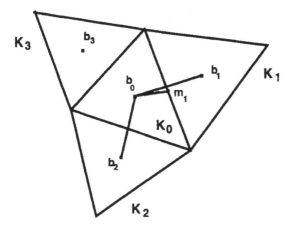

Fig. 4.1: Illustration of limiting.

for some nonnegative coefficients α_1, α_2 which depend only on m_1 and the geometry, we can write, for any linear function u_h,

$$u_h(m_1) - u_h(b_0) = \alpha_1 \left(u_h(b_1) - u_h(b_0)\right) + \alpha_2 \left(u_h(b_2) - u_h(b_0)\right),$$

and since

$$\bar{u}_{K_i} = \frac{1}{|K_i|} \int_{K_i} u_h = u_h(b_i), \qquad i = 0, 1, 2, 3,$$

we have that

$$\tilde{u}_h(m_1, K_0) \equiv u_h(m_1) - \bar{u}_{K_0} = \alpha_1 \left(\bar{u}_{K_1} - \bar{u}_{K_0}\right) + \alpha_2 \left(\bar{u}_{K_2} - \bar{u}_{K_0}\right) \equiv \Delta \bar{u}(m_1, K_0)$$

Now, we are ready to describe the slope limiting. Let us consider a piecewise linear function u_h, and let $m_i, i = 1, 2, 3$ be the three mid-points of the edges of the triangle K_0. We then can write, for $(x, y) \in K_0$,

$$u_h(x, y) = \sum_{i=1}^{3} u_h(m_i)\varphi_i(x, y) = \bar{u}_{K_0} + \sum_{i=1}^{3} \tilde{u}_h(m_i, K_0)\varphi_i(x, y).$$

To compute $\Lambda\Pi_h u_h$, we first compute the quantities

$$\Delta_i = \bar{m}(\tilde{u}_h(m_i, K_0), \nu \Delta\bar{u}(m_i, K_0)),$$

where $\bar{m}$ is the TVB modified *minmod* function and $\nu > 1$. We take $\nu = 1.5$ in our numerical runs. Then, if $\sum_{i=1}^{3} \Delta_i = 0$, we simply set

$$\Lambda\Pi_h u_h(x, y) = \bar{u}_{K_0} + \sum_{i=1}^{3} \Delta_i \varphi_i(x, y).$$

If $\sum_{i=1}^{3} \Delta_i \neq 0$, we compute

$$pos = \sum_{i=1}^{3} \max(0, \Delta_i), \qquad neg = \sum_{i=1}^{3} \max(0, -\Delta_i),$$

and set

$$\theta^+ = \min\left(1, \frac{neg}{pos}\right), \qquad \theta^- = \min\left(1, \frac{pos}{neg}\right).$$

Then, we define

$$\Lambda \Pi_h u_h(x, y) = \bar{u}_{K_0} + \sum_{i=1}^{3} \hat{\Delta}_i \, \varphi_i(x, y),$$

where

$$\hat{\Delta}_i = \theta^+ \max(0, \Delta_i) - \theta^- \max(0, -\Delta_i).$$

It is very easy to see that this slope limiting operator satisfies the three properties listed above.

For systems, we perform the limiting in the local characteristic variables. To limit Δ_i, we proceed as in the rectangular case, the only difference being that we work with the following Jacobian

$$\frac{\partial}{\partial u} f(\bar{u}_{K_0}) \cdot \frac{m_i - b_0}{|m_i - b_0|}.$$

4.4 Computational results: Transient, nonsmooth solutions

In this section we present several numerical results obtained with the P^1 and P^2 (second and third order accurate) RKDG methods with either rectangles or triangles in the triangulation. These are standard test problems for Euler equations of compressible gas dynamics.

The double-Mach reflection problem Double Mach reflection of a strong shock. This problem was studied extensively in Woodward and Colella [66] and later by many others. We use exactly the same setup as in [66], namely a Mach 10 shock initially makes a 60° angle with a reflecting wall. The undisturbed air ahead of the shock has a density of 1.4 and a pressure of 1.

For the rectangle based triangulation, we use a rectangular computational domain $[0, 4] \times [0, 1]$, as in [66]. The reflecting wall lies at the bottom of the computational domain for $\frac{1}{6} \leq x \leq 4$. Initially a right-moving Mach 10 shock is positioned at $x = \frac{1}{6}, y = 0$ and makes a 60° angle with the x-axis. For the bottom boundary, the exact post-shock condition is imposed for the part from $x = 0$ to $x = \frac{1}{6}$, to mimic an angled wedge. Reflective boundary condition is used for the rest. At the top boundary of our computational domain, the flow values are set to describe the exact motion of the Mach

10 shock. Inflow/outflow boundary conditions are used for the left and right boundaries. As in [66], only the results in $[0, 3] \times [0, 1]$ are displayed.

For the triangle based triangulation, we have the freedom to treat irregular domains and thus use a true wedged computational domain. Reflective boundary conditions are then used for all the bottom boundary, including the sloped portion. Other boundary conditions are the same as in the rectangle case.

Uniform rectangles are used in the rectangle based triangulations. Four different meshes are used: 240×60 rectangles ($\Delta x = \Delta y = \frac{1}{60}$); 480×120 rectangles ($\Delta x = \Delta y = \frac{1}{120}$); 960×240 rectangles ($\Delta x = \Delta y = \frac{1}{240}$); and 1920×480 rectangles ($\Delta x = \Delta y = \frac{1}{480}$). The density is plotted in Figure 4.2 for the P^1 case and in 4.3 for the P^2 case.

To better appreciate the difference between the P^1 and P^2 results in these pictures, we show a "blowed up" portion around the double Mach region in Figure 4.4 and show one-dimensional cuts along the line $y = 0.4$ in Figures 4.5 and 4.6. In Figure 4.4, w can see that P^2 with $\Delta x = \Delta y = \frac{1}{240}$ has qualitatively the same resolution as P^1 with $\Delta x = \Delta y = \frac{1}{480}$, for the fine details of the complicated structure in this region. P^2 with $\Delta x = \Delta y = \frac{1}{480}$ gives a much better resolution for these structures than P^1 with the same number of rectangles.

Moreover, from Figure 4.5, we clearly see that the difference between the results obtained by using P^1 and P^2, on the same mesh, increases dramatically as the mesh size decreases. This indicates that the use of polynomials of high degree might be beneficial for capturing the above mentioned structures. From Figure 4.6, we see that the results obtained with P^1 are qualitatively similar to those obtained with P^2 in a coarser mesh; the similarity increases as the meshsize decreases. The conclusion here is that, if one is interested in the above mentioned fine structures, then one can use the third order scheme P^2 with only half of the mesh points in each direction as in P^1. This translates into a reduction of a factor of 8 in space-time grid points for 2D time dependent problems, and will more than off-set the increase of cost per mesh point and the smaller CFL number by using the higher order P^2 method. This saving will be even more significant for 3D.

The optimal strategy, of course, is to use adaptivity and concentrate triangles around the interesting region, and/or change the order of the scheme in different regions.

The forward-facing step problem Flow past a forward facing step. This problem was again studied extensively in Woodward and Colella [66] and later by many others. The set up of the problem is the following: A right going Mach 3 uniform flow enters a wind tunnel of 1 unit wide and 3 units long. The step is 0.2 units high and is located 0.6 units from the left-hand end of the tunnel. The problem is initialized by a uniform, right-going Mach 3 flow. Reflective boundary conditions are applied along the walls of the tunnel and in-flow and out-flow boundary conditions are applied at the entrance (left-hand end) and the exit (right-hand end), respectively.

The corner of the step is a singularity, which we study carefully in our numerical experiments. Unlike in [66] and many other papers, we do not modify our scheme near the corner in any way. It is well known that this leads to an errorneous entropy layer at the downstream bottom wall, as well as a spurious Mach stem at the bottom wall. However, these artifacts decrease when the mesh is refined. In Figure 4.7, second order P^1 results using rectangle triangulations are shown, for a grid refinement study using $\Delta x = \Delta y = \frac{1}{40}$, $\Delta x = \Delta y = \frac{1}{80}$, $\Delta x = \Delta y = \frac{1}{160}$, and $\Delta x = \Delta y = \frac{1}{320}$ as mesh sizes. We can clearly see the improved resolution (especially at the upper slip line from the triple point) and decreased artifacts caused by the corner, with increased mesh points. In Figure 4.8, third order P^2 results using the same meshes are shown.

In order to verify that the erroneous entropy layer at the downstream bottom wall and the spurious Mach stem at the bottom wall are both artifacts caused by the corner singularity, we use our triangle code to locally refine near the corner progressively; we use the meshes displayed in Figure 4.9. In Figure 4.10, we plot the density obtained by the P^1 triangle code, with triangles (roughly the resolution of $\Delta x = \Delta y = \frac{1}{40}$, except around the corner). In Figure 4.11, we plot the entropy around the corner for the same runs. We can see that, with more triangles concentrated near the corner, the artifacts gradually decrease. Results with P^2 codes in Figures 4.12 and 4.13 show a similar trend.

4.5 Computational results: Steady state, smooth solutions

In this section, we present some of the numerical results of Bassi and Rebay [2] in two dimensions and Warburton, Lomtev, Kirby and Karniadakis [65] in three dimensions.

The purpose of the numerical results of Bassi and Rebay [2] we are presenting is to assess (i) the effect of the quality of the approximation of curved boundaries and of (ii) the effect of the degree of the polynomials on the quality of the approximate solution. The test problem we consider here is the two-dimensional steady-state, subsonic flow around a disk at Mach number $M_\infty = 0.38$. Since the solution is smooth and can be computed analytically, the quality of the approximation can be easily assessed.

In the figures 4.14, 4.15, 4.16, and 4.17, details of the meshes around the disk are shown together with the approximate solution given by the RKDG method using piecewise linear elements. These meshes approximate the circle with a polygonal. It can be seen that the approximate solution are of very low quality even for the most refined grid. This is an effect caused by the kinks of the polygonal approximating the circle.

This statement can be easily verified by taking a look to the figures 4.18, 4.19, 4.20, and 4.21. In these pictures the approximate solutions with piecewise linear, quadratic, and cubic elements are shown; the meshes have been modified to render *exactly* the circle. It is clear that the improvement in the quality of the approximation is enormous. Thus, a high-quality approximation of the boundaries has a dramatic improvement on the quality of the approximations.

Also, it can be seen that the higher the degree of the polynomials, the better the quality of the approximations, in particular from figures 4.18 and 4.19. In [2], Bassi and Rebay show that the RKDG method using polynomilas of degree k are $(k+1)$-th order accurate for $k = 1, 2, 3$. As a consequence, a RKDG method using polynomials of a higher degree is more efficient than a RKDG method using polynomials of lower degree.

In [65], Warburton, Lomtev, Kirby and Karniadakis present the same test problem in a three dimensional setting. In Figure 4.22, we can see the three-dimensional mesh and the density isosurfaces. We can also see how, while the mesh is being kept fixed and the degree of the polynomials k is increased from 1 to 9, the maximum error on the entropy goes exponentialy to zero. (In the picture, a so-called 'mode' is equal to $k + 1$).

4.6 Concluding remarks

In this section, we have extended the RKDG methods to multidimensional systems. We have described in full detail the algorithms and displayed numerical results showing the performance of the methods for the Euler equations of gas dynamics.

The flexibility of the RKDG method to handle nontrivial geometries and to work with different elements has been displayed. Moreover, it has been shown that the use of polynomials of high degree not only does not degrade the resolution of strong shocks, but enhances the resolution of the contact discontinuities and renders the scheme more efficient on smooth regions.

Next, we extend the RKDG methods to convection-dominated problems.

214

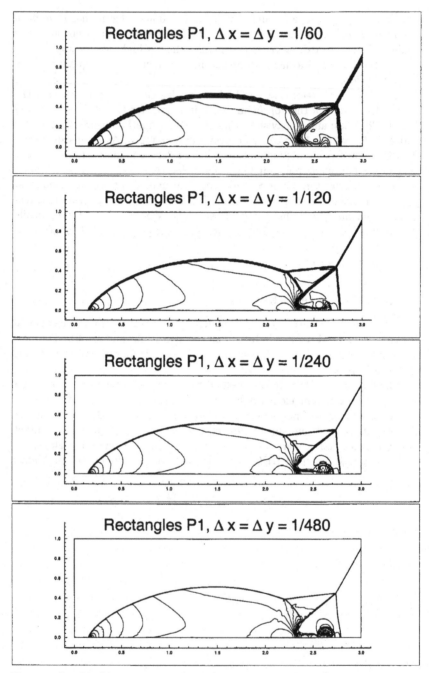

Fig. 4.2: Double Mach reflection problem. Second order P^1 results. Density ρ. 30 equally spaced contour lines from $\rho = 1.3965$ to $\rho = 22.682$. Mesh refinement study. From top to bottom: $\Delta x = \Delta y = \frac{1}{60}, \frac{1}{120}, \frac{1}{240}$, and $\frac{1}{480}$.

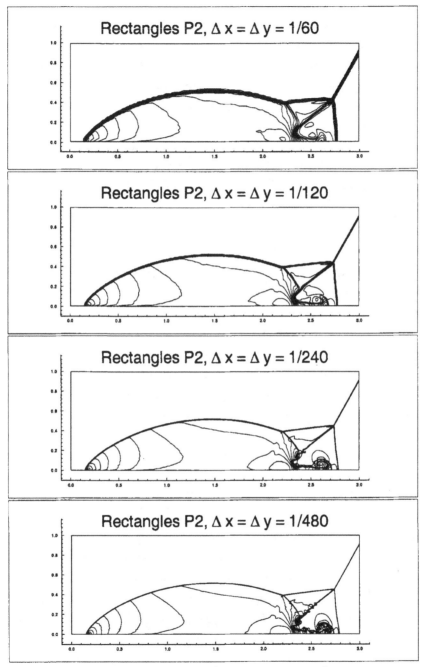

Fig. 4.3: Double Mach reflection problem. Third order P^2 results. Density ρ. 30 equally spaced contour lines from $\rho = 1.3965$ to $\rho = 22.682$. Mesh refinement study. From top to bottom: $\Delta x = \Delta y = \frac{1}{60}, \frac{1}{120}, \frac{1}{240}$, and $\frac{1}{480}$.

216

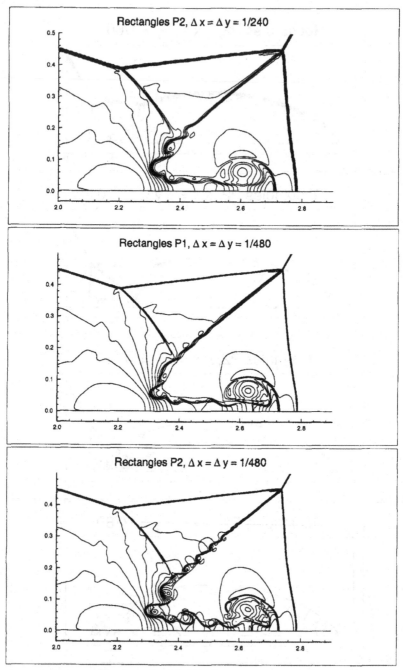

Fig. 4.4: Double Mach reflection problem. Blowed-up region around the double Mach stems. Density ρ. Third order P^2 with $\Delta x = \Delta y = \frac{1}{240}$ (top); second order P^1 with $\Delta x = \Delta y = \frac{1}{480}$ (middle); and third order P^2 with $\Delta x = \Delta y = \frac{1}{480}$ (bottom).

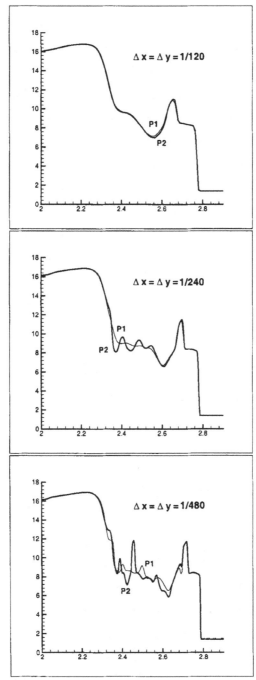

Fig. 4.5: Double Mach reflection problem. Cut $y = 0.4$ of the blowed-up region. Density ρ. Comparison of second order P^1 with third order P^2 on the same mesh

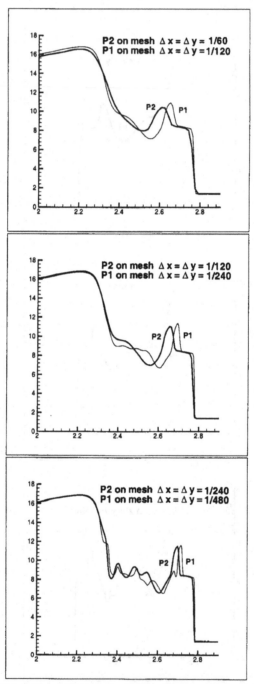

Fig. 4.6: Double Mach reflection problem. Cut $y = 0.4$ of the blowed-up region. Density ρ. Comparison of second order P^1 with third order P^2 on a coarser mesh

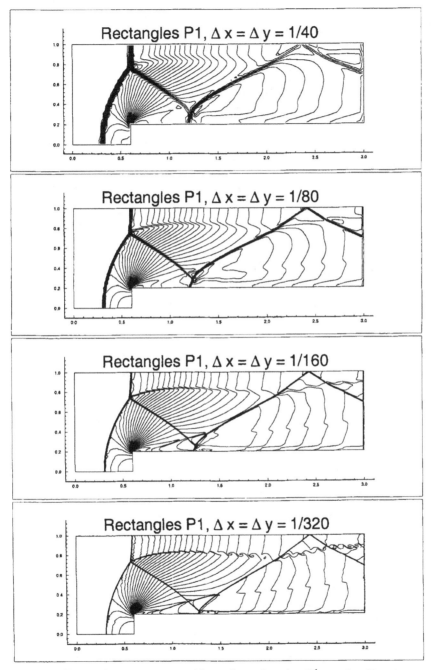

Fig. 4.7: Forward facing step problem. Second order P^1 results. Density ρ. 30 equally spaced contour lines from $\rho = 0.090338$ to $\rho = 6.2365$. Mesh refinement study. From top to bottom: $\Delta x = \Delta y = \frac{1}{40}, \frac{1}{80}, \frac{1}{160}$, and $\frac{1}{320}$.

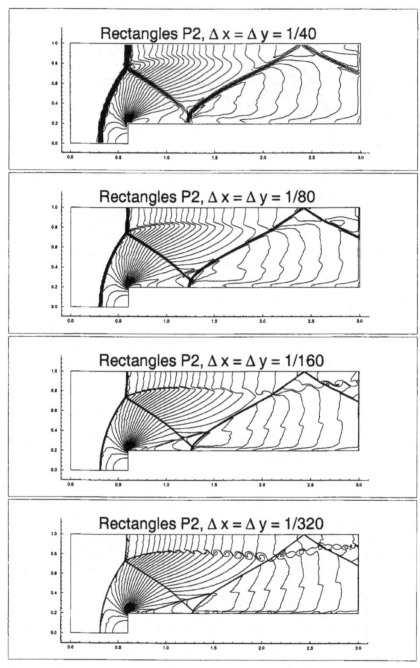

Fig. 4.8: Forward facing step problem. Third order P^2 results. Density ρ. 30 equally spaced contour lines from $\rho = 0.090338$ to $\rho = 6.2365$. Mesh refinement study. From top to bottom: $\Delta x = \Delta y = \frac{1}{40}, \frac{1}{80}, \frac{1}{160}$, and $\frac{1}{320}$.

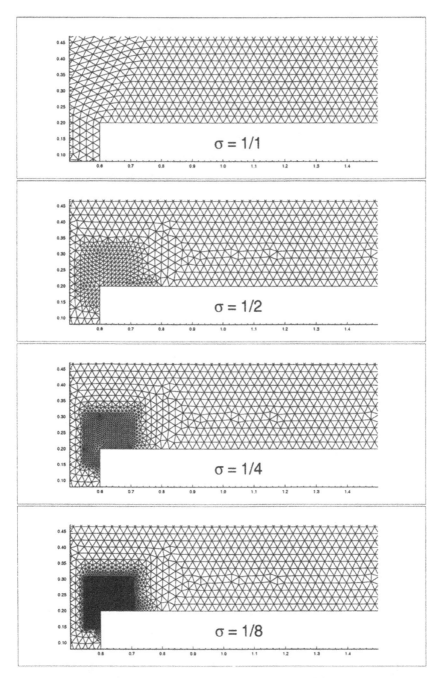

Fig. 4.9: Forward facing step problem. Detail of the triangulations associated with the different values of σ. The parameter σ is the ratio between the typical size of the triangles near the corner and that elsewhere.

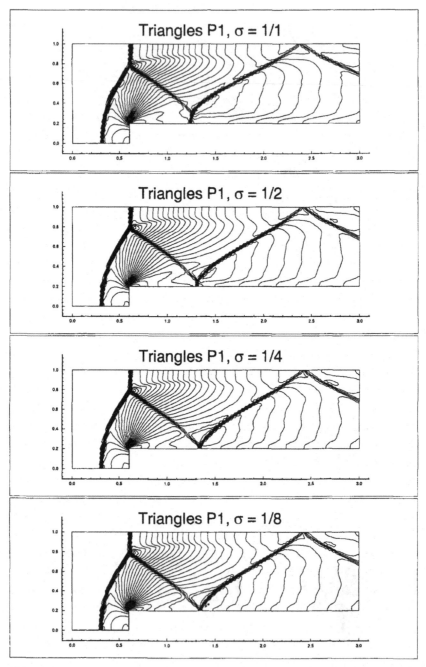

Fig. 4.10: Forward facing step problem. Second order P^1 results. Density ρ. 30 equally spaced contour lines from $\rho = 0.090338$ to $\rho = 6.2365$. Triangle code. Progressive refinement near the corner

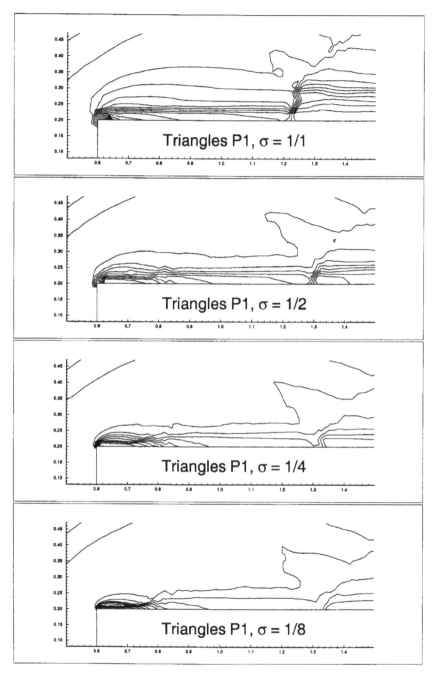

Fig. 4.11: Forward facing step problem. Second order P^1 results. Entropy level curves around the corner. Triangle code. Progressive refinement near the corner

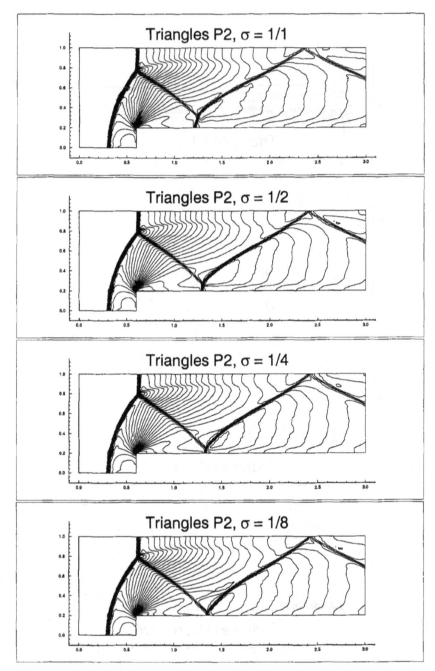

Fig. 4.12: Forward facing step problem. Third order P^2 results. Density ρ. 30 equally spaced contour lines from $\rho = 0.090338$ to $\rho = 6.2365$. Triangle code. Progressive refinement near the corner

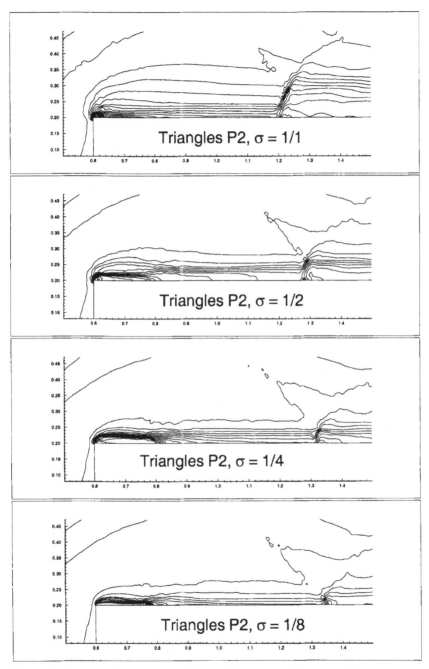

Fig. 4.13: Forward facing step problem. Third order P^1 results. Entropy level curves around the corner. Triangle code. Progressive refinement near the corner

226

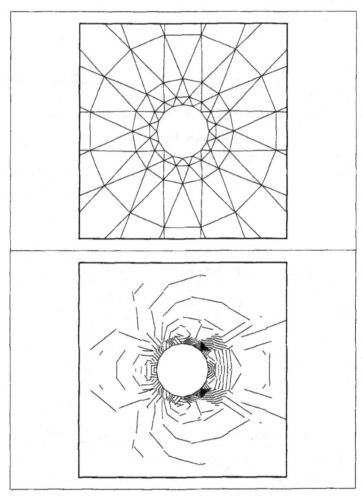

Fig. 4.14: Grid "16 × 8" with a piecewise linear approximation of the circle (top) and the corresponding solution (Mach isolines) using P^1 elements (bottom).

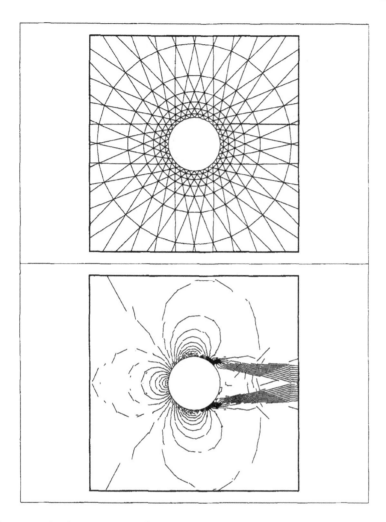

Fig. 4.15: Grid "32 × 8" with a piecewise linear approximation of the circle (top) and the corresponding solution (Mach isolines) using P^1 elements (bottom).

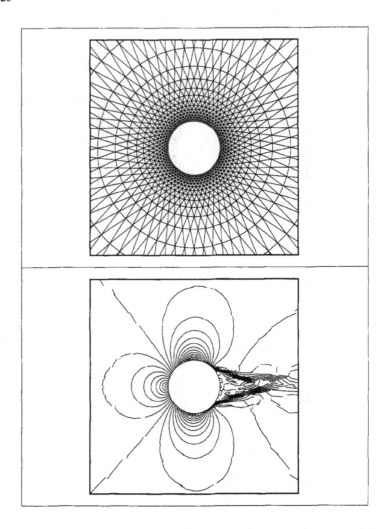

Fig. 4.16: Grid "64 × 16" with a piecewise linear approximation of the circle (top) and the corresponding solution (Mach isolines) using P^1 elements (bottom).

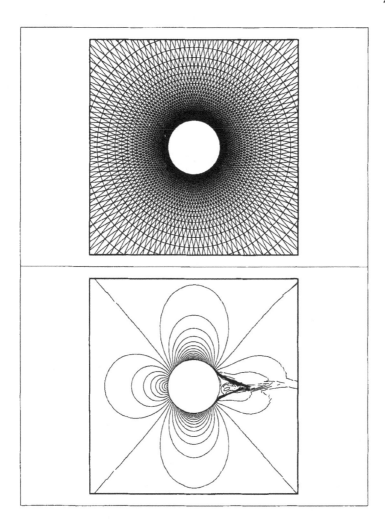

Fig. 4.17: Grid "128 × 32" a piecewise linear approximation of the circle (top) and the corresponding solution (Mach isolines) using P^1 elements (bottom).

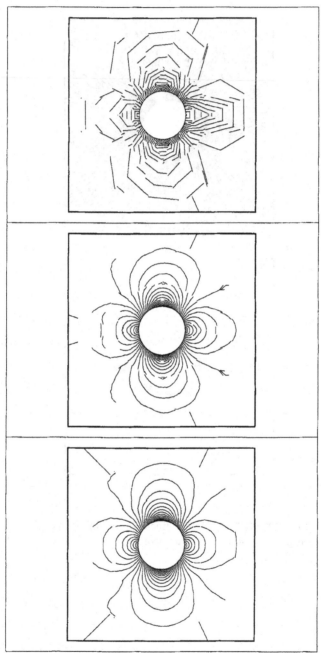

Fig. 4.18: Grid "16 × 4" with exact rendering of the circle and the corresponding P^1 (top), P^2(middle), and P^3 (bottom) approximations (Mach isolines).

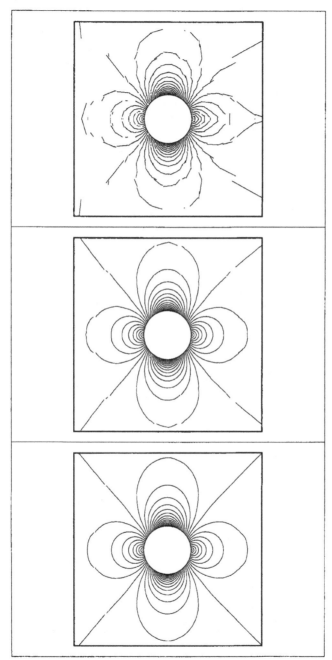

Fig. 4.19: Grid "32×8" with exact rendering of the circle and the corresponding P^1 (top), P^2(middle), and P^3 (bottom) approximations (Mach isolines).

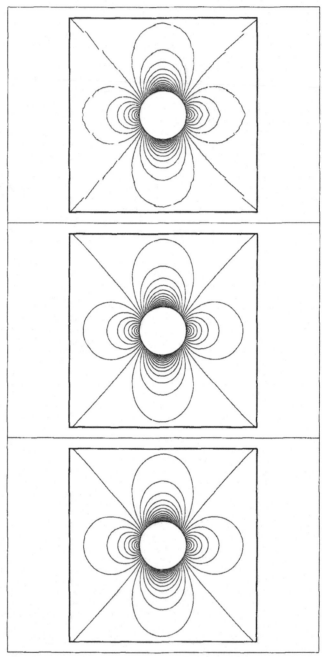

Fig. 4.20: Grid "64 × 16" with exact rendering of the circle and the corresponding P^1 (top), P^2 (middle), and P^3 (bottom) approximations (Mach isolines).

Fig. 4.21: Grid "128 × 32" with exact rendering of the circle and the corresponding P^1 (top), P^2(middle), and P^3 (bottom) approximations (Mach isolines).

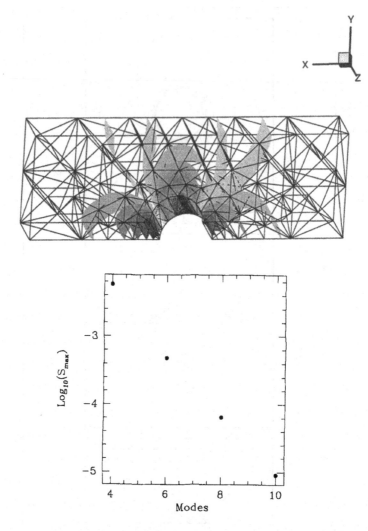

Fig. 4.22: Three-dimensional flow over a semicircular bump. Mesh and density isosurfaces (top) and history of convergence with *p*-refinement of the maximum entropy generated (bottom). The degree of the polynomial plus one is plotted on the 'modes' axis.

5 Convection-diffusion problems: The LDG method

5.1 Introduction

In this chapter, which follows the work by Cockburn and Shu [18], we restrict ourselves to the semidiscrete LDG methods for convection-diffusion problems with periodic boundary conditions. Our aim is to clearly display the most distinctive features of the LDG methods in a setting as simple as possible; the extension of the method to the fully discrete case is straightforward. In §2, we introduce the LDG methods for the simple one-dimensional case $d = 1$ in which

$$\mathbf{F}(u, Du) = f(u) - a(u)\,\partial_x u,$$

u is a scalar and $a(u) \geq 0$ and show some preliminary numerical results displaying the performance of the method. In this simple setting, the main ideas of how to device the method and how to analyze it can be clearly displayed in a simple way. Thus, the L^2-stability of the method is proven in the general nonlinear case and the rate of convergence of $(\Delta x)^k$ in the $L^\infty(0, T; L^2)$-norm for polynomials of degree $k \geq 0$ in the linear case is obtained; this estimate is sharp. In §3, we extend these results to the case in which u is a scalar and

$$\mathbf{F}_i(u, Du) = f_i(u) - \sum_{1 \leq j \leq d} a_{ij}(u)\,\partial_{x_j} u,$$

where a_{ij} defines a positive semidefinite matrix. Again, the L^2-stability of the method is proven for the general nonlinear case and the rate of convergence of $(\Delta x)^k$ in the $L^\infty(0, T; L^2)$-norm for polynomials of degree $k \geq 0$ and arbitrary triangulations is proven in the linear case. In this case, the multidimensionality of the problem and the arbitrariness of the grids increase the technicality of the analysis of the method which, nevertheless, uses the same ideas of the one-dimensional case. In §4, the extension of the LDG method to multidimensional systems is briefly described some numerical results for the compressible Navier-Stokes equations from the paper by Bassi and Rebay [3] and from the paper by Lomtev and Karniadakis [46] are presented.

5.2 The LDG methods for the one-dimensional case

In this section, we present and analyze the LDG methods for the following simple model problem:

$$\partial_t u + \partial_x \left(f(u) - a(u)\,\partial_x u \right) = 0 \quad \text{in } (0, T) \times (0, 1), \qquad (5.1)$$

$$u(t = 0) = u_0, \quad \text{on } (0, 1), \qquad (5.2)$$

with periodic boundary conditions.

General formulation and main properties To define the LDG method, we introduce the new variable $q = \sqrt{a(u)}\, \partial_x u$ and rewrite the problem (5.1), (5.2) as follows:

$$\partial_t u + \partial_x \left(f(u) - \sqrt{a(u)}\, q \right) = 0 \quad \text{in } (0, T) \times (0, 1), \tag{5.3}$$

$$q - \partial_x g(u) = 0 \quad \text{in } (0, T) \times (0, 1), \tag{5.4}$$

$$u(t = 0) = u_0, \quad \text{on } (0, 1), \tag{5.5}$$

where $g(u) = \int^u \sqrt{a(s)}\, ds$. The LDG method for (5.1), (5.2) is now obtained by simply discretizing the above system with the Discontinuous Galerkin method.

To do that, we follow [15] and [14]. We define the flux $\mathbf{h} = (h_u, h_q)^t$ as follows:

$$\mathbf{h}(u, q) = \left(f(u) - \sqrt{a(u)}\, q, \, -g(u) \right)^t. \tag{5.6}$$

For each partition of the interval $(0, 1)$, $\{ x_{j+1/2} \}_{j=0}^N$, we set $I_j = (x_{j-1/2}, x_{j+1/2})$, and $\Delta x_j = x_{j+1/2} - x_{j-1/2}$ for $j = 1, \ldots, N$; we denote the quantity $\max_{1 \leq j \leq N} \Delta x_j$ by Δx. We seek an approximation $\mathbf{w}_h = (u_h, q_h)^t$ to $\mathbf{w} = (u, q)^t$ such that for each time $t \in [0, T]$, both $u_h(t)$ and $q_h(t)$ belong to the finite dimensional space

$$V_h = V_h^k = \{ v \in L^1(0, 1) : v|_{I_j} \in P^k(I_j), \; j = 1, \ldots, N \}, \tag{5.7}$$

where $P^k(I)$ denotes the space of polynomials in I of degree at most k. In order to determine the approximate solution (u_h, q_h), we first note that by multiplying (5.3), (5.4), and (5.5) by arbitrary, smooth functions v_u, v_q, and v_i, respectively, and integrating over I_j, we get, after a simple formal integration by parts in (5.3) and (5.4),

$$\int_{I_j} \partial_t u(x, t)\, v_u(x)\, dx - \int_{I_j} h_u(\mathbf{w}(x, t))\, \partial_x v_u(x)\, dx$$
$$+ h_u(\mathbf{w}(x_{j+1/2}, t))\, v_u(x_{j+1/2}^-) - h_u(\mathbf{w}(x_{j-1/2}, t))\, v_u(x_{j-1/2}^+) = 0, \tag{5.8}$$

$$\int_{I_j} q(x, t)\, v_q(x)\, dx - \int_{I_j} h_q(\mathbf{w}(x, t))\, \partial_x v_q(x)\, dx$$
$$+ h_q(\mathbf{w}(x_{j+1/2}, t))\, v_q(x_{j+1/2}^-) - h_q(\mathbf{w}(x_{j-1/2}, t))\, v_q(x_{j-1/2}^+) = 0, \tag{5.9}$$

$$\int_{I_j} u(x, 0)\, v_i(x)\, dx = \int_{I_j} u_0(x)\, v_i(x)\, dx. \tag{5.10}$$

Next, we replace the smooth functions v_u, v_q, and v_i by test functions $v_{h,u}$, $v_{h,q}$, and $v_{h,i}$, respectively, in the finite element space V_h and the exact solution $\mathbf{w} = (u,q)^t$ by the approximate solution $\mathbf{w}_h = (u_h, q_h)^t$. Since this function is discontinuous in each of its components, we must also replace the nonlinear flux $\mathbf{h}(\mathbf{w}(x_{j+1/2}, t))$ by a numerical flux $\hat{\mathbf{h}}(\mathbf{w})_{j+1/2}(t) = (\hat{h}_u(\mathbf{w}_h)_{j+1/2}(t), \hat{h}_q(\mathbf{w}_h)_{j+1/2}(t))$ that will be suitably chosen later. Thus, the approximate solution given by the LDG method is defined as the solution of the following weak formulation:

$$\forall\, v_{h,u} \in P^k(I_j):$$
$$\int_{I_j} \partial_t\, u_h(x,t)\, v_{h,u}(x)\, dx - \int_{I_j} h_u(\mathbf{w}_h(x,t))\, \partial_x\, v_{h,u}(x)\, dx$$
$$+\hat{h}_u(\mathbf{w}_h)_{j+1/2}(t)\, v_{h,u}(x_{j+1/2}^-) - \hat{h}_u(\mathbf{w}_h)_{j-1/2}(t)\, v_{h,u}(x_{j-1/2}^+) = 0, \quad (5.11)$$
$$\forall\, v_{h,q} \in P^k(I_j):$$
$$\int_{I_j} q_h(x,t)\, v_{h,q}(x)\, dx - \int_{I_j} h_q(\mathbf{w}_h(x,t))\, \partial_x\, v_{h,q}(x)\, dx$$
$$+\hat{h}_q(\mathbf{w}_h)_{j+1/2}(t)\, v_{h,q}(x_{j+1/2}^-) - \hat{h}_q(\mathbf{w}_h)_{j-1/2}(t)\, v_{h,q}(x_{j-1/2}^+) = 0, \quad (5.12)$$
$$\forall\, v_{h,i} \in P^k(I_j):$$
$$\int_{I_j} u_h(x,0)\, v_{h,i}(x)\, dx = \int_{I_j} u_0(x)\, v_{h,i}(x)\, dx. \quad (5.13)$$

It only remains to choose the numerical flux $\hat{\mathbf{h}}(\mathbf{w}_h)_{j+1/2}(t)$. We use the notation:

$$[p] = p^+ - p^-, \quad \text{and} \quad \bar{p} = \frac{1}{2}(p^+ + p^-), \text{and } p_{j+1/2}^{\pm} = p(x_{j+1/2}^{\pm}).$$

To be consistent with the type of numerical fluxes used in the RKDG methods, we consider numerical fluxes of the form

$$\hat{\mathbf{h}}(\mathbf{w}_h)_{j+1/2}(t) \equiv \hat{\mathbf{h}}(\mathbf{w}_h(x_{j+1/2}^-, t), \mathbf{w}_h(x_{j+1/2}^+, t)),$$

that (i) are locally Lipschitz and consistent with the flux $\mathbf{h}$, (ii) allow for a local resolution of q_h in terms of u_h, (iii) reduce to an E-flux (see Osher [51]) when $a(\cdot) \equiv 0$, and that (iv) enforce the L^2-stability of the method.

To reflect the convection-diffusion nature of the problem under consideration, we write our numerical flux as the sum of a convective flux and a diffusive flux:

$$\hat{h}(\mathbf{w}^-, \mathbf{w}^+) = \hat{h}_{conv}(\mathbf{w}^-, \mathbf{w}^+) + \hat{h}_{diff}(\mathbf{w}^-, \mathbf{w}^+). \qquad (5.14)$$

The convective flux is given by

$$\hat{h}_{conv}(\mathbf{w}^-, \mathbf{w}^+) = (\hat{f}(u^-, u^+), 0)^t, \qquad (5.15)$$

where $\hat{f}(u^-, u^+)$ is any locally Lipschitz E-flux consistent with the nonlinearity f, and the diffusive flux is given by

$$\hat{h}_{diff}(\mathbf{w}^-, \mathbf{w}^+) = \left(-\frac{[g(u)]}{[u]}\, \bar{q},\ -\overline{g(u)} \right)^t - \mathbb{C}_{diff}[\mathbf{w}], \qquad (5.16)$$

where

$$\mathbb{C}_{diff} = \begin{pmatrix} 0 & c_{12} \\ -c_{12} & 0 \end{pmatrix}, \qquad (5.17)$$

$$c_{12} = c_{12}(\mathbf{w}^-, \mathbf{w}^+) \quad \text{is locally Lipschitz}, \qquad (5.18)$$

$$c_{12} \equiv 0 \quad \text{when } a(\cdot) \equiv 0. \qquad (5.19)$$

We claim that this flux satisfies the properties (i) to (iv).

Let us prove our claim. That the flux $\hat{h}$ is consistent with the flux h easily follows from their definitions. That $\hat{h}$ is locally Lipschitz follows from the fact that $\hat{f}(\cdot, \cdot)$ is locally Lipschitz and from (5.17); we assume that $f(\cdot)$ and $a(\cdot)$ are locally Lipschitz functions, of course. Property (i) is hence satisfied.

That the approximate solution q_h can be resolved element by element in terms of u_h by using (5.12) follows from the fact that, by (5.16), the flux $\hat{h}_q = -\overline{g(u)} - c_{12}[u]$ is independent of q_h. Property (ii) is hence satisfied.

Property (iii) is also satisfied by (5.19) and by the construction of the convective flux.

To see that the property (iv) is satisfied, let us first rewrite the flux $\hat{h}$ in the following way:

$$\hat{h}(\mathbf{w}^-, \mathbf{w}^+) = \left(\frac{[\varphi(u)]}{[u]} - \frac{[g(u)]}{[u]}\, \bar{q},\ -\overline{g(u)} \right)^t - \mathbb{C}[\mathbf{w}],$$

where

$$\mathbb{C} = \begin{pmatrix} c_{11} & c_{12} \\ -c_{12} & 0 \end{pmatrix}, \quad c_{11} = \frac{1}{[u]}\left(\frac{[\varphi(u)]}{[u]} - \hat{f}(u^-, u^+)\right). \quad (5.20)$$

with $\varphi(u)$ defined by $\varphi(u) = \int^u f(s)\,ds$. Since $\hat{f}(\cdot, \cdot)$ is an E-flux,

$$c_{11} = \frac{1}{[u]^2}\int_{u^-}^{u^+}\left(f(s) - \hat{f}(u^-, u^+)\right)ds \geq 0,$$

and so, by (5.17), the matrix $\mathbb{C}$ is semipositive definite. The property (iv) follows from this fact and from the following result.

Theorem 5.1 *We have,*

$$\tfrac{1}{2}\int_0^1 u_h^2(x, T)\,dx + \int_0^T\int_0^1 q_h^2(x, t)\,dx\,dt + \Theta_{T,\mathbb{C}}([\mathbf{w}_h]) \leq \tfrac{1}{2}\int_0^1 u_0^2(x)\,dx,$$

where

$$\Theta_{T,\mathbb{C}}([\mathbf{w}_h]) = \int_0^T \sum_{1 \leq j \leq N}\left\{[\mathbf{w}_h(t)]^t\mathbb{C}\,[\mathbf{w}_h(t)]\right\}_{j+1/2}dt.$$

For a proof, see [18]. Thus, this shows that the flux $\hat{\mathbf{h}}$ under consideration does satisfy the properties (i) to (iv)- as claimed.

Now, we turn to the question of the quality of the approximate solution defined by the LDG method. In the linear case $f' \equiv c$ and $a(\cdot) \equiv a$, from the above stability result and from the the approximation properties of the finite element space V_h, we can prove the following error estimate. We denote the $L^2(0,1)$-norm of the ℓ-th derivative of u by $|u|_\ell$.

Theorem 5.2 *Let e be the approximation error $\mathbf{w} - \mathbf{w}_h$. Then we have,*

$$\left\{\int_0^1 |e_u(x,T)|^2\,dx + \int_0^T\int_0^1 |e_q(x,t)|^2\,dx\,dt + \Theta_{T,\mathbb{C}}([\mathbf{e}])\right\}^{1/2} \leq C\,(\Delta x)^k,$$

where $C = C(k, |u|_{k+1}, |u|_{k+2})$. In the purely hyperbolic case $a = 0$, the constant C is of order $(\Delta x)^{1/2}$. In the purely parabolic case $c = 0$, the constant C is of order Δx for even values of k for uniform grids and for $\mathbb{C}$ identically zero.

For a proof, see [18]. The above error estimate gives a suboptimal order of convergence, but it is sharp for the LDG methods. Indeed, Bassi *et al* [4] report an order of convergence of order $k+1$ for even values of k and of order k for odd values of k for a steady state, purely elliptic problem for uniform grids and for $\mathbb{C}$ identically zero. The numerical results for a purely parabolic problem that will be displayed later lead to the same conclusions; see Table 5 in the section §2.b.

The error estimate is also sharp in that the optimal order of convergence of $k+1/2$ is recovered in the purely hyperbolic case, as expected. This improvement of the order of convergence is a reflection of the *semipositive definiteness* of the matrix $\mathbb{C}$, which enhances the stability properties of the LDG method. Indeed, since in the purely hyperbolic case

$$\Theta_{T,\mathbb{C}}([\mathbf{w}_h]) = \int_0^T \sum_{1 \le j \le N} \left\{ [u_h(t)]^t \, c_{11} \, [u_h(t)] \right\}_{j+1/2} dt,$$

the method enforces a control of the jumps of the variable u_h, as shown in Proposition lemenergy. This additional control is reflected in the improvement of the order of accuracy from k in the general case to $k+1/2$ in the purely hyperbolic case.

However, this can only happen in the purely hyperbolic case for the LDG methods. Indeed, since $c_{11} = 0$ for $c = 0$, the control of the jumps of u_h is not enforced in the purely parabolic case. As indicated by the numerical experiments of Bassi *et al.* [4] and those of section §2.b below, this can result in the effective degradation of the order of convergence. To remedy this situation, the control of the jumps of u_h in the purely parabolic case can be easily enforced by letting c_{11} be strictly positive if $|c| + |a| > 0$. Unfortunately, this is not enough to guarantee an improvement of the accuracy: an additional control on the jumps of q_h is required! This can be easily achieved by allowing the matrix $\mathbb{C}$ to be *symmetric and positive definite* when $a > 0$. In this case, the order of convergence of $k + 1/2$ can be easily obtained for the general convection-diffusion case. However, this would force the matrix entry c_{22} to be nonzero and the property (ii) of local resolvability of q_h in terms of u_h would not be satisfied anymore. As a consequence, the high parallelizability of the LDG would be lost.

The above result shows how strongly the order of convergence of the LDG methods depend on the choice of the matrix $\mathbb{C}$. In fact, the numerical results of section §2.b in uniform grids indicate that with yet another choice of the matrix $\mathbb{C}$, see (5.21), the LDG method converges with the optimal order of $k + 1$ in the general case. The analysis of this phenomenon constitutes the subject of ongoing work.

5.3 Numerical results in the one-dimensional case

In this section we present some numerical results for the schemes discussed in this paper. We will only provide results for the following one dimensional, linear convection diffusion equation

$$\partial_t u + c \partial_x u - a \partial_x^2 u = 0 \quad \text{in } (0,T) \times (0,2\pi),$$
$$u(t=0,x) = \sin(x), \quad \text{on } (0,2\pi),$$

where c and $a \geq 0$ are both constants; periodic boundary conditions are used. The exact solution is $u(t,x) = e^{-at} \sin(x - ct)$. We compute the solution up to $T = 2$, and use the LDG method with $\mathbb{C}$ defined by

$$\mathbb{C} = \begin{pmatrix} \frac{|c|}{2} & -\frac{\sqrt{a}}{2} \\ \frac{\sqrt{a}}{2} & 0 \end{pmatrix}. \tag{5.21}$$

We notice that, for this choice of fluxes, the approximation to the convective term cu_x is the standard upwinding, and that the approximation to the diffusion term $a \partial_x^2 u$ is the standard three point central difference, for the P^0 case. On the other hand, if one uses a central flux corresponding to $c_{12} = -c_{21} = 0$, one gets a spread-out five point central difference approximation to the diffusion term $a \partial_x^2 u$.

The LDG methods based on P^k, with $k = 1, 2, 3, 4$ are tested. Elements with equal size are used. Time discretization is by the third-order accurate TVD Runge-Kutta method [58], with a sufficiently small time step so that error in time is negligible comparing with spatial errors. We list the L_∞ errors and numerical orders of accuracy, for u_h, as well as for its derivatives suitably scaled $\Delta x^m \partial_x^m u_h$ for $1 \leq m \leq k$, at the center of of each element. This gives the complete description of the error for u_h over the whole domain, as u_h in each element is a polynomial of degree k. We also list the L_∞ errors and numerical orders of accuracy for q_h at the element center.

In all the convection-diffusion runs with $a > 0$, accuracy of at least $(k+1)$-th order is obtained, for both u_h and q_h, when P^k elements are used. See Tables 1 to 3. The P^4 case for the purely convection equation $a = 0$ seems to be not in the asymptotic regime yet with $N = 40$ elements (further refinement with $N = 80$ suffers from round-off effects due to our choice of non-orthogonal basis functions), Table 4. However, the absolute values of the errors are comparable with the convection dominated case in Table 3.

Table 1. The heat equation $a = 1$, $c = 0$. L_∞ errors and numerical order of accuracy, measured at the center of each element, for $\Delta x^m \partial_x^m u_h$ for $0 \le m \le k$, and for q_h.

k	variable	$N = 10$	$N = 20$		$N = 40$	
		error	error	order	error	order
1	u	4.55E-4	5.79E-5	2.97	7.27E-6	2.99
	$\Delta x\, \partial_x u$	9.01E-3	2.22E-3	2.02	5.56E-4	2.00
	q	4.17E-5	2.48E-6	4.07	1.53E-7	4.02
2	u	1.43E-4	1.76E-5	3.02	2.19E-6	3.01
	$\Delta x\, \partial_x u$	7.87E-4	1.03E-4	2.93	1.31E-5	2.98
	$(\Delta x)^2\, \partial_x^2 u$	1.64E-3	2.09E-4	2.98	2.62E-5	2.99
	q	1.42E-4	1.76E-5	3.01	2.19E-6	3.01
3	u	1.54E-5	9.66E-7	4.00	6.11E-8	3.98
	$\Delta x\, \partial_x u$	3.77E-5	2.36E-6	3.99	1.47E-7	4.00
	$(\Delta x)^2\, \partial_x^2 u$	1.90E-4	1.17E-5	4.02	7.34E-7	3.99
	$(\Delta x)^3\, \partial_x^3 u$	2.51E-4	1.56E-5	4.00	9.80E-7	4.00
	q	1.48E-5	9.66E-7	3.93	6.11E-8	3.98
4	u	2.02E-7	5.51E-9	5.20	1.63E-10	5.07
	$\Delta x\, \partial_x u$	1.65E-6	5.14E-8	5.00	1.61E-9	5.00
	$(\Delta x)^2\, \partial_x^2 u$	6.34E-6	2.04E-7	4.96	6.40E-9	4.99
	$(\Delta x)^3\, \partial_x^3 u$	2.92E-5	9.47E-7	4.95	2.99E-8	4.99
	$(\Delta x)^4\, \partial_x^4 u$	3.03E-5	9.55E-7	4.98	2.99E-8	5.00
	q	2.10E-7	5.51E-9	5.25	1.63E-10	5.07

Table 2. The convection diffusion equation $a = 1$, $c = 1$. L_∞ errors and numerical order of accuracy, measured at the center of each element, for $\Delta x^m \partial_x^m u_h$ for $0 \le m \le k$, and for q_h.

k	variable	$N = 10$ error	$N = 20$ error	$N = 20$ order	$N = 40$ error	$N = 40$ order
1	u	6.47E-4	1.25E-4	2.37	1.59E-5	2.97
	$\Delta x\, \partial_x u$	9.61E-3	2.24E-3	2.10	5.56E-4	2.01
	q	2.96E-3	1.20E-4	4.63	1.47E-5	3.02
2	u	1.42E-4	1.76E-5	3.02	2.18E-6	3.01
	$\Delta x\, \partial_x u$	7.93E-4	1.04E-4	2.93	1.31E-5	2.99
	$(\Delta x)^2\, \partial_x^2 u$	1.61E-3	2.09E-4	2.94	2.62E-5	3.00
	q	1.26E-4	1.63E-5	2.94	2.12E-6	2.95
3	u	1.53E-5	9.75E-7	3.98	6.12E-8	3.99
	$\Delta x\, \partial_x u$	3.84E-5	2.34E-6	4.04	1.47E-7	3.99
	$(\Delta x)^2\, \partial_x^2 u$	1.89E-4	1.18E-5	4.00	7.36E-7	4.00
	$(\Delta x)^3\, \partial_x^3 u$	2.52E-4	1.56E-5	4.01	9.81E-7	3.99
	q	1.57E-5	9.93E-7	3.98	6.17E-8	4.01
4	u	2.04E-7	5.50E-9	5.22	1.64E-10	5.07
	$\Delta x\, \partial_x u$	1.68E-6	5.19E-8	5.01	1.61E-9	5.01
	$(\Delta x)^2\, \partial_x^2 u$	6.36E-6	2.05E-7	4.96	6.42E-8	5.00
	$(\Delta x)^3\, \partial_x^3 u$	2.99E-5	9.57E-7	4.97	2.99E-8	5.00
	$(\Delta x)^4\, \partial_x^4 u$	2.94E-5	9.55E-7	4.95	3.00E-8	4.99
	q	1.96E-7	5.35E-9	5.19	1.61E-10	5.06

Table 3. The convection dominated convection diffusion equation $a = 0.01$, $c = 1$. L_∞ errors and numerical order of accuracy, measured at the center of each element, for $\Delta x^m \partial_x^m u_h$ for $0 \le m \le k$, and for q_h.

k	variable	$N = 10$ error	$N = 20$ error	$N = 20$ order	$N = 40$ error	$N = 40$ order
1	u	7.14E-3	9.30E-4	2.94	1.17E-4	2.98
	$\Delta x\, \partial_x u$	6.04E-2	1.58E-2	1.93	4.02E-3	1.98
	q	8.68E-4	1.09E-4	3.00	1.31E-5	3.05
2	u	9.59E-4	1.25E-4	2.94	1.58E-5	2.99
	$\Delta x\, \partial_x u$	5.88E-3	7.55E-4	2.96	9.47E-5	3.00
	$(\Delta x)^2\, \partial_x^2 u$	1.20E-2	1.50E-3	3.00	1.90E-4	2.98
	q	8.99E-5	1.11E-5	3.01	1.10E-6	3.34
3	u	1.11E-4	7.07E-6	3.97	4.43E-7	4.00
	$\Delta x\, \partial_x u$	2.52E-4	1.71E-5	3.88	1.07E-6	4.00
	$(\Delta x)^2\, \partial_x^2 u$	1.37E-3	8.54E-5	4.00	5.33E-6	4.00
	$(\Delta x)^3\, \partial_x^3 u$	1.75E-3	1.13E-4	3.95	7.11E-6	3.99
	q	1.18E-5	7.28E-7	4.02	4.75E-8	3.94
4	u	1.85E-6	4.02E-8	5.53	1.19E-9	5.08
	$\Delta x\, \partial_x u$	1.29E-5	3.76E-7	5.10	1.16E-8	5.01
	$(\Delta x)^2\, \partial_x^2 u$	5.19E-5	1.48E-6	5.13	4.65E-8	4.99
	$(\Delta x)^3\, \partial_x^3 u$	2.21E-4	6.93E-6	4.99	2.17E-7	5.00
	$(\Delta x)^4\, \partial_x^4 u$	2.25E-4	6.89E-6	5.03	2.17E-7	4.99
	q	3.58E-7	3.06E-9	6.87	5.05E-11	5.92

Table 4. The convection equation $a = 0$, $c = 1$. L_∞ errors and numerical order of accuracy, measured at the center of each element, for $\Delta x^m \partial_x^m u_h$ for $0 \le m \le k$.

k	variable	$N = 10$	$N = 20$		$N = 40$	
		error	error	order	error	order
1	u	7.24E-3	9.46E-4	2.94	1.20E-4	2.98
	$\Delta x\,\partial_x u$	6.09E-2	1.60E-2	1.92	4.09E-3	1.97
2	u	9.96E-4	1.28E-4	2.96	1.61E-5	2.99
	$\Delta x\,\partial_x u$	6.00E-3	7.71E-4	2.96	9.67E-5	3.00
	$(\Delta x)^2\,\partial_x^2 u$	1.23E-2	1.54E-3	3.00	1.94E-4	2.99
3	u	1.26E-4	7.50E-6	4.07	4.54E-7	4.05
	$\Delta x\,\partial_x u$	1.63E-4	2.00E-5	3.03	1.07E-6	4.21
	$(\Delta x)^2\,\partial_x^2 u$	1.52E-3	9.03E-5	4.07	5.45E-6	4.05
	$(\Delta x)^3\,\partial_x^3 u$	1.35E-3	1.24E-4	3.45	7.19E-6	4.10
4	u	3.55E-6	8.59E-8	5.37	3.28E-10	8.03
	$\Delta x\,\partial_x u$	1.89E-5	1.27E-7	7.22	1.54E-8	3.05
	$(\Delta x)^2\,\partial_x^2 u$	8.49E-5	2.28E-6	5.22	2.33E-8	6.61
	$(\Delta x)^3\,\partial_x^3 u$	2.36E-4	5.77E-6	5.36	2.34E-7	4.62
	$(\Delta x)^4\,\partial_x^4 u$	2.80E-4	8.93E-6	4.97	1.70E-7	5.72

Finally, to show that the order of accuracy could really degenerate to k for P^k, as was already observed in [4], we rerun the heat equation case $a = 1, c = 0$ with the central flux

$$\mathbb{C} = \begin{pmatrix} 0 & 0 \\ 0 & 0 \end{pmatrix}. \tag{5.22}$$

This time we can see that the global order of accuracy in L_∞ is only k when P^k is used with an odd value of k.

Table 5. The heat equation $a = 1$, $c = 0$. L_∞ errors and numerical order of accuracy, measured at the center of each element, for $\Delta x^m \partial_x^m u_h$ for $0 \leq m \leq k$, and for q_h, using the central flux.

k	variable	$N = 10$ error	$N = 20$ error	$N = 20$ order	$N = 40$ error	$N = 40$ order
1	u	3.59E-3	8.92E-4	2.01	2.25E-4	1.98
	$\Delta x\, \partial_x u$	2.10E-2	1.06E-2	0.98	5.31E-3	1.00
	q	2.39E-3	6.19E-4	1.95	1.56E-4	1.99
2	u	6.91E-5	4.12E-6	4.07	2.57E-7	4.00
	$\Delta x\, \partial_x u$	7.66E-4	1.03E-4	2.90	1.30E-5	2.98
	$(\Delta x)^2\, \partial_x^2 u$	2.98E-4	1.68E-5	4.15	1.03E-6	4.02
	q	6.52E-5	4.11E-6	3.99	2.57E-7	4.00
3	u	1.62E-5	1.01E-6	4.00	6.41E-8	3.98
	$\Delta x\, \partial_x u$	1.06E-4	1.32E-5	3.01	1.64E-6	3.00
	$(\Delta x)^2\, \partial_x^2 u$	1.99E-4	1.22E-5	4.03	7.70E-7	3.99
	$(\Delta x)^3\, \partial_x^3 u$	6.81E-4	8.68E-5	2.97	1.09E-5	2.99
	q	1.54E-5	1.01E-6	3.93	6.41E-8	3.98
4	u	8.25E-8	1.31E-9	5.97	2.11E-11	5.96
	$\Delta x\, \partial_x u$	1.62E-6	5.12E-8	4.98	1.60E-9	5.00
	$(\Delta x)^2\, \partial_x^2 u$	1.61E-6	2.41E-8	6.06	3.78E-10	6.00
	$(\Delta x)^3\, \partial_x^3 u$	2.90E-5	9.46E-7	4.94	2.99E-8	4.99
	$(\Delta x)^4\, \partial_x^4 u$	5.23E-6	7.59E-8	6.11	1.18E-9	6.01
	q	7.85E-8	1.31E-9	5.90	2.11E-11	5.96

5.4 The LDG methods for the multi-dimensional case

In this section, we consider the LDG methods for the following convection-diffusion model problem

$$\partial_t u + \sum_{1 \leq i \leq d} \partial_{x_i} \left(f_i(u) - \sum_{1 \leq j \leq d} a_{ij}(u) \, \partial_{x_j} u \right) = 0 \quad \text{in } (0,T) \times (0,1)^d \quad (5.23)$$

$$u(t=0) = u_0, \quad \text{on } (0,1)^d, \qquad (5.24)$$

with periodic boundary conditions. Essentially, the one-dimensional case and the multidimensional case can be studied in exactly the same way. However, there are two important differences that deserve explicit discussion. The first is the treatment of the matrix of entries $a_{ij}(u)$, which is assumed to be *symmetric, semipositive definite* and the introduction of the variables q_ℓ, and the second is the treatment of arbitrary meshes. 4 To define the LDG method, we first notice that, since the matrix $a_{ij}(u)$ is assumed to be symmetric and semipositive definite, there exists a symmetric matrix $b_{ij}(u)$ such that

$$a_{ij}(u) = \sum_{1 \leq \ell \leq d} b_{i\ell}(u) \, b_{\ell j}(u). \qquad (5.25)$$

Then we define the new scalar variables $q_\ell = \sum_{1 \leq j \leq d} b_{\ell j}(u) \, \partial_{x_j} u$ and rewrite the problem (5.23), (5.24) as follows:

$$\partial_t u + \sum_{1 \leq i \leq d} \partial_{x_i} \left(f_i(u) - \sum_{1 \leq \ell \leq d} b_{i\ell}(u) \, q_\ell \right) = 0 \quad \text{in } (0,T) \times (0,1)^d, (5.26)$$

$$q_\ell - \sum_{1 \leq j \leq d} \partial_{x_j} g_{\ell j}(u) = 0, \quad \ell = 1, \ldots d, \quad \text{in } (0,T) \times (0,1)^d, \qquad (5.27)$$

$$u(t=0) = u_0, \quad \text{on } (0,1)^d, \qquad (5.28)$$

where $g_{\ell j}(u) = \int^u b_{\ell j}(s) \, ds$. The LDG method is now obtained by discretizing the above equations by the Discontinuous Galerkin method.

We follow what was done in §2. So, we set $\mathbf{w} = (u, \mathbf{q})^t = (u, q_1, \cdots, q_d)^t$ and, for each $i = 1, \cdots, d$, introduce the flux

$$\mathbf{h}_i(\mathbf{w}) = \left(f_i(u) - \sum_{1 \leq \ell \leq d} b_{i\ell}(u) \, q_\ell, -g_{1i}(u), \cdots, -g_{di}(u) \right)^t. \quad (5.29)$$

We consider triangulations of $(0,1)^d$, $\mathbb{T}_{\Delta x} = \{K\}$, made of non-overlapping polyhedra. We require that for any two elements K and K', $\overline{K} \cap \overline{K}'$ is either a face e of both K and K' with nonzero $(d-1)$-Lebesgue measure $|e|$, or has Hausdorff dimension less than $d-1$. We denote by $\mathbb{E}_{\Delta x}$ the set of all faces e of the border of K for all $K \in \mathbb{T}_{\Delta x}$. The diameter of K is denoted by Δx_K and the maximum Δx_K, for $K \in \mathbb{T}_{\Delta x}$ is denoted by Δx. We require, for the sake of simplicity, that the triangulations $\mathbb{T}_{\Delta x}$ be regular, that is, there is a constant independent of Δx such that

$$\frac{\Delta x_K}{\rho_K} \leq \sigma \quad \forall K \in \mathbb{T}_{\Delta x},$$

where ρ_K denotes the diameter of the maximum ball included in K.

We seek an approximation $\mathbf{w}_h = (u_h, \mathbf{q}_h)^t = (u_h, q_{h1}, \cdots, q_{hd})^t$ to $\mathbf{w}$ such that for each time $t \in [0, T]$, each of the components of $\mathbf{w}_h$ belong to the finite element space

$$V_h = V_h^k = \{ v \in L^1((0,1)^d) : v|_K \in P^k(K) \ \forall \ K \in \mathbb{T}_{\Delta x} \}, \qquad (5.30)$$

where $P^k(K)$ denotes the space of polynomials of total degree at most k. In order to determine the approximate solution $\mathbf{w}_h$, we proceed exactly as in the one-dimensional case. This time, however, the integrals are made on each element K of the triangulation $\mathbb{T}_{\Delta x}$. We obtain the following weak formulation on each element K of the triangulation $\mathbb{T}_{\Delta x}$:

$$\int_K \partial_t u_h(x,t)\, v_{h,u}(x)\, dx - \sum_{1 \leq i \leq d} \int_K h_{iu}(\mathbf{w}_h(x,t))\, \partial_{x_i} v_{h,u}(x)\, dx$$
$$+ \int_{\partial K} \hat{h}_u(\mathbf{w}_h, \mathbf{n}_{\partial K})(x,t)\, v_{h,u}(x)\, d\,\Gamma(x) = 0, \qquad \forall\, v_{h,u} \in P^k(K), (5.31)$$

$$\text{For } \ell = 1, \cdots, d:$$
$$\int_K q_{h\ell}(x,t)\, v_{h,q_\ell}(x)\, dx - \sum_{1 \leq j \leq d} \int_K h_{j\,q_\ell}(\mathbf{w}_h(x,t))\, \partial_{x_j} v_{h,q_\ell}(x)\, dx$$
$$+ \int_{\partial K} \hat{h}_{q_\ell}(\mathbf{w}_h, \mathbf{n}_{\partial K})(x,t)\, v_{h,q_\ell}(x)\, d\,\Gamma(x) = 0, \quad \forall\, v_{h,q_\ell} \in P^k(K), (5.32)$$
$$\int_K u_h(x,0)\, v_{h,i}(x)\, dx = \int_K u_0(x)\, v_{h,i}(x)\, dx, \qquad \forall\, v_{h,i} \in P^k(K), (5.33)$$

where $\mathbf{n}_{\partial K}$ denotes the outward unit normal to the element K at $x \in \partial K$. It remains to choose the numerical flux $(\hat{h}_u, \hat{h}_{q_1}, \cdots, \hat{h}_{q_d})^t \equiv \hat{\mathbf{h}} \equiv \hat{\mathbf{h}}(\mathbf{w}_h, \mathbf{n}_{\partial K})(x,t)$.

As in the one-dimensional case, we require that the fluxes $\hat{\mathbf{h}}$ be of the form

$$\hat{\mathbf{h}}(\mathbf{w}_h, \mathbf{n}_{\partial K})(x) \equiv \hat{\mathbf{h}}(\mathbf{w}_h(x^{int_K}, t), \mathbf{w}_h(x^{ext_K}, t); \mathbf{n}_{\partial K}),$$

where $\mathbf{w}_h(x^{int_K})$ is the limit at x taken from the interior of K and $\mathbf{w}_h(x^{ext_K})$ the limit at x from the exterior of K, and consider fluxes that (i) are locally Lipschitz, conservative, that is,

$$\hat{\mathbf{h}}(\mathbf{w}_h(x^{int_K}), \mathbf{w}_h(x^{ext_K}); \mathbf{n}_{\partial K}) + \hat{\mathbf{h}}(\mathbf{w}_h(x^{ext_K}), \mathbf{w}_h(x^{int_K}); -\mathbf{n}_{\partial K}) = 0,$$

and consistent with the flux

$$\sum_{1 \leq i \leq d} \mathbf{h}_i\, n_{\partial K, i},$$

(ii) allow for a local resolution of each component of $\mathbf{q}_h$ in terms of u_h *only*, (iii) reduce to an E-flux when $a(\cdot) \equiv 0$, and that (iv) enforce the L^2-stability of the method.

Again, we write our numerical flux as the sum of a convective flux and a diffusive flux:

$$\hat{\mathbf{h}} = \hat{\mathbf{h}}_{conv} + \hat{\mathbf{h}}_{diff},$$

where the convective flux is given by

$$\hat{\mathbf{h}}_{conv}(\mathbf{w}^-, \mathbf{w}^+; \mathbf{n}) = (\hat{f}(u^-, u^+; \mathbf{n}), 0)^t,$$

where $\hat{f}(u^-, u^+; \mathbf{n})$ is any locally Lipschitz E-flux which is conservative and consistent with the nonlinearity

$$\sum_{1 \leq i \leq d} f_i(u)\, n_i,$$

and the diffusive flux $\hat{\mathbf{h}}_{diff}(\mathbf{w}^-, \mathbf{w}^+; \mathbf{n})$ is given by

$$\left(- \sum_{1 \leq i, \ell \leq d} \frac{[g_{i\ell}(u)]}{[u]} \overline{q_\ell}\, n_i, \ -\sum_{1 \leq i \leq d} \overline{g_{i1}(u)}\, n_i, \cdots, \ -\sum_{1 \leq i \leq d} \overline{g_{id}(u)}\, n_i \right)^t - \mathbb{C}_{diff}[\mathbf{w}],$$

where

$$\mathbb{C}_{diff} = \begin{pmatrix} 0 & c_{12} & c_{13} & \cdots & c_{1d} \\ -c_{12} & 0 & 0 & \cdots & 0 \\ -c_{13} & 0 & 0 & \cdots & 0 \\ \vdots & \vdots & \vdots & \ddots & \vdots \\ -c_{1d} & 0 & 0 & \cdots & 0 \end{pmatrix},$$

$$c_{1j} = c_{1j}(\mathbf{w}^-, \mathbf{w}^+) \quad \text{is locally Lipschitz for } j = 1, \cdots, d,$$

$$c_{1j} \equiv 0 \quad \text{when } a(\cdot) \equiv 0 \quad \text{for } j = 1, \cdots, d.$$

We claim that this flux satisfies the properties (i) to (iv).

To prove that properties (i) to (iii) are satisfied is now a simple exercise. To see that the property (iv) is satisfied, we first rewrite the flux $\hat{\mathbf{h}}$ in the following way:

$$\left(- \sum_{1 \le i, \ell \le d} \frac{[g_{i\ell}(u)]}{[u]} \overline{q_\ell}\, n_i, \ - \sum_{1 \le i \le d} \overline{g_{i1}(u)}\, n_i, \cdots, \ - \sum_{1 \le i \le d} \overline{g_{id}(u)}\, n_i \right)^t - \mathbb{C}[\mathbf{w}],$$

where

$$\mathbb{C} = \begin{pmatrix} c_{11} & c_{12} & c_{13} & \cdots & c_{1d} \\ -c_{12} & 0 & 0 & \cdots & 0 \\ -c_{13} & 0 & 0 & \cdots & 0 \\ \vdots & \vdots & \vdots & \ddots & \vdots \\ -c_{1d} & 0 & 0 & \cdots & 0 \end{pmatrix},$$

$$c_{11} = \frac{1}{[u]}\left(\sum_{1 \le i \le d} \frac{[\varphi_i(u)]}{[u]}\, n_i - \hat{f}(u^-, u^+; \mathbf{n}) \right),$$

where $\varphi_i(u) = \int^u f_i(s)\, ds$. Since $\hat{f}(\cdot, \cdot; \mathbf{n})$ is an E-flux,

$$c_{11} = \frac{1}{[u]^2} \int_{u^-}^{u^+} \left(\sum_{1 \le i \le d} f_i(s)\, n_i - \hat{f}(u^-, u^+; \mathbf{n}) \right) ds \ge 0,$$

and so the matrix $\mathbb{C}$ is semipositive definite. The property (iv) follows from this fact and from the following result.

Theorem 5.3 *We have,*

$$\frac{1}{2} \int_{(0,1)^d} u_h^2(x, T)\, dx + \int_0^T \int_{(0,1)^d} |\, q_h(x, t)\,|^2\, dx\, dt + \Theta_{T, \mathbb{C}}([\mathbf{w}_h]) \le \frac{1}{2} \int_{(0,1)^d} u_0^2(x)\, dx,$$

where

$$\Theta_{T,\mathbb{C}}([\mathbf{w}_h]) = \int_0^T \sum_{e \in \mathbb{E}_{\Delta x}} \int_e [\mathbf{w}_h(x,t)]^t \mathbb{C} \, [\mathbf{w}_h(x,t)] \, d\,\Gamma(x)\, dt.$$

We can also prove the following error estimate. We denote the integral over $(0,1)^d$ of the sum of the squares of all the derivatives of order $(k+1)$ of u by $|u|^2_{k+1}$.

Theorem 5.4 *Let* **e** *be the approximation error* $\mathbf{w} - \mathbf{w}_h$. *Then we have, for arbitrary, regular grids,*

$$\left\{ \int_{(0,1)^d} |\mathbf{e}_u(x,T)|^2 \, dx + \int_0^T \int_{(0,1)^d} |\mathbf{e}_q(x,t)|^2 \, dx\, dt + \Theta_{T,\mathbb{C}}([\mathbf{e}]) \right\}^{1/2} \leq C\,(\Delta x)^k,$$

where $C = C(k, |u|_{k+1}, |u|_{k+2})$. *In the purely hyperbolic case* $a_{ij} = 0$, *the constant* C *is of order* $(\Delta x)^{1/2}$. *In the purely parabolic case* $c = 0$, *the constant* C *is of order* Δx *for even values of* k *and of order* 1 *otherwise for Cartesian products of uniform grids and for* $\mathbb{C}$ *identically zero provided that the local spaces* Q^k *are used instead of the spaces* P^k, *where* Q^k *is the space of tensor products of one dimensional polynomials of degree* k.

5.5 Extension to multidimensional systems

In this chapter, we have considered the so-called LDG methods for convection-diffusion problems. For scalar problems in multidimensions, we have shown that they are L^2-stable and that in the linear case, they are of order k if polynomials of order k are used. We have also shown that this estimate is sharp and have displayed the strong dependence of the order of convergence of the LDG methods on the choice of the numerical fluxes.

The main advantage of these methods is their extremely high parallelizability and their high-order accuracy which render them suitable for computations of convection-dominated flows. Indeed, although the LDG method have a large amount of degrees of freedom per element, and hence more computations per element are necessary, its extremely local domain of dependency allows a very efficient parallelization that by far compensates for the extra amount of local computations.

The LDG methods for multidimensional systems, like for example the compressible Navier-Stokes equations and the equations of the hydrodynamic model for semiconductor device simulation, can be easily defined by simply applying the procedure described for the multidimensional scalar case to each component of **u**. In practice, especially for viscous terms which are not symmetric but still semipositive definite, such as for the compressible Navier-Stokes equations, we can use $\mathbf{q} = (\partial_{x_1} u, ..., \partial_{x_d} u)$ as the auxilary variables. Although with this choice, the L^2-stability result will not be available theoretically, this would not cause any problem in practical implementations.

5.6 Some numerical results

Next, we present some numerical results from the papers by Bassi and Rebay [3] and Lomtev and Karniadakis [46].

• **Smooth, steady state solutions.** We start by displaying the convergence of the method for a p-refinement done by Lomtev and Karniadakis [46]. In Figure 5.23, we can see how the maximum errors in density, momentum, and energy decrease exponentially to zero as the degree k of the approximating polynomials increases while the grid is kept fixed; details about the exact solution can be found in [46].

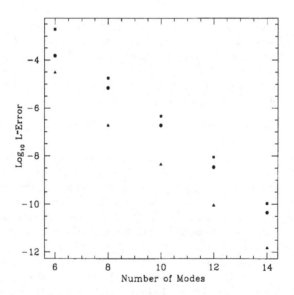

Fig. 5.23: Maximum errors of the density (triangles), momemtum (circles) and energy (squares) as a function of the degree of the approximating polynomial plus one (called "number of modes" in the picture).

Now, let us consider the laminar, transonic flow around the NACA0012 airfoil at an angle of attack of ten degrees, freestream Mach number $M =$

0.8, and Reynolds number (based on the freestream velocity and the airfoil chord) equal to 73; the wall temperature is set equal to the freestream total temperature. Bassy and Rebay [3] have computed the solution of this problem with polynomials of degree 1, 2, and 3 and Lomtev and Karniadakis [46] have tried the same test problem with polynomials of degree 2, 4, and 6 in a mesh of 592 elements which is about four times less elements than the mesh used by Bassi and Rebay [3]. In Figure 5.25, taken from [46], we display the pressure and drag coefficient distributions computed by Bassi and Rebay [3] with polynomials on degree 3 and the ones computed by Lomtev and Karniadakis [46] computed with polynomials of degree 6. We can see good agreement of both computations. In Figure 5.24, taken from [46], we see the mesh and the Mach isolines obtained with polynomials of degree two and four; note the improvement of the solution.

Next, we show a result from the paper by Bassi and Rebay [3]. We consider the laminar, subsonic flow around the NACA0012 airfoil at an angle of attack of zero degrees, freestream Mach number $M = 0.5$, and Reynolds number equal to 5000. In figure 5.26, we can see the Mach isolines corresponding to linear, quadratic, and cubic elements. In the figures 5.27, 5.28, and 5.29 details of the results with cubic elements are shown. Note how the boundary layer is captured withing a few layers of elements and how its separation at the trailing edge of the airfoil has been clearly resolved. Bassi and Rebay [3] report that these results are comparable to common structured and unstructures finite volume methods on much finer grids- a result consistent with the computational results we have displayed in these notes.

Finally, we present a not-yet-published result kindly provided by Lomtev and Karniadakis about the simulation of an expansion pipe flow. The smaller cylinder has a diameter of 1 and the larger cylinder has a diameter of 2. In Figure 5.30, we display the velocity profile and some streamlines for a Reynolds number equal to 50 and Mach number 0.2. The computation was made with polynomials of degree 5 and a mesh of 600 tetrahedra; of course the tetrahedra have curved faces to accomodate the exact boundaries. In Figure 5.31, we display a comparison between computational and experimental results. As a function of the Reynolds number, two quantities are plotted. The first is the distance between the step and the center of the vertex (lower brach) and the second is the distance from the step to the separation point (upper branch). The computational results are obtained by the method under consideration with polynomials of degree 5 for the compressible Navier Stokes equations, and by a standard Galerkin formulation in terms of velocity-pressure (NEKTAR), by Sherwin and Karniadakis [56], or in terms of velocity-vorticity (IVVA), by Trujillo [61], for the *incompressible* Navier Stokes equations; results produced by the code called PRISM are also included, see Newmann [48]. The experimental data was taken from Macagno and Tung [49]. The agreement between computations and experiments is remarkable.

• **Unsteady solutions.** To end this chapter, we present the computation of an unsteady solution by Lomtev and Karniadakis [46]. The test problem is the classical problem of a flow around a cylinder in two space dimensions. The Reynolds number is 10, 000 and the Mach number 0.2.

In Figure 5.32, the streamlines are shown for a computation made on a grid of 680 triangles (with curved sides fitting the cylinder) and polynomials whose degree could vary from element to element; the maximum degree was 5. In Figure 5.33, details of the mesh and the density around the cylinder are shown. Note how the method is able to capture the shear layer instability observed experimentally. For more details, see [46].

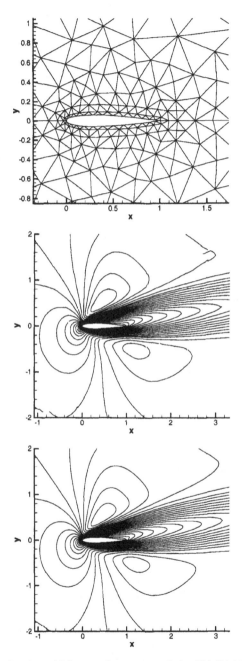

Fig. 5.24: Mesh (top) and Mach isolines around the NACA0012 airfoil, ($Re =$ 73, $M = 0.8$, angle of attack of ten degrees) for quadratic (middle) and quartic (bottom) elements.

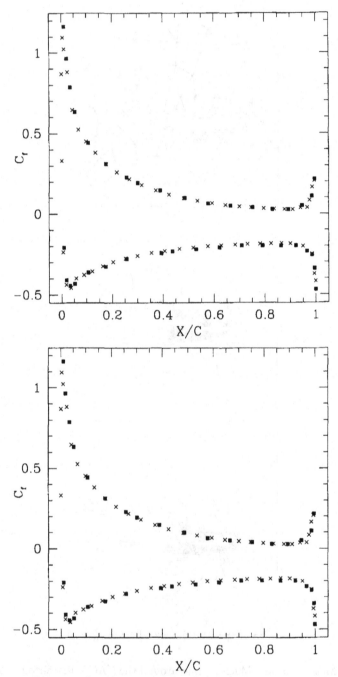

Fig. 5.25: Pressure (top) and drag(bottom) coefficient distributions. The squares were obtained by Bassi and Rebay [3] with cubics and the crosses by Lomtev and Karniadakis [46] with polynomials of degree 6.

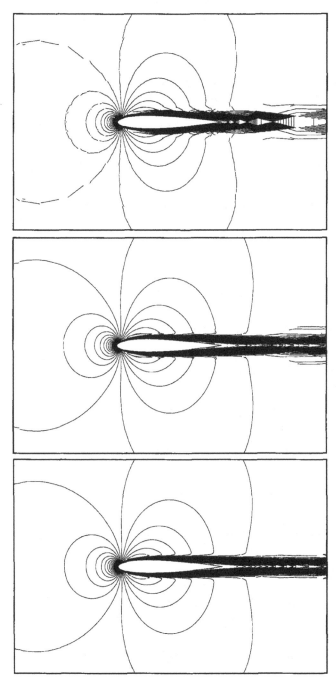

Fig. 5.26: Mach isolines around the NACA0012 airfoil, ($Re = 5000, M = 0.5$, zero angle of attack) for the linear (top), quadratic (middle), and cubic (bottom) elements.

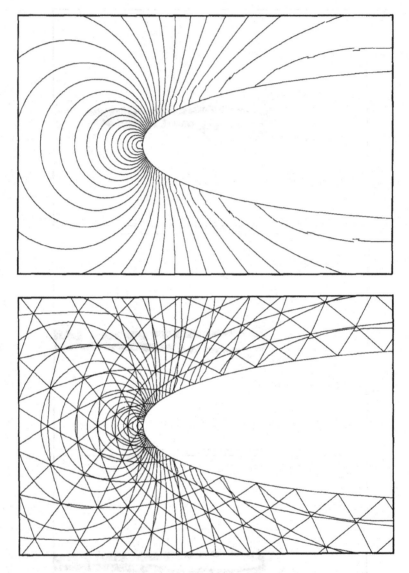

Fig. 5.27: Pressure isolines around the NACA0012 airfoil, ($Re = 5000, M = 0.5$, zero angle of attack) for the for cubic elements without (top) and with (bottom) the corresponding grid.

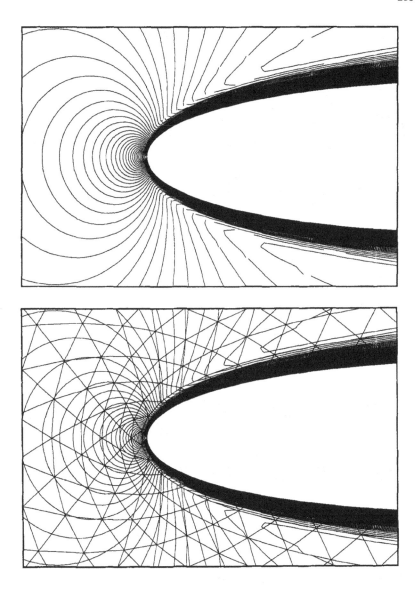

Fig. 5.28: Mach isolines around the leading edge of the NACA0012 airfoil, ($Re = 5000$, $M = 0.5$, zero angle of attack) for the for cubic elements without (top) and with (bottom) the corresponding grid.

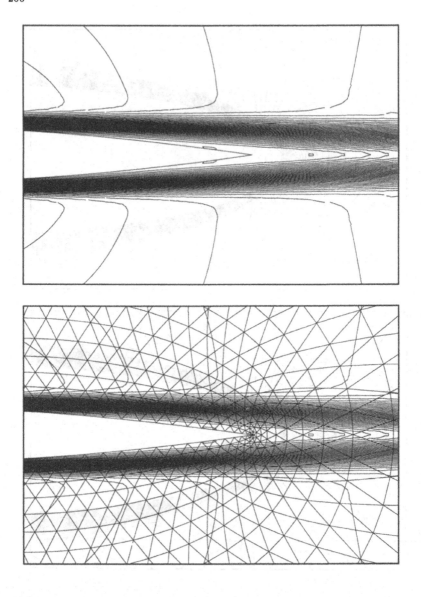

Fig. 5.29: Mach isolines around the trailing edge of the NACA0012 airfoil, ($Re = 5000, M = 0.5$, zero angle of attack) for the for cubic elements without (top) and with (bottom) the corresponding grid.

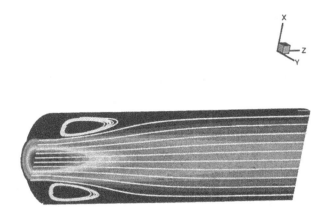

Fig. 5.30: Expansion pipe flow at Reynolds number 50 and Mach number 0.2. Velocity profile and streamlines computed with a mesh of 600 elements and polynomials of degree 5.

262

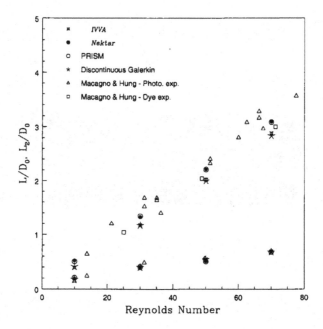

Fig. 5.31: Expansion pipe flow: Comparison between computational and experimental results.

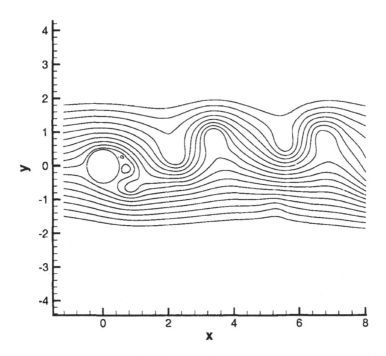

Fig. 5.32: Flow around a cylinder with Reynolds number 10,000 and Mach number 0.2. Streamlines. A mesh of 680 elements was used with polynomials that could change degree from element to element; the maximum degree was 5.

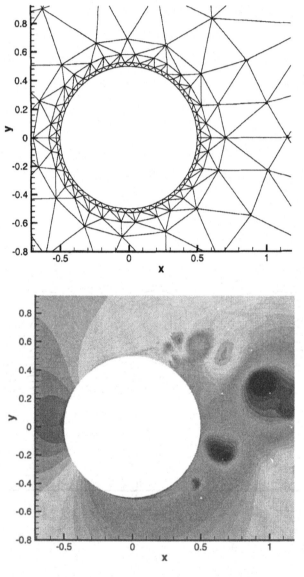

Fig. 5.33: Flow around a cylinder with Reynolds number 10,000 and Mach number 0.2. Detail of the mesh (top) and density (bottom) around the cylinder.

References

1. H.L. Atkins and C.-W. Shu. Quadrature-free implementation of discontinuous Galerkin methods for hyperbolic equations. *ICASE Report 96-51*, 1996. submitted to AIAA J.

2. F. Bassi and S. Rebay. High-order accurate discontinuous finite element solution of the 2d Euler equations. *J. Comput. Phys.* to appear.

3. F. Bassi and S. Rebay. A high-order accurate discontinuous finite element method for the numerical solution of the compressible Navier-Stokes equations. *J. Comput. Phys*, 131:267–279, 1997.

4. F. Bassi, S. Rebay, M. Savini, G. Mariotti, and S. Pedinotti. A high-order accurate discontinuous finite element method for inviscid and viscous turbomachinery flows. *Proceedings of the Second European Conference ASME on Turbomachinery Fluid Dynamics and Thermodynamics*, 1995.

5. K.S. Bey and J.T. Oden. A Runge-Kutta discontinuous Galerkin finite element method for high speed flows. *info AIAA 10th Computational Fluid Dynamics Conference, Honolulu, Hawaii, June 24-27*, 1991.

6. R. Biswas, K.D. Devine, and J. Flaherty. Parallel, adaptive finite element methods for conservation laws. *Applied Numerical Mathematics*, 14:255–283, 1994.

7. G. Chavent and B. Cockburn. The local projection p^0 p^1-discontinuous-Galerkin finite element method for scalar conservation laws. *M^2AN*, 23:565–592, 1989.

8. G. Chavent and G. Salzano. A finite element method for the 1d water flooding problem with gravity. *J. Comput. Phys*, 45:307–344, 1982.

9. Z. Chen, B. Cockburn, C. Gardner, and J. Jerome. Quantum hydrodynamic simulation of hysteresis in the resonant tunneling diode. *J. Comput. Phys*, 117:274–280, 1995.

10. Z. Chen, B. Cockburn, J. Jerome, and C.-W. Shu. Mixed-RKDG finite element method for the drift-diffusion semiconductor device equations. *VLSI Design*, 3:145–158, 1995.

11. P. Ciarlet. *The finite element method for elliptic problems*. North Holland, 1975.

12. B. Cockburn and P.-A. Gremaud. A priori error estimates for numerical methods for scalar conservation laws. part i: The general approach. *Math. Comp.*, 65:533–573, 1996.

13. B. Cockburn, S. Hou, and C.W. Shu. Tvb Runge-Kutta local projection discontinuous Galerkin finite element method for conservation laws iv: The multidimensional case. *Math. Comp.*, 54:545–581, 1990.

14. B. Cockburn, S.Y. Lin, and C.W. Shu. Tvb Runge-Kutta local projection discontinuous Galerkin finite element method for conservation laws iii: One dimensional systems. *J. Comput. Phys*, 84:90–113, 1989.

15. B. Cockburn and C.W. Shu. Tvb Runge-Kutta local projection discontinuous Galerkin finite element method for scalar conservation laws ii: General framework. *Math. Comp.*, 52:411–435, 1989.

16. B. Cockburn and C.W. Shu. The p^1-Rkdg method for two-dimensional Euler equations of gas dynamics. *ICASE Report No.91-32*, 1991.

17. B. Cockburn and C.W. Shu. The Runge-Kutta local projection p^1-discontinuous Galerkin method for scalar conservation laws. *M^2AN*, 25:337–361, 1991.

18. B. Cockburn and C.W. Shu. The local discontinuous Galerkin finite element method for convection-diffusion systems. *SIAM J. Numer. Anal.*. to appear.

19. B. Cockburn and C.W. Shu. The Runge-Kutta discontinuous Galerkin finite element method for conservation laws v: Multidimensional systems. *J. Comput. Phys.*. to appear.

20. H.L. deCougny, K.D. Devine, J.E. Flaherty, R.M. Loy, C. Ozturan, and M.S. Shephard. High-order accurate discontinuous finite element solution of the 2d Euler equations. *Applied Numerical Mathematics*, 16:157–182, 1994.

21. K.D. Devine, J.E. Flaherty, R.M. Loy, and S.R. Wheat. Parallel partitioning strategies for the adaptive solution of conservation laws. *Rensselaer Polytechnic Institute Report No. 94-1*, 1994.

22. K.D. Devine, J.E. Flaherty, S.R. Wheat, and A.B. Maccabe. A massively parallel adaptive finite element method with dynamic load balancing. *SAND 93-0936C*, 1993.

23. K. Eriksson and C. Johnson. Adaptive finite element methods for parabolic problems i: A linear model problem. *SIAM J. Numer. Anal.*, 28:43–77, 1991.

24. K. Eriksson and C. Johnson. Adaptive finite element methods for parabolic problems ii: Optimal error estimates in $l_\infty l_2$ and $l_\infty l_\infty$. *SIAM J. Numer. Anal.*, 32:706–740, 1995.

25. K. Eriksson and C. Johnson. Adaptive finite element methods for parabolic problems iv: A nonlinear model problem. *SIAM J. Numer. Anal.*, 32:1729–1749, 1995.

26. K. Eriksson and C. Johnson. Adaptive finite element methods for parabolic problems v: Long time integration. *SIAM J. Numer. Anal.*, 32:1750–1762, 1995.

27. K. Eriksson, C. Johnson, and V. Thomée. Time discretization of parabolic problems by the discontinuous Galerkin method. *RAIRO, Anal. Numér.*, 19:611–643, 1985.

28. J. Goodman and R. LeVeque. On the accuracy of stable schemes for 2d scalar conservation laws. *Math. Comp.*, 45:15–21, 1985.

29. T. Hughes and A. Brook. Streamline upwind-Petrov-Galerkin formulations for convection dominated flows with particular emphasis on the incompressible navier-stokes equations. *Comp. Meth. in App. Mech. and Eng.*, 32:199–259, 1982.

30. T. Hughes, L.P. Franca, M. Mallet, and A. Misukami. A new finite element formulation for computational fluid dynamics, i. *Comp. Meth. in App. Mech. and Eng.*, 54:223–234, 1986.

31. T. Hughes, L.P. Franca, M. Mallet, and A. Misukami. A new finite element formulation for computational fluid dynamics, ii. *Comp. Meth. in App. Mech. and Eng.*, 54:341–355, 1986.

32. T. Hughes, L.P. Franca, M. Mallet, and A. Misukami. A new finite element formulation for computational fluid dynamics, iii. *Comp. Meth. in App. Mech. and Eng.*, 58:305–328, 1986.

33. T. Hughes, L.P. Franca, M. Mallet, and A. Misukami. A new finite element formulation for computational fluid dynamics, iv. *Comp. Meth. in App. Mech. and Eng.*, 58:329–336, 1986.

34. T. Hughes and M. Mallet. A high-precision finite element method for shock-tube calculations. *Finite Element in Fluids*, 6:339–, 1985.

35. P. Jamet. Galerkin-type approximations which are discontinuous in time for parabolic equations in a variable domain. *SIAM J. Numer. Anal.*, 15:912–928, 1978.

36. G. Jiang and C.-W. Shu. On cell entropy inequality for discontinuous Galerkin methods. *Math. Comp.*, 62:531–538, 1994.

37. C. Johnson and J. Pitkaranta. An analysis of the discontinuous Galerkin method for a scalar hyperbolic equation. *Math. Comp.*, 46:1–26, 1986.

38. C. Johnson and J. Saranen. Streamline diffusion methods for problems in fluid mechanics. *Math. Comp.*, 47:1–18, 1986.

39. C. Johnson and A. Szepessy. On the convergence of a finite element method for a non-linear hyperbolic conservation law. *Math. Comp.*, 49:427–444, 1987.

40. C. Johnson, A. Szepessy, and P. Hansbo. On the convergence of shock capturing streamline diffusion finite element methods for hyperbolic conservation laws. *Math. Comp.*, 54:107–129, 1990.

41. P. LeSaint and P.A. Raviart. On a finite element method for solving the neutron transport equation. *Mathematical aspects of finite elements in partial differential equations (C. de Boor, Ed.), Academic Press*, pages 89–145, 1974.

42. W. B. Lindquist. Construction of solutions for two-dimensional riemann problems. *Comp. & Maths. with Appls.*, 12:615–630, 1986.

43. W. B. Lindquist. The scalar Riemann problem in two spatial dimensions: piecewise smoothness of solutions and its breakdown. *SIAM J. Numer. Anal.*, 17:1178–1197, 1986.

44. I. Lomtev and G.E. Karniadakis. A discontinuous spectral/hp element Galerkin method for the Navier-Stokes equations on unstructured grids. *Proc. IMACS WC'97*, Berlin, Germany, 1997.

45. I. Lomtev and G.E. Karniadakis Simulations of viscous supersonic flows on unstructured h-p meshes. *AIAA 97-0754, 35th Aerospace Sciences Meeting*, Reno, 1997.

46. I. Lomtev and G.E. Karniadakis A Discontinuous Galerkin Method for the Navier-Stokes equations. *Int. J. Num. Meth. Fluids*, submitted.

47. I. Lomtev, C.B. Quillen and G.E. Karniadakis. Spectral/hp methods for viscous compressible flows on unstructured 2D meshes. *J. Comp. Phys.*, in press.

48. D. Newmann. A Computational Study of Fluid/Structure Interactions: Flow-Induced Vibrations of a Flexible Cable Ph.D., Princeton, 1996.

49. E.O. Macagno and T. Hung. Computational and experimental study of a captive annular eddy. *J.F.M.*, 28:43 –, 1967.

50. X. Makridakis and I. Babuška. On the stability of the discontinuous Galerkin method for the heat equation. *SIAM J. Numer. Anal.*, 34:389–401, 1997.

51. S. Osher. Riemann solvers, the entropy condition and difference approximations. *SIAM J. Numer. Anal.*, 21:217–235, 1984.

52. C. Ozturan, H.L. deCougny, M.S. Shephard, and J.E. Flaherty. Parallel adaptive mesh refinement and redistribution on distributed memory computers. *Comput. Methods in Appl. Mech. and Engrg.*, 119:123–137, 1994.

53. T. Peterson. A note on the convergence of the discontinuous Galerkin method for a scalar hyperbolic equation. *SIAM J. Numer. Anal.*, 28:133–140, 1991.

54. W.H. Reed and T.R. Hill. Triangular mesh methods for the neutron transport equation. *Los Alamos Scientific Laboratory Report LA-UR-73-479*, 1973.

55. G.R. Richter. An optimal-order error estimate for the discontinuous Galerkin method. *Math. Comp.*, 50:75–88, 1988.

56. S.J. Sherwin and G. Karniadakis Tetrahedral hp finite elements: Algorithms and flow simulations. *J. Comput. Phys*, 124:314–45, 1996.

57. C.-W. Shu and S. Osher. Efficient implementation of essentially non-oscillatory shock-capturing schemes. *J. Comput. Phys*, 77:439–471, 1988.

58. C.-W. Shu and S. Osher. Efficient implementation of essentially non-oscillatory shock capturing schemes, ii. *J. Comput. Phys*, 83:32–78, 1989.

59. C.W. Shu. TVB uniformly high order schemes for conservation laws. *Math. Comp.*, 49:105–121, 1987.

60. C.W. Shu. TVD time discretizations. *SIAM J. Sci. Stat. Comput.*, 9:1073–1084, 1988.

61. J.R. Trujillo. Effective High-Order Vorticity-Velocity Formulation. Ph.D., Princeton, 1997.

62. B. van Leer. Towards the ultimate conservation difference scheme, ii. *J. Comput. Phys*, 14:361–376, 1974.

63. B. van Leer. Towards the ultimate conservation difference scheme, v. *J. Comput. Phys*, 32:1–136, 1979.

64. D. Wagner. The Riemann problem in two space dimensions for a single conservation law. *SIAM J. Math. Anal.*, 14:534–559, 1983.

65. T.C. Warburton, I. Lomtev, R.M. Kirby and G.E. Karniadakis. A discontinuous Galerkin method for the Navier-Stokes equations on hybrid grids. *Center for Fluid Mechanics # 97-14*, Division of Applied Mathematics, Brown University, 1997.

66. P. Woodward and P. Colella. The numerical simulation of two-dimensional fluid flow with strong shocks. *J. Comput. Phys*, 54:115–173, 1984.

67. T. Zhang and G.Q.. Chen. Some fundamental concepts about systems of two spatial dimensional conservation laws. *Acta Math. Sci. (English Ed.)*, 6:463–474, 1986.

68. T. Zhang and Y.X. Zheng. Two dimensional Riemann problems for a single conservation law. *Trans. Amer. Math. Soc.*, 312:589–619,1989.

Chapter 3

Adaptive Finite Element Methods for Conservation Laws

Claes Johnson

Mathematics Department
Chalmers University of Technology
412 96 Göteborg, Sweden
E-mail: claes@math.chalmers.se

ABSTRACT

The purpose of this lecture is to give an introductory overview of our recent work together with coworkers on computational methods for conservation laws, which are reliable in the sense that the computational error may be controled on a given tolerance level in a given norm, and efficient in the sense that the desired error control may be achieved at (nearly) minimal computational cost. To satisfy the desired criteria of reliability and efficiency, the computational methods are adaptive with feed back from the computational process. The adaptive methods are based on a posteriori error estimates, where the error is estimated in terms of the mesh size, the residual of the computed solution, and certain stability factors measuring relevant stability properties of the solution being approximated through the solution of an associated linearized dual problem. The a posteriori error estimates give stopping criteria guaranteeing the desired error control, and also serve as part of the modification criteria for adaptively choosing the computational mesh. We prove analytically that the stability factors in the basic model cases of shocks and rarefactions in one dimension, are of moderate size. We also present results from numerical computations of dual solutions and stability factors.

Contents

1 Introduction

In this chapter we give an introductory overview of our recent work together with E. Burman, P. Hansbo, M. Larson, R. Sandboge, and A. Szepessy, on computational methods for conservation laws, which are *reliable* in the sense that the computational error may be controled on a given tolerance level in a given norm, and *efficient* in the sense that the desired error control may be achieved at (nearly) minimal computational cost. To satisfy the desired criteria of reliability and efficiency, the computational methods are *adaptive* with *feed back* from the computational process. Key references to this work includes [23], [19]-[21], [32], [33], [37], [2], [3], [34]. The presented work is part of a larger program on adaptive computational methods for differential equations based on generalized Galerkin methods outlined in [8], presented in [9]-[18], [27], [28], [30], [31], [35], and further developed in the series of books [4]-[7], and realized in the related software *Femlab* (http://www.md.chalmers.se/Math/Research/Femlab/), and *Mechlab* ([29]. We refer the reader to these sources for a more complete account of our work.

Conservation laws are systems of nonlinear partial differential equations on *conservation form*:

$$\frac{\partial u}{\partial t} + \sum_{j=1}^{3} \frac{\partial}{\partial x_j} f_j(u, x, t) = S(u), \qquad (1.34)$$

where $u = u(x, t)$ is a vector function depending on the Euclidean space coordinate $x \in \mathbb{R}^3$ and time variable $t > 0$, the f_j are given flux vectors depending on (u, x, t) and include both convective and diffusive effects, and $S(u, x, t)$ is a given source vector. The *Navier-Stokes equations* for incompressible and compressible fluid flow take this form with u representing mass, momentum and energy, and (1.34) expressing conservation or balance laws for u, and the source term $S(u)$ representing exterior forces. In generalizations including chemical reactions and multi-phase flow, u includes the different mass fractions and $S(u)$ includes reaction or interaction terms. The inviscid form of the Navier-Stokes equations with diffusive viscosities equal to zero, is referred to as the *Euler equations*.

We consider the problem of reliable and efficient numerical computation of solutions of conservation laws, including the Euler and Navier-Stokes equations, which we may refer to as the problem of *computability* of solutions of conservation laws. We say that a certain solution u is *computable* with a given amount of computational work if

$$\|u - U\| \le TOL, \qquad (1.35)$$

where $\|\cdot\|$ is a given norm, TOL is a a given tolerance, and U is an approximation of u computed with the given work. This problem is intimately connected to the mathematical problem of existence, uniqueness and stability of the solution u. Clearly, a solution must exist to be computable, and because of the approximate nature of computed solutions, the concept of *stability* measuring the distance between solutions with nearby data, is central. There are results indicating that the viscid Navier-Stokes equations have unique solutions even for large time and data, but quantitative a priori information on stability is very sparse. It appears that as the viscosity decreases, in general the complexity of solutions increases and their stability deteriorates, corresponding to the development of fluctuating turbulent solutions with very fine-scale features. Accordingly, the work to compute solutions within a given tolerance, in general increases with decreasing viscosity. For the Euler equation limit with zero viscosity, existence and uniqueness may be seriously questioned, as well as of course the computability of solutions. It may be argued that real physical processes always involve some viscous effects, and thus it appears that the natural object of study is the Navier-Stokes equations.

Paradoxically, the development of computational methods for conservation laws initiated in the 50s, has largely been based on an opposite conception of the Euler equations as being more accessible to computation than the Navier-Stokes equations. Accordingly, the main interest has been focussed on inviscid equations, and a particular issue arising in this case, namely the existence of certain discontinuous solutions to the inviscid Euler equations for compressible flow corresponding to *shocks* or *contact discontinuities*. The analogous solutions of the viscous Navier-Stokes equations are smooth with layers replacing the discontinuities, of width decreasing with the viscosity. To mathematically handle such discontinuous solutions of the inviscid equations, the concept of *weak solutions* of (1.34) was introduced, defined using the conservation form allowing the derivatives to be applied to smooth test functions in a duality setting. It was then noted that certain discontinuous weak solutions have an unphysical nature, and *physically admissible weak solutions*, or *entropy solutions*, were identified as satisfying certain additional *entropy conditions* or *entropy inequalities* related to certain *entropy functions* of the solution. The entropy inequalities are automatically satisfied by smooth solutions, but not necessarily by discontinuous weak solutions. For scalar conservation laws, enough entropy functions are known to guarantee uniqueness, but for the inviscid Euler equations of compressible flow, essentially only one entropy function is known, the physical entropy, which does not appear to be enough to guarantee uniqueness. The question of existence and uniqueness of solutions of the inviscid Euler equations is therefore still unresolved, and it appears that the indicated traditional program based on weak solutions of the inviscid equations satisfying entropy conditions, can not be carried to the desired goal. Instead, it appears that Euler solutions (in three dimensions) in fact may be effectively uncomputable, and as already indicated it appears to be necessary to shift the focus to viscid forms of the equations, which may be expected to have unique solutions, albeit their complexity and the work to compute solutions in general may increase with decreasing viscosity. The increasing complexity requires turbulence modeling introducing viscous effects which reduce the complexity. The design of turbulence models is a main open problem, where today computational methods may offer new possibilitites.

A corresponding situation is met e.g. in elasto-plasticity, where the traditional model of perfect plasticity, which was motivated by its simplicity, in fact may be degenerate with non-unique solutions, which must be regularized by introducing viscous effects. The simplicity of the inviscid models were appealing in the early years of computing when computational power was very limited, and could then give useful information in certain very special cases, but with the massive computational power available today, it is natural to shift focus to more complete and non-degenerate viscid models.

The traditional computational methodology for conservation laws, with focus on the inviscid case, has been dominated by *finite difference methods* initiated in the 40s by Friedrichs and von Neumann and *finite volume methods* based on the work by Godunov in the 50's. The basic interest and work has been directed to the following topics:

- weak consistency
- entropy consistency
- shock capturing
- artificial viscosity
- a priori convergence results.

A computational method for a conservation law is *weakly consistent* if a limit of computed solutions as the mesh size tends to zero, is a weak solution of the conservation law. If the limit also satisfies the entropy conditions, then the method is *entropy consistent. Shock capturing* concerns oscillation-free (monotone) approximation of discontinuous solutions such as shocks or contact discontinuities. The traditional route to guarantee weak consistency is to use finite difference methods on so called *conservation form*, mimicing the form of the conservation law on uniform computational grids. The *artificial viscosity* is the effective viscosity used in the computation, which usually is nonzero even solving inviscid problems. Entropy consistency and monotone shock capturing may be guaranteed by introducing enough artificial viscosity, but too much will decrease the accuracy and produce smeared numerical layers. A basic problem is to find just the right amount of artificial viscosity to guarantee entropy consistency while keeping numerical layers as sharp as possible. A basic traditional result states that first order artificial viscosity (proportional to the mesh size) is enough, but too much because the resulting method is only first order accurate, and in particular contact discontinuities will be seriously smeared. The traditional convergence results based on the work of Kruzkov from the 60s are of a priori type proving typically convergence of first order schemes with first order artificial viscosity for scalar conservation laws using compactness and monotonicity methods. Very few a priori convergence results are available for systems of conservation laws (even in one space dimension), and/or higher order schemes, reflecting the lack of mathematical results on existence, uniqueness and stability of solutions, and generality of numerics.

Our work is based on *General Galerkin finite element methods*, referred to as G^2-methods, including *least squares stabilization, residual-based artificial viscosity,* and *space-time finite elements* as prime ingredients, which offer a rich spectrum of computational techniques with a common mathematical basis, and which may be applied to a large variety of differential equations, such as systems of conservation laws including models for compressible flow. The richness of this approach makes it possible to view many existing finite difference and finite volume methods as particular cases, but also opens entirely new possibilities of computational solution of conservation laws including in particular the goals of reliability and efficiency, which were not addressed in the traditional setting of finite difference and finite volume methods. We seek to realize these goals by using adaptive methods based on *a posteriori error estimates*, which take the following basic form:

$$\|u - U\| \leq S_c \| \frac{h^2}{\hat{\epsilon}} R(U)\|, \tag{1.36}$$

where the error $u - U$ in the given norm $\| \cdot \|$, e.g. an $L_p(L_q)$-norm in time-space, is estimated in terms of the mesh size $h = h(x)$, the artificial viscosity $\hat{\epsilon} = \hat{\epsilon}(U, x)$, the residual of the computed solution $R(U)$, and certain computable stability factors $S_c = S_c(u)$ defined through a continuous linearized dual problem and expressing relevant stability properties of the solution u being computed. The adaptive methods use

$$S_c \| \frac{h^2}{\hat{\epsilon}} R(U)\| \leq TOL \tag{1.37}$$

as *stopping criterion* guaranteeing the desired error control (1.35) through (1.36), and involve *modification strategies* aimed at realizing (1.37) at minimal computational cost, often based on *equi-distribution* of element contributions to the total error estimate. The *a priori* analog of (1.36) takes the basic form

$$\|u - U\| \leq S_{c,h} \|h^2 D^2 u\|, \tag{1.38}$$

where $S_{c,h}$ is an analog of S_c defined through a discrete dual problem, and D^2 denotes second derivatives in space. Conceptually, this estimate is obtained from (1.36) by using that $R(U) \approx \hat{\epsilon} D^2 U$ (including only the viscous term of the conservation law), and replacing U by u. The size of the stability factors S_c and $S_{c,h}$ of course is decisive for computability: if these factors are very large, then the solution u is uncomputable, while if they are of moderate size, then the computational work is comparable to that of direct interpolation of the solution u (if the solver is efficient with work proportional to the number of unknowns).

In the new setting of G^2-methods, the classical, largely unresolved, crucial questions concerning weak consistency, entropy consistency and the design of artificial viscosities, in particular for higher order methods for systems in several dimensions, may be given conclusive answers. For example, weak consistency is guaranteed simply by using Galerkin-type methods, and entropy consistency by requiring $\hat{\epsilon} \geq h^2 R(U)$, coupling the artificial viscosity to the residual $R(U)$ of the computed solution. Thus, the new approach offers not only answers to certain new questions of prime importance, but also offers new answers to some of the basic old questions. In the new setting, the quantity $\frac{h^2}{\epsilon} R(U)$ obviously plays a key role.

We focus on the following issues in the setting of systems of conservation laws in several dimensions:

- design of G^2-methods: least squares modification, artificial viscosity, space-time meshes
- a posteriori error estimates, adaptive methods
- stability properties of the linearized equations, including computation of the stability factors S_c for bench-mark problems.

For maximal simplicity of presentation, we use the context of a scalar conservation law in one space dimension as much as possible, and present the extensions to systems of conservation laws in several dimensions more concisely. We first quickly review some basic facts concerning conservation laws, using the context of a scalar conservation law in one space dimension, including the basic types of solutions (shocks, and rarefactions), and the concepts of weak solution and entropy solution. This material overlaps some with the background material presented in the chapter by E. Tadmor, but the emphasis in this chapter is somewhat different because the numerics is different. We then turn to the main topic, the G^2-method, starting with the *characteristic Galerkin method*, we prove a posteriori error estimates of the form (1.36), and give theoretical information on the crucial stability issues in basic cases. We illustrate with new computational results from [3] focussing on the aspects of stability factors and dual solutions. We then extend the scope to the compressible Euler and Navier-Stokes equations. For additional numerical results using the adaptive methods, we refer to the references cited above.

2 Scalar conservation laws in one dimension

We start considering a scalar conservation law in one space dimension:

$$\dot{u} + (f(u))' - \epsilon u'' = 0, \quad x \in \mathbb{R}, \ t > 0,$$
$$u(x, 0) = u_0(x), \quad x \in \mathbb{R}, \tag{2.39}$$

where the convective flux $f(u)$ is a given smooth function of u, u_0 is a given initial value, ϵ is a constant positive viscosity, $\dot{v} = \frac{\partial v}{\partial t}$ and $v' = \frac{\partial v}{\partial x}$. Since $(f(u))' = A(u)u'$, where $A = \frac{df}{du}$ denotes the derivative of f, the conservation law (2.39) can be written

$$\dot{u} + A(u)u' - \epsilon u'' = 0, \tag{2.40}$$

which has the form of a convection-diffusion equation $\dot{u} + \beta u' - \epsilon u'' = 0$, with the convection velocity $\beta(x,t) = A(u(x,t))$ depending on the solution $u = u(x,t)$.

We consider (2.39) posed on $\mathbb{R}$, and we also need to impose suitable boundary conditions at $x = \pm\infty$ for $t > 0$, such as the homogenous Dirichlet boundary condition $u(x,t) \to 0, x \to \pm\infty$. We may also pose the conservation law on a bounded interval, again with e.g. Dirichlet boundary conditions.

We focus mainly on the case with the viscosity ϵ being small, which offers the richest interaction between convective and diffusive fluxes $f(u)$ and $-\epsilon u'$, respectively. The *inviscid form* of (2.39) with $\epsilon = 0$, reads:

$$\dot{u} + (f(u))' = 0, \quad x \in \mathbb{R}, \ t > 0,$$
$$u(x,0) = u_0(x), \quad x \in \mathbb{R}. \tag{2.41}$$

For scalar problems in one space dimension, inviscid solutions are limits of viscid solutions as ϵ tends to zero, and may have a simple form in basic cases, including shocks, rarefactions and contact discontinuities. For systems in several dimensions, the complexity of solutions in general increases with decreasing ϵ, when eventually the flow becomes turbulent.

With $f(u) = u^2/2$, the problem (2.39) is referred to as *Burgers' equation*:

$$\dot{u} + uu' - \epsilon u'' = 0,$$

after the American mathematician Burgers, who used this model in a study of turbulence in fluid flow. We now present a similar basic flow model with $f(u) = u(1-u)$.

2.1 A model for traffic flow

We consider a model for traffic flow along a highway represented by $\mathbb{R}$, where $\rho(x,t)$ denotes the density ($0 \le \rho \le 1$) and $u(x,t)$ the velocity of cars at $x \in \mathbb{R}$ at time t. Conservation of mass, assuming that no cars enter or leave the highway, is expressed through the conservation law

$$\dot{\rho} + (u\rho)' = 0 \quad \text{in} \quad Q = \mathbb{R} \times (0,T), \tag{2.42}$$

which is complemented by a constitutive relation between u and ρ. A simple model is given by

$$u = 1 - \rho, \tag{2.43}$$

stating that the velocity varies linearly with the density from 1 if the road is empty, to zero if the road is packed. A more elaborate model is given by

$$u = 1 - \rho - \epsilon \rho'/\rho, \tag{2.44}$$

where $\epsilon > 0$, describing a driver who reduces speed if the density is increasing ahead, and vice versa. The simple model (2.43) is obtained setting $\epsilon = 0$. Combining (2.42) and (2.44), we obtain the following model for traffic flow

$$\dot{\rho} + ((1 - \rho)\rho)' - \epsilon\rho'' = 0. \qquad (2.45)$$

This equation is very similar to Burgers' equation and the above discussion directly applies. In particular, (2.45) also has shocks, corresponding to the sudden stops met in driving, and rarefaction waves corresponding to the gradual initiation of the traffic flow when a traffic light switches from red to green, see the figure below illustrating a situation with the red light at $x = 0$ turning green at $t = 0$.

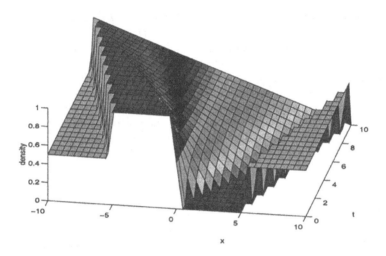

An unphysical discontinuous solution in the case $\epsilon = 0$, corresponds to the situation when the driver closest to the traffic light looking in his rear mirror notices that the density of cars is equal to one behind, and therefore decides to maintain zero speed even when the traffic light changes to green. One may say that this corresponds to a *stupid driver* who does not understand the entropy inequality, while the model (2.44) corresponds to an *intelligent driver* who will respect this condition, and thus in particular will get started as the first car at the switch to green.

We note that the unphysical solution is unstable: even a very slight "honk" should make the stupid driver aware of the green traffic light and make him get started into a rarefaction, which is very different from the unphysical solution after some time. The physical solution on the other hand is stable: a small perturbation of initial data will not change the solution dramatically. This is another aspect of the entropy condition characterizing physical solutions: Weak solutions which do not respect the entropy condition, may be unstable to very small disturbances, which give them an "unphysical" nature. We will return below to aspects of stability of shocks and rarefactions.

2.2 The Euler equations for a compressible inviscid perfect gas

The Euler equations in one dimension for a compressible inviscid perfect gas in conservation variables (ρ, m, e), take the form of the following system of conservations laws

$$
\begin{aligned}
\dot{\rho} + (u\rho)' &= 0, & x \in \mathbb{R}, \ t > 0, \\
\dot{m} + (um)' + p' &= 0, & x \in \mathbb{R}, \ t > 0, \\
\dot{e} + (ue)' + (pu)' &= 0, & x \in \mathbb{R}, \ t > 0,
\end{aligned}
\tag{2.46}
$$

together with boundary and initial conditions, where ρ is the density, $m = \rho u$ the momentum density with u the velocity, e the energy density, and p the pressure given by the perfect gas law $p = (\gamma - 1)(e - m^2/(2\rho))$, where $\gamma > 1$ is a constant. The equations express conservation of mass, momentum and energy, where the momentum equation corresponds to Newton's law with p' representing the net force density on a fluid element from the pressure, and in the energy equation $(pu)'$ represents the work density from the pressure acting on a fluid element. Here $u\rho$, um and ue are the convective fluxes of the mass, momentum and energy, and p' and $(pu)'$ are fluxes related to the pressure. The physical entropy η is given by $\eta = -\rho \log(p\rho^{-\gamma})$ with corresponding entropy flux $u\eta$.

2.3 Characteristics

We may attempt to solve the inviscid conservation law (2.41) using the *method of characteristics*, and in particular we could hope this way to obtain approximate solutions of the viscid equation (2.39) if ϵ is small. The *characteristics* of (2.41) are curves $x = x(\bar{x}, \bar{t})$, such that

$$
\frac{dx}{d\bar{t}} = A(u(x, \bar{t})), \text{ for } \bar{t} > 0, \quad x(\bar{x}, 0) = \bar{x},
$$

where u is a differentiable solution of (2.41) satisfying $\dot{u} + A(u)u' = 0$. Since by the chain rule,

$$
\frac{d}{d\bar{t}} u(x(\bar{x}, \bar{t}), \bar{t}) = \frac{du}{dx}\frac{dx}{d\bar{t}} + \frac{du}{d\bar{t}} = A(u)u' + \dot{u} = 0, \ \bar{t} > 0,
$$

it follows that u is constant along a characteristic. Thus, $\frac{dx}{d\bar{t}}$ is also constant along a characteristic, which therefore must be a straight line. This means that the solution of (2.41) is obtained by drawing from each point $(\bar{x}, 0)$ a straight line with slope $A(u_0(\bar{x}))$ and noting that the solution u is constant equal to $u_0(\bar{x})$ along that line. We may express the nature of the solution of (2.41) by stating that the level curves of a solution are straight lines coinciding

with characteristics. This solution method works as longs as characteristics
do not cross.

Considering now for definiteness the case of Burgers' equation with $A(u) = u$, we next note that if u_0 is decreasing, then two characteristics starting at
two distinct points $\bar{x}_1$ and $\bar{x}_2$ may actually cross for $\bar{t} > 0$, and then we
have the conflicting statement that u at the intersection should have both
the value $u_0(\bar{x}_1)$ and $u_0(\bar{x}_2)$, which are different. We shall see that this corresponds to the formation of a shock wave, which is a discontinuous solution
corresponding to a propagating sharp front. The "bang" from a supersonic
airplane, or the "bang" in the water pipe system of a house from closing a
faucet, are examples of shock waves.

2.4 Basic solutions: rarefaction and shock waves

We now recall the basic types of solutions in the context of the inviscid
Burgers' equation. We have seen that if u_0 is increasing and continuous, then
the method of characteristics works, and we obtain a continuous solution. If
u_0 is increasing and discontinuous, with two constant states in the basic case,
then the method of characteristics can be extended to give a solution referred
to as a *rarefaction wave*, which is a continuous solution connecting at each
time level the two constant states by a linear transition of width proportional
to $t > 0$. We give the details shortly. If u_0 is decreasing and discontinuous
with two constant states, then the corresponding solution is a shock wave
consisting of two constant states for all $t > 0$. A shock may also develop
after some time starting with decreasing continuous data. This corresponds
to the development of a breaking wave on a shore. We now consider the case
of rarefaction wave and shock wave separately in more detail.

Rarefaction wave Consider the inviscid Burgers' equation with the increasing discontinuous initial data $u_0(x) = 0$ for $x < 0$, and $u_0(x) = 1$ for $x > 0$.
The exact solution is the rarefaction wave given by

$$
\begin{array}{lll}
u(x,t) = 0 & \text{for} & x < 0, \\
u(x,t) = \frac{x}{t} & \text{for} & 0 \leq \frac{x}{t} \leq 1, \\
u(x,t) = 1 & \text{for} & 1 < \frac{x}{t}.
\end{array}
\tag{2.47}
$$

This is a continuous function for $t > 0$, differentiable off the lines $x = 0$
and $x = t$, which satisfies (2.41) in fact for all (x,t) for $t > 0$, and thus is
a solution in a classical sense. In a rarefaction wave, an initial discontinuity
separating two constant states develops into a continuous linear transition
from one state to the other of width t in space, corresponding to "fan-like"
level curves in space-time, see below. The solution may be found using the
method of characteristics for $x < 0$ and $1 < \frac{x}{t}$ and is completed with the
fan-like transition which gives a solution which is continuous for $t > 0$.

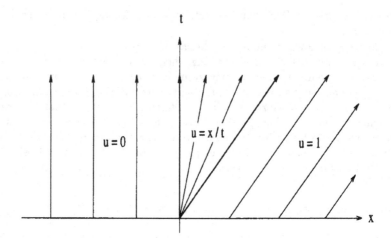

Shock We consider now the decreasing discontinuous initial data $u_0(x) = 1$ for $x < 0$, and $u_0(x) = 0$ for $x > 0$. The limit of solutions of the viscid Burgers' equation as ϵ tends to zero, turns out to be the following discontinuous function corresponding to a shock wave moving with speed $\frac{1}{2}$,

$$u(x,t) = 1 \quad \text{for} \quad x < \tfrac{t}{2},$$
$$u(x,t) = 0 \quad \text{for} \quad x > \tfrac{t}{2}. \tag{2.48}$$

The corresponding solutions for $\epsilon > 0$ are smooth approximations of the discontinuity of width ϵ, with u' being negative and of size ϵ^{-1} in the transition zone from 1 to 0.

In the present case the initial discontinuity is maintained, in contrast to the above case of a rarefaction wave, where the initial discontinuity is turned into a continuous linear transition from one state to the other over a distance in space of length t. The method of characteristics breaks down in the shock wave case with the two characteristics at the two sides of the shock with slopes 1 and 0 respectively, converging into the shock with slope $1/2$, see below.

2.5 Weak solutions and the Rankine-Hugoniot condition

We say that the bounded function $u(x,t)$ is a *weak solution* of the equation $\dot{u} + (f(u))' = 0$ for $x \in \mathbb{R}$ and $t > 0$ if for all smooth functions $\varphi(x,t)$ which vanish for $t = 0$ and for $|x| + t$ large,

$$\int_{\mathbb{R}} \int_0^\infty -(u\dot{\varphi} + f(u)\varphi')dx\,dt = 0. \tag{2.49}$$

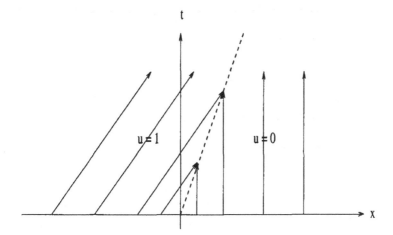

This relation is formally obtained multiplying $\dot{u} + (f(u))' = 0$ by φ and integrating by parts putting the derivatives on φ. Note that by the vanishing of φ, no boundary terms arise from the integration by parts. The concept of weak solution is the minimal requirement we can put on a function to satisfy the conservation law $\dot{u} + (f(u))' = 0$. This is analogous to the variational (or weak formulation) of the Poisson equation, where "half" of the derivatives are put on the test functions. In the present case we go one step further and put all derivatives on the test function φ. It is the conservation form of the equation which makes this possible.

Let us thus seek to find a discontinuous weak solution $u(x,t)$ defined by $u(x,t) = u_+$ if $x > st$ and $u(x,t) = u_-$ if $x < st$, where u_+ and u_- are two constant states, and s is a constant, corresponding to a discontinuity propagating with speed s. Inserting this function into (2.49) and integrating by parts gives

$$\int_\Gamma ([u]n_t + [f(u)]n_x)\varphi \, ds = 0,$$

where Γ is the straight line $\{(x,t) : x = st, t \geq 0\}$ with normal (n_x, n_t) parallel to $(-1, s)$, and $[u] = u^+ - u^-$ and $[f(u)] = f(u^+) - f(u^-)$ are the jumps of u and $f(u)$ across Γ. Since $s = -n_t/n_x$ this gives the following relation, referred to as the *Rankine-Hugoniot condition*, relating the shock speed s to the jumps in u and $f(u)$:

$$s = \frac{[f(u)]}{[u]}. \tag{2.50}$$

For Burgers' equation with $f(u) = u^2/2$, we have

$$s = (u_+ + u_-)/2. \tag{2.51}$$

In the above example, we thus find that $s = \frac{1}{2}$. The Rankine-Hugoniot condition expresses the conservation of the quantity u expressed in the conservation law $\dot{u} + (f(u))' = 0$ in the case of a discontinuous solution.

2.6 Weak solutions may be non-unique

Consider the inviscid Burgers' equation with the rarefaction wave initial data $u_0(x) = 0$ for $x < 0$ and $u_0(x) = 1$ for $x > 0$. The function

$$
\begin{aligned}
u(x,t) &= 0 \quad \text{for} \quad x < \tfrac{t}{2}, \\
u(x,t) &= 1 \quad \text{for} \quad x > \tfrac{t}{2},
\end{aligned}
\tag{2.52}
$$

is a weak solution, with a discontinuity $\{x,t\} : x = st\}$ moving with speed $s = \frac{1}{2}$. However, this solution is different from the rarefaction wave solution (2.47), which since it is a classical solution, also is a weak solution. Thus, we have in this case two different weak solutions, and thus we have an example of non-uniqueness of weak solutions. From a physical point of view, only the rarefaction wave solution is an acceptable solution. This is because the function (2.52) does not satisfy certain entropy inequalities, which characterize physically admissible weak solutions, while the rarefaction wave solution does so. On the other hand, the discontinuous shock wave solution does satisfy the entropy inequalities, and thus the shock is a physically admissible solution. We now turn to a discussion of the entropy inequalities satisfied by physically admissible solutions.

We have already remarked that exact solutions of the viscous conservation law always do satisfy the entropy inequalities, while weak solutions of the inviscid law may fail to do so. This turns out to be of mayor concern for the design of numerical methods for conservation laws in the case of small viscosity. In fact, to guarantee that approximate numerical solutions properly satisfy the entropy inequalities, which is necessary to avoid that the numerical solution seeks to approximate a non-physical solution, the numerical method must effectively introduce a sufficient amount of numerical or "artificial" viscosity. This is because the given viscosity may be too small compared to the mesh size used to guarantee satisfaction of the entropy inequalities. A basic problem in the design of numerical methods for conservation laws with small viscosity, is thus the design of the artificial viscosity of the numerical method. We present below a solution to this basic problem, which generalizes to systems of conservation laws.

2.7 Entropy inequalities

The entropy inequalities are related to the Second Law of Thermodynamics stating that in a closed physical system, the entropy cannot decrease. Intuitively, the physical concept of "entropy" is related to "disorder" or "loss of information" and thus the The Second Law states that disorder cannot decrease. To see how the Second Law in this form applies to the shock wave case above, we note that the characteristics in this case "converge into" the shock, which may be viewed as a "loss of information", or increase of entropy, which is physically admissible. On the other hand, for the unphysical solution (2.52), the characteristics seem to "emerge from" the discontinuity, which may be viewed as some kind of "creation of information", which would contradict the Second Law. Thus, in brief we may say that a discontinuous weak solution with the characteristics converging into the discontinuity, is a physical solution while it is unphysical if the characteristics are diverging out from the discontinuity.

Let us now turn to the entropy inequalitites which express the Second Law in quantitative analytical form. For definiteness we consider Burgers' equation in viscid form. We first notice that multiplying by u, integrating in time and space, we obtain the following basic energy estimate:

$$\int_{\mathbb{R}} |u(x,t)|^2 \, dx + D_\epsilon(u) = \int_{\mathbb{R}} |u(x,0)|^2 \, dx, \qquad (2.53)$$

where

$$D_\epsilon(u) = 2 \int_0^t \int_{\mathbb{R}} \epsilon(u')^2 \, dx \, ds. \qquad (2.54)$$

In particular, it follows that for $t > 0$

$$\|u(\cdot, t)\| \leq \|u_0\|, \qquad (2.55)$$

with $\| \cdot \|$ denoting the $L_2(\mathbb{R})$-norm.

We now compare the size of $D_\epsilon(u)$ in the case of a rarefaction wave and a shock wave. We find that $D_\epsilon(u) \propto \epsilon$ in the case of a rarefaction wave and $D_\epsilon(u) \propto [u]$ in the case of a shock wave with jump $[u]$. Thus, in the rarefaction wave case $D_\epsilon(u) \to 0$ as $\epsilon \to 0$, while in the shock case we have $D_\epsilon(u) \to [u] \neq 0$ as $\epsilon \to 0$. Thus, we will have equality in (2.55) in the case of a rarefaction wave, and strict inequality in the case of a shock wave. This may be viewed as expressing a "loss of information" or "entropy production" in the case of a shock wave.

We now present a sharper variant of the preceeding analysis with the aim of connecting to the above discussion of the direction of characteristics at a shock. We then assume that the solutions u of the viscid Burgers' equation are bounded for $\epsilon > 0$, and tend pointwise to some bounded limit, again denoted by u, as ϵ tends to zero. Multiplying Burgers' equation first by smooth test function φ with compact support (vanishing for $t = 0$ and $t + |x|$ large), and integrating by parts, we obtain

$$-\int_0^t \int_{\mathbb{R}} u\dot{\varphi} \, dx \, ds - \tfrac{1}{2} \int_0^t \int_{\mathbb{R}} u^2 \varphi' \, dx \, ds$$
$$= -\int_0^t \int_{\mathbb{R}} \epsilon u' \varphi' \, dx \, ds.$$

Since by (2.53)

$$2 \int_0^t \int_{\mathbb{R}} \epsilon (u')^2 \, dx \, ds \leq \|u_0\|^2, \tag{2.56}$$

we have using Cauchy's inequality and the smoothness of φ,

$$\int_0^t \int_{\mathbb{R}} \epsilon u' \varphi' \, dx \, ds \leq \|u_0\| \sqrt{\epsilon} \left(\int_0^t \int_{\mathbb{R}} (\varphi')^2 dx \, ds \right)^{1/2} \to 0 \quad \text{as } \epsilon \to 0. \tag{2.57}$$

Thus, we conclude that the limit u satisfies (2.49) with $f(u) = u^2/2$ and hence is a weak solution of the inviscid Burgers' equation.

Next, multiplying by $u\varphi$, where φ is a smooth function with compact support, now assumed to be also non-negative, we obtain integrating by parts

$$- \int_0^t \int_{\mathbb{R}} \tfrac{u^2}{2} \dot{\varphi} \, dx \, ds - \int_0^t \int_{\mathbb{R}} \tfrac{u^3}{3} \varphi' \, dx \, ds$$
$$+ \int_0^t \int_{\mathbb{R}} \epsilon (u')^2 \varphi \, dx \, ds = - \int_0^t \int_{\mathbb{R}} \epsilon u u' \varphi' \, dx \, ds.$$

Arguing as above, using also the boundedness of u, we see that the right hand side tends to zero. Using also the positivity of φ to see that the third term on the right hand side is positive, we conclude that the limit u satisfies

$$- \int_0^t \int_{\mathbb{R}} \frac{u^2}{2} \dot{\varphi} \, dx \, ds - \int_0^t \int_{\mathbb{R}} \frac{u^3}{3} \varphi' \, dx \, ds \leq 0, \tag{2.58}$$

for all smooth non-negative functions φ with compact support, with strict inequality if $D_\epsilon(u)$ tends to some non-zero limit. This is an entropy inequality stated in weak form, with the derivatives on the test function φ. In strong form obtained formally integrating by parts and using the positivity of φ, the entropy inequality may be stated

$$\frac{d}{dt} \left(\frac{-u^2}{2} \right) + \frac{d}{dx} \left(\frac{-u^3}{3} \right) \geq 0, \tag{2.59}$$

For smooth solutions of the inviscid Burgers' equation $\dot{u} + uu' = 0$, this relation is obtained (with equality) simply by multiplying by u and integrating. The above derivation shows that a limit of solutions of the viscid Burgers' equation, satisfies this relation, possibly with strict inequality (e.g. in the case of a shock). Integrating (2.59) in space and time with homogenous Dirichlet boundary conditions, in which case the x-derivative term integrates to zero, we obtain a global form of the entropy inequality which coincides with the basic L_2 estimate (2.55). In the present case the entropy of u is equal to $-u^2/2$.

Suppose now that the limit u is discontinuous with two states separated by the line $x = st$: $u(x,t) = u_-$ for $x < st$ and $u(x,t)$ for $x > st$. The Rankine-Hugoniot condition (2.51) expressing that this function is a weak solution, is $s = \frac{1}{2}(u_- + u_+)$. Inserting next the function u into the entropy inequality (2.58), and using partial integration as in the derivation of the Rankine-Hugoniot condition, we find that

$$s[\frac{u^2}{2}] - [\frac{u^3}{3}] \geq 0. \tag{2.60}$$

We thus find by a simple computation that

$$0 \leq \frac{1}{2}(u_- + u_+)\frac{1}{2}[u^2] - \frac{1}{3}[u^3] = (u_- - u_+)\frac{1}{12}(u_- - u_+)^2. \tag{2.61}$$

We conclude that the entropy inequality in this case may be stated $u_- \geq u_+$, or equivalently,

$$u_- \geq s \geq u_+. \tag{2.62}$$

This version of the entropy inequality states that the characteristics of a physically admissible discontinuous weak solution of the inviscid Burgers equation "converge into" the shock.

2.8 General convex entropies

We will refer to $\eta(u) = \frac{u^2}{2}$ as the "mathematial entropy" of u corresponding to the "physical entropy" $-\frac{u^2}{2}$ under consideration, the difference being the sign. The global form of the entropy inequality, which coincides with the basic L_2-estimate (2.55), may then be stated

$$\int_{\mathbb{R}} \eta(u(x,t))dx \leq \int_{\mathbb{R}} \eta(u(x,0))dx, \quad \text{for } t > 0, \tag{2.63}$$

Below, we will work with the concept of mathematical entropy, following the tradition in the mathematical literature on conservation laws, and simply refer to this concept as entropy. With this terminology, the Second Law states that the entropy cannot increase.

More generally, for scalar conservation laws any convex function $\eta(u)$, may be used as an *entropy*. The function $\eta(u) = u^2/2$ is then the basic example of an entropy. Arguing as above multiplying (2.39) by $\eta'(u)\varphi$, and using that $\eta'' \geq 0$, we obtain the following entropy inequality stated in strong form

$$\frac{\partial}{\partial t}(\eta(u)) + \frac{\partial}{\partial x}(q(u)) \leq 0, \tag{2.64}$$

where $q(u)$ is the *entropy flux* satisfying the compatibility condition

$$q' = f'\eta'. \tag{2.65}$$

We note that in the present case, we may find for any choice of $\eta(u)$, an entropy flux $q(u)$ satisfying the compatibility condition (2.65), and as a consequence we obtain a family of entropy inequalites corresponding to the family of convex functions $\eta(u)$. This means that a physically admissible solution in the present scalar case satisfies (2.64) for a whole family of different entropies. A basic mathematical result for scalar conservation laws guarantess uniqueness of weak solutions satisfying all the entropy inequalities related to convex entropies.

For the system of conservation laws modeling inviscid compressible flow the situation is different: in this case only one entropy is known (up to simple modifications corresponding to adding a constant and multiplying by a constant), which is the one physical entropy given by Nature. It is not known if the corresponding single entropy inequality is enough to guarantee uniqueness of weak solutions. Accordingly, the mathematical theory for systems of inviscid conservation laws remains severely incomplete.

To sum up, we have shown that a limit of viscous solutions of (2.39), as ϵ tends to zero, satisfies the entropy inequality (2.64) for any convex entropy $\eta(u)$. A basic problem in the analysis of numerical methods will be to establish the same property for limits of numerical solutions. We shall see below that the basic mechanism to build this property into the numerical method, is to add enough artificial viscosity. On the other hand, from accuracy point of view we should not use more artificial viscosity than necessary. A basic problem is thus to find the minimal amount of artificial viscosity needed to guarantee the proper satisfaction of the entropy inequalities. We will present a solution to this problem below.

3 G^2: Generalized Galerkin methods

We now turn to the G^2-method based on Galerkin's method combined with (i) weighted least-squares modification, (ii) residual-based artificial viscosity and (iii) space-time piecewise polynomial approximation. The G^2-method offers a rich spectrum of computational methods depending on the choice of the space-time mesh. These methods may be classified as *Eulerian* with the space-time mesh oriented along the space and time coordinate axis, *Lagrangean* with the space-time mesh oriented along particle paths in space-time, and Arbitrary Lagrangean-Eulerian or *ALE* methods with the space-time mesh oriented according to some other feature such as space-time gradients of the solution. We also refer to Lagrangean variants as *characteristic Galerkin* or chG-methods, or more precisely as chG(p,q)-methods with piecewise polynomials of order p in space, and piecewise polynomials of order q along space-time particle paths, and similarly to ALE-methods with the space-time mesh oriented according to space-time gradients of the solution as *oriented Galerkin* orG(p,q)-methods. We also refer to the Eulerian variants as *Streamline diffusion*, or Sd-methods. In all these variants the space-time mesh is usually organized in *space-time slabs* between discrete time levels, and the space mesh may be changed across the discrete time levels to avoid mesh distortion and allow mesh adaption. When the space mesh is changed across a discrete time level, a *projection* from the previous mesh to the new mesh at that time level, is performed. The projection is built into the Galerkin method through a jump term and corresponds to a modified L_2 projection with a residual-based artificial viscosity. The discrete solution between the discrete time levels may be viewed as an approximate *transport* step, and the whole process may be viewed as a method of the basic form *projection-transport*.

The traditional finite difference methods are of Eulerian type with the first order Lax-Friedrichs' scheme from the 50s as a prototype on conservation form and with artificial viscosity proportional to the mesh size. The next generation of classical schemes originates from Godunov's method in 1d, which is of the form projection-transport with piecewise constant (discontinuous) approximation and a Riemann solver for the transport step. The multidimensional finite volume schemes developed in recent decades, use discontinuous polynomial approximation with numerical fluxes often constructed using 1d Riemann solvers. These methods may alternatively be viewed as particular G^2 methods with discontinuous piecewise polynomials, or *discontinuous Galerkin* or dG-methods.

An outline of the remainder of this chapter is as follows: We first present the chG(1,0) method, continue with an Eulerian Sd-method, which we refer to as cG(1)dG(0) with continuous piecewise linears in space and discontinuous piecewise constants in time. We also present a full dG(q)-method, based on discontinuous piecewise polynomial approximation in space-time of degree q. We start with scalar conservation laws in 1d, and conclude with the multi-d systems given by the Euler and Navier-Stokes equations for compressible flow. We present in particular computations of dual solutions and stability factors.

3.1   The characteristic Galerkin method chG(1,0)

For the numerical solution of the conservation law (2.39), we now consider the characteristic Galerkin method chG(1,0), which is the particular variant of the General Galerkin G^2-method with the space-time mesh oriented according to characteristics, and with piecewise linear approximation in space and piecewise constant approximation along characteristics. We complement (2.39) with homogenous Dirichlet boundary conditions, assuming that $u_0(x)$ vanishes for $|x|$ large.

To define chG(1,0), let $0 = t_0 < t_1 < ...$, be an increasing sequence of time levels, let $S_n = \mathbb{R} \times I_n$ where $I_n = (t_{n-1}, t_n)$, be the corresponding space-time slabs. Associate to each slab S_n a set V_n of continuous piecewise linear functions on $\mathbb{R}$ with mesh function $h_n(x)$, and assume that the functions in V_n vanish for $|x|$ large. For a given velocity β defined on S_n, let

$$W_{kn}^{\beta} = \{v \in \mathcal{C}(S_n) : \dot{v} + \beta v' = 0 \quad \text{in } S_n, \quad v_{n-1}^+ \in V_n\}, \tag{3.66}$$

where $\mathcal{C}(S_n)$ is the set of continuous functions on S_n, and

$$v_{n-1}^+(x) = \lim_{s \to 0^+} v(x, t_{n-1} + s).$$

Finally, set $W_k^\beta = \prod_n W_{kn}^\beta$. A function v in W_k^β is constant along the characteristics on each slab S_n defined by the mesh tilting velocity β, is in general discontinuous in time across the discrete time levels t_n, and the limit from above v_{n-1}^+ at the initial time level t_{n-1} of each slab S_n is a piecewise linear function in V_n. To guarantee existence of the characteristics, we assume that β is Lipschitz continuous in x, and that the time step $k_n = t_n - t_{n-1}$ is sufficiently small. Below we shall see that it is natural to require that $\bar{h}_n^2/k_n$ tends to zero as $\bar{h}_n$ tends to zero, where $\bar{h}_n$ is the maximum of $h_n(x)$, which requires k_n not to be too small. Normally, the condition on k_n to avoid intersection of characteristics, is of the form $k_n \leq C h_n(x)$, so that no conflict arises.

The chG(1,0) method for (2.39) may now be formulated as follows: find $U \in W_k^\beta$, where $\beta = A(U)$, such that for $n = 1, 2, ..., \forall v \in W_{kn}^\beta$,

$$(\dot{U} + A(U)U', v)_n + (\hat{\epsilon}\nabla U, \nabla v)_n + ([U_{n-1}], v_{n-1}^+) = 0, \qquad (3.67)$$

where $[U_{n-1}] = U_{n-1}^+ - U_{n-1}^-$ is the jump at time t_{n-1} with $v_{n-1}^-(x) = \lim_{s \to 0+} v(x, t_{n-1} - s)$ the limit from below,

$$\hat{\epsilon} = \max(\epsilon, \gamma_1 h^{2-\kappa} R(U), \gamma_2 h^{\frac{3}{2}}), \qquad (3.68)$$

$(\cdot, \cdot)_n = (\cdot, \cdot)_{L_2(S_n)}$, $R(U) = |[U_{n-1}]|/k_n$ on S_n (extended from t_{n-1} to S_n as constant in time), the γ_i are positive constants of moderate size, and κ is a small positive constant. This method can also be applied to the inviscid form (2.41) of (2.39) simply be setting $\epsilon = 0$ in the definition of $\hat{\epsilon}$.

The artificial viscosity $\hat{\epsilon}$ defined by (3.68) has the same form as that used in [5] for linear convection-diffusion problems. Roughly speaking, $\hat{\epsilon}$ may be viewed to be designed to guarantee that limits of chG(1,0)-solutions satisfy all entropy inequalities, while being as small as possible in order to maintain high accuracy. The crucial design feature is that

$$\frac{h^2 R(U)}{\hat{\epsilon}} \leq h^\kappa/\gamma_1 \to 0 \quad \text{as } \bar{h} \to 0 \qquad (3.69)$$

where $\bar{h} = \max_n \bar{h}_n$, the mathematical justification of which we present below. The artificial vicosity $\hat{\epsilon}$ defined by (3.68) is locally much smaller than the *classical artificial viscosity* of the simple form $\hat{\epsilon} \sim h$, which is the artificial viscosity of the Lax-Friedrichs' scheme, and which degrades the accuracy to first order everywhere. In a shock, $\hat{\epsilon}$ defined by (3.68) may be of order h, because there R may be of size $1/h$, but elsewhere $\hat{\epsilon}$ will be smaller, and the accuracy is improved.

By a fixed point argument one can prove that the nonlinear problem (3.67) for a given n, has solution if the time step k_n is so small that intersection of characteristics is avoided. Notice that the mesh tilting velocity $A(U)$ depends on the solution U. Notice further that the basic energy estimate for U implies that $|U'| \leq C\hat{\epsilon}^{-1}$, which guarantees that the time steps do not have to be chosen smaller than $\hat{\epsilon}$.

Since by the construction, $\dot{U} + A(U)U' = 0$ on each slab S_n, the chG(1,0) method (3.67) reduces to finding $U \in W_k^\beta$ with $\beta = A(U)$, such that for $n = 1, 2, ...,$

$$(U_{n-1}^+, v_{n-1}^+) + (\hat{\epsilon}U', v')_n = (U_{n-1}^-, v_{n-1}^+) \quad \forall v \in W_{kn}^\beta. \tag{3.70}$$

This method may be viewed to be of the form "projection + transport". To see this, we may consider the following variant with the diffusion term acting on the discrete time level t_{n-1}, instead of the slab S_n: find $U \in W_k^\beta$ with $\beta = A(U)$, such that for $n = 1, 2, ...,$,

$$(U_{n-1}^+, v) + (\hat{\epsilon}U', v')_n^+ k_n = (U_{n-1}^-, v), \quad \forall v \in V_n, \tag{3.71}$$

where we use the notation

$$(v, w)_n^+ = \int_\mathbb{R} v_{n-1}^+ w_{n-1}^+ \, dx,$$

and a factor k_n came from the change of integration from S_n to $\mathbb{R}$. This method may be formulated as finding for $n = 1, 2, ..$, the function $U_n \equiv U_n^-$, such that

$$U_n = T_n \tilde{P}_n U_{n-1}, \tag{3.72}$$

where $\tilde{P}_n$ is a modified $L_2(\mathbb{R})$ projection into V_n with artificial viscosity, defined by

$$(\tilde{P}_n w, v) + (\hat{\epsilon}(\tilde{P}_n w)', v')k_n = (w, v), \quad \forall v \in V_n, \tag{3.73}$$

where $\hat{\epsilon} = \max(\epsilon, \gamma_1 h^{2-\kappa}|w - \tilde{P}_n w|/k_n, \gamma_2 h^{\frac{3}{2}})$, and $T_n w$ is the solution at time t_n of the inviscid equation $\dot{u} + A(u)u' = 0$ on the slab S_n, with given initial data equal to w at time t_{n-1}. Other variants of this method are obtained by choosing different projections, and more generally using approximations of the transport step.

A very simple variant of (3.71) is obtained using the same uniform mesh underlying all V_n, and using a lumped mass $L_2(\mathbb{R})$ projection, see [4]. This method effectively adds classical artificial viscosity of order h from the lumping of the mass matrix in the projection performed at each time step, which dominates the artificial viscosity $\hat{\epsilon}$ given by (3.68). The simplified method may be summarized as: on time level t_{n-1}, compute $U_{n-1}^+ \in V_n$ at the nodes, as a mean value of U_{n-1}^- weighted with the proper hat function. Then solve the equation $\dot{u} + A(u)u' = 0$ on the slab S_n with initial data U_{n-1}^+ at time t_{n-1} using the method of characteristics, drawing from each node of T_n a line with slope $A(U_{n-1}^+)$, to get the solution U_n^-, and repeat. The artificial viscosity from the lumped L_2 projection is quite a bit of over-kill, and thus the accuracy of the lumped version is not as good as that of the more elaborate chG(1,0) method with the residual dependent artificial viscosity $\hat{\epsilon}$.

Following [4], we can generalize chG(1,0) to chG(q,r) basing V_n on continuous piecewise polynomials of order q, and using discontinuous piecewise polynomial approximation of order r along characteristics.

3.2 The Sd-method

Choosing the mesh tilting velocity $\beta = 0$ and stabilizing using a weighted least squares modification, we obtain the following Eulerian variant, which we refer to as the cG(1)dG(0) Sd-method or Sd(1,0): find $U \in W_k^0$, such that for $n = 1, 2, ..., \forall v \in W_{kn}^0$,

$$(\dot{U} + A(U)U', v + \delta(\dot{v} + A(U)v'))_n + (\hat{\epsilon}\nabla U, \nabla v)_n + ([U_{n-1}], v_{n-1}^+) = 0,$$
(3.74)

where

$$\delta = \frac{1}{2}(k_n^{-2} + h_n^{-2}A(U)^2)^{-1/2},$$

$\hat{\epsilon} = \max(\epsilon, \gamma_1 h^{2-\kappa} R(U), \gamma_2 h^{\frac{3}{2}})$, and now $R(U) = |\dot{U} + A(U)U'| + |[U_{n-1}]|/k_n$ on S_n. The accuracy of this method is inferior to that of chG(1,0), because of the piecewise constant approximation in time, which does not in general fit with the solution feature of being constant along space-time characteristics. In particular, the Sd-method with piecewise constants in time, introduces artificial viscosity of order h, through the δ term. This method can directly be generalized to Sd(q,r)-methods based on cG(q)dG(r), which have higher accuracy and less viscosity, but are more computationally intensive.

3.3 The oriented Galerkin method

Orienting the mesh in the Sd-method (3.74) along characteristics replacing W_k^0 by $W_k^{A(U)}$, we have effectively $\dot{v} + A(U)v' = 0$ for $v \in W_k^{A(U)}$, so that we may choose $\delta = 0$, and thus the Sd(q,r)-method with the mesh orientation along characteristics, reduces to chG(q,r).

One can also choose the mesh tilting velocity β according to some other principle than following characteristics or setting $\beta = 0$. For example, one may seek to orient β to be perpendicular to the space-time gradient of the solution, with the direction of piecewise constant approximation along space-time level curves. This gives a different orientation of the mesh at shocks, which are propagated at a speed different from that of the two converging characteristics at the shock. We refer to such a method as an oriented Galerkin method (orG-method).

3.4 An analysis of chG(1,0)

We will analyze chG(1,0) in the form (3.71) through the following sequence of steps. We start with the inviscid case with $\epsilon = 0$. We first show that if a bounded sequence of chG(1,0)-solutions $\{U\}$ converges pointwise to a limit function u as the mesh size h tends to zero, then the limit u is (i) a weak solution, (ii) satisfies all entropy inequalities, that is, we prove (i) weak consistency and (ii) entropy consistency. By the uniqueness of a weak solution satisfying all entropy inequalities in the present scalar case, we thus can assure that a limit of chG(1,0)-solutions must be equal to the exact solution. Next, we prove L_2 a posteriori error estimates of the form (3.78), and we also indicate how to turn this estimate into an a priori estimate using an energy norm stability estimate to give the needed a priori information on the regularity of the chG(1,0) solution. Finally, we derive analytical estimates of the stability factors for the dual linearized problems in model cases associated to smooth solutions, shocks and rarefactions, showing that these factors are of moderate size. We conclude that the basic types of solutions of scalar conservation laws are computable in L_2 norms with essentially optimal work compared to interpolation.

We will assume below that the quantity $\alpha \equiv \max_n \bar{h}_n^2/k_n$ tends to zero as $\bar{h}$ tends to zero. This requires the time steps k_n not to be too small, but is not in conflict with the requirement of choosing sufficiently small time steps in order to avoid crossing of characteristics.

A basic energy stability estimate Choosing $v = U$ in (3.71), integrating by parts and summing over n, we obtain the following familiar basic energy estimate for $N \geq 1$,

$$\|U_N^-\|^2 + \sum_{n=1}^N \|[U_{n-1}]\|^2 + 2\sum_{n=1}^N (\hat{\epsilon}U', U')_n^+ k_n \leq \|u_0\|^2. \tag{3.75}$$

A limit of chG(1,0)-solutions is a weak solution To prove that a pointwise limit of a bounded sequence of chG-solutions $\{U\}$, is a weak solution of (2.41), we let φ be an arbitrary smooth test function with compact support and choose in (3.71) the function $v = \pi_h\varphi \in W_{kn}^\beta$ to interpolate φ_{n-1} at the nodes of V_n. Integrating by parts, replacing v by φ in the first term using the fact that $\dot{U} + A(U)U' = 0$, and summing over n, we then obtain

$$0 = \int_{\mathbb{R}} \int_0^{t_N} -(U\dot{\varphi} + f(U)\varphi')dx\,dt$$

$$+ \sum_{n=1}^N (\hat{\epsilon}U', (\pi_h\varphi)')_n^+ k_n + \sum_{n=1}^N ([U_{n-1}], (\pi_h\varphi)_{n-1} - \varphi_{n-1}).$$

By the basic energy estimate (3.75), it follows as above using that φ is smooth (with compact support), that the second term on the right hand side tends to zero. Further, for the third term, we have by a standard interpolation estimate bringing in a factor h_n^2, recalling that $\alpha = \max_n \bar{h}_n^2/k_n$, and letting $\bar{k} = \max_n k_n$,

$$\sum_n |([U_{n-1}], (\pi_h\varphi)_{n-1} - \varphi_{n-1})| \leq C_i C_\varphi (\sum_n \|[U_{n-1}]\|^2)^{1/2} \alpha(\bar{k}\sum_n k_n)^{1/2},$$

where C_i is an interpolation constant and C_φ a constant depending on φ, which using (3.75) surely tends to zero as $\bar{h}$ and $\bar{k}$ tend to 0. This proves that a limit of chG(1,0) solutions is a weak solution of (2.41).

A limit of chG(1,0)-solutions satisfies all entropy inequalities To prove entropy consistency, let now $\eta(u)$ be a smooth strictly convex entropy with corresponding entropy flux $q(u)$. In (3.71) we then choose $v = \pi_h(\eta'(U)\varphi) \in W_{kn}^\beta$, where φ now is a non-negative smooth function with compact support, and obtain integrating by parts using again the fact that in the integration over the slab S_n the interpolation operator π_h can be removed, while changing from $\pi_h(\eta'(U)\varphi)$ to $\eta'(U)\varphi$ in the jump term on the discrete time levels brings in the factor $\psi_{n-1} = \eta'(U_{n-1}^+)\varphi_{n-1} - \pi_n(\eta'(U_{n-1}^+)\varphi_{n-1})$, with π_n the usual nodal interpolation operator into V_n,

$$0 = -\int_0^{t_N} \int_{\mathbb{R}} \eta(U)\dot\varphi \, dx \, dt - \int_0^{t_N} \int_{\mathbb{R}} q(U)\varphi' \, dx \, dt$$

$$+ \sum_{n=1}^N (\hat\epsilon U', (\pi_n\eta'(U)\varphi)')_n^+ k_n - \sum_{n=1}^N ([U_{n-1}], \psi_{n-1}^+)$$

$$+ \sum_{n=1}^N (\eta(U_{n-1}^-) - \eta(U_{n-1}^+) - \eta'(U_{n-1}^+)(U_{n-1}^- - U_{n-1}^+), \varphi_{n-1})$$

$$= -I - II + III - IV + V,$$

with the obvious notation. Here, the terms I and II will generate the desired terms in the entropy inequality. By the convexity of η, and the positivity of φ, the term V is non-negative. We will now show, using in a crucial way the design of the artificial viscosity $\hat\epsilon$, that III will generate a positive term through which we will be able to dominate IV. We then first split III into

$$\sum_n (\hat\epsilon U', (\pi_n\eta'(U)\varphi)')_n^+ k_n = \sum_n (\hat\epsilon U', (\pi_n\eta'(U))'\varphi)_n^+ k_n$$

$$+ \sum_n (\hat\epsilon U', (\pi_n\eta'(U)\varphi)' - (\pi_n\eta'(U))'\varphi)_n^+ k_n = III_1 + III_2.$$

Now, it is straight forward to show that III_2 will vanish in the limit, using the fact that $(\pi_n\eta'(U)\varphi)' = (\pi_n\eta'(U))'\varphi$ if φ is constant, from which follows that in III_2 we may replace φ by $\varphi - \bar\varphi$ with $\bar\varphi$ a piecewise constant approximation. Thus, we may focus our interest on III_1, and we then make the crucial observation that

$$(\hat\epsilon U', (\pi_n\eta'(U))'\varphi)_n^+ k_n \geq c(\hat\epsilon U', U'\varphi)_n^+ k_n$$

where c is a lower bound of η'', recalling the strict convexity of η. This follows because on an interval $I_j = (x_{j-1}, x_j)$ underlying V_n of length h_{nj} with $W = U_{n-1}^+$ as above,

$$W'(\pi_n \eta'(W))' = h_{nj}^{-2}(W(x_j) - W(x_{j-1}))(\eta'(W(x_j)) - \eta'(W(x_{j-1})))$$
$$\geq c h_{nj}^{-2}(W(x_j) - W(x_{j-1}))^2 = c(W')^2.$$

We conclude the following crucial positivity property of the artificial viscosity term III:

$$III_1 \geq c \sum_n (\hat{\epsilon} U', U' \varphi)_{n-1}^+ k_n.$$

We now return to the critical term IV with the objective of bounding this term using the positive term III_1. To this end, we note that by a standard interpolation estimate, we have

$$|([U_{n-1}], \psi_{n-1})| \leq C_i \int_{\mathbb{R}} h_n^2 \frac{\|[U_{n-1}]\|}{k_n} (|\eta'''(W)|(W')^2 \varphi_{n-1}$$
$$+ C_\varphi |\eta''(W) W'| + C_\varphi |\eta'(W)|) k_n \, dx, \tag{3.76}$$

which follows by computing the derivative $(\eta'(W) \varphi_{n-1})''$ element by element, noting in particular that $W'' = 0$ on each element. Now, if $\bar{h}$ is sufficiently small, then by the design property (3.69), using the strict convexity of η to see that $|\eta'''|/\eta''$ is bounded from above,

$$\int_{\mathbb{R}} h_n^2 \frac{\|[U_{n-1}]\|}{k_n} C_i |\eta'''(W)|(W')^2 \varphi_{n-1} k_n \leq c(\hat{\epsilon} U', U' \varphi)_n^+ k_n. \tag{3.77}$$

The remaining terms on the left hand side of (3.76) may be dominated similarly (with a good margin) using also (3.75). Thus, the crucial term V can be dominated by the positive artificial viscosity term III_1 together with (3.75), and we conclude that a limit of chG(1,0)-solutions, satisfies

$$-\int_0^t \int_{\mathbb{R}} \eta(w) \dot{\varphi} \, dx \, dt - \int_0^t \int_{\mathbb{R}} q(w) \varphi' \, dx \, dt \leq 0,$$

which is the weak form of the entropy inequality for given entropy $\eta(u)$.

What is the minimal amount of artificial viscosity? The above analysis shows that the design property $\hat{\epsilon} \geq C h^{2-\kappa} R$ with κ small positive, is sufficient to guarantee entropy consistency for chG(1,0), the critical inequality being (3.77) containing the ingredients h_n^2, $R(U) = \|[U_{n-1}]\|/k_n$, $\hat{\epsilon}$, and $((U_{n-1}^+)')^2$. The analysis also indicates that to guarantee entropy consistency, we need $h^2 R / \hat{\epsilon}$ to be sufficiently small.

The design of the artificial viscosity for the Sd-method with non-oriented space-time mesh is similar: $\hat{\epsilon} \sim h^2 R(U)$, with $R(U)$ the full residual including also a contribution $|\dot{U} + (f(U))'|$ from the space-time slab.

The essence of the argument showing entropy consistency can be exhibited in the setting of a Galerkin method for a nonlinear problem $L(u) = 0$, for which the entropy inequality for a given entropy $\eta(u)$ is obtained multiplying by $\eta'(u)\varphi$, where φ is a smooth non-negative test function with compact support. To prove entropy consistency we choose $v = \pi_h(\eta'(U)\varphi) \in V_h$ in the Galerkin equations $(L(U), v) + (\hat{\epsilon}U', v') = 0$ for all $v \in V_h$, where V_h is the finite element space underlying the Galerkin method, π_h is an interpolation opertator into V_h, and $\hat{\epsilon}$ is the artificial viscosity. Replacing then $\pi_h(\eta'(U)\varphi)$ by $\eta'(U)\varphi$ in the first term, brings in the crucial compensating term

$$E = (L(U), \eta'(U)\varphi - \pi_h(\eta'(U)\varphi)),$$

which will have to be dominated by the positive term $(\hat{\epsilon}U', \eta''U'\varphi)$ resulting from the artificial viscosity term. By standard interpolation estimates computing the derivative $(\eta'(U)\varphi)'$, we have, writing out only the most critical term with no differentiation of φ,

$$E \leq (C_i h^2 R(U)U', U'\varphi)$$

where $R(U) = |L(U)|$ is the residual. If now $\hat{\epsilon} = Ch^2 R(U)$ with C large enough, then this term will be dominated by the indicated positive term, and the desired entropy inequality will follow.

3.5 Adaptive algorithms

Adaptive algorithms for chG(1,0) are based on a posteriori error estimates of the form

$$\|u - U\| \leq C_i S_c \|\frac{h^2}{\hat{\epsilon}} R(U)\| + E(\hat{\epsilon} - \epsilon), \qquad (3.78)$$

where $E(\hat{\epsilon} - \epsilon)$ accounts for the perturbation effect of the artificial viscosity with $E(0) = 0$, and S_c is a strong stability factor. The norm $\| \cdot \|$ may be an $L_q(L_p)$-norm in time-space. Similar estimates may be derived for the Sd and orG-methods. The modification criterion of the adaptive algorithm may include modification of (i) mesh size and (ii) $\hat{\epsilon}$. In particular, we may seek to adaptively achieve $\hat{\epsilon} = \epsilon$, in which case the second term in (3.78) disappears. This may be attempted if ϵ is not too small, and corresponds to pointwise resolution of the exact solution. The mesh modification may be based on equidistribution, seeking to make the element contribution to the norm $\|h^2 \hat{\epsilon}^{-1} R(U)\|$ equal. In orG variants also the mesh orientation may be modified for example according to the space-time gradients of computed solutions. For numerical results illustrating the effect of different orientations in space-time, see [22].

3.6 A posteriori error estimates

As an illustration of the general approach, we now prove an a posteriori error estimate for the following variant of the chG(1,0) method for (2.39) with $\epsilon > 0$ and $A(U) = U$ considering Burgers' equation: find $U = U(x,t)$ such that, for $n = 1, 2, \ldots, U|_{S_n}$ satisfies

$$\dot{U} + UU' - \epsilon U'' = 0, \quad (x,t) \in \mathbb{R} \times I_n,$$
$$U(x,t) \to 0, \quad x \to \pm\infty, \ t \in I_n, \qquad (3.79)$$
$$U_{n-1}^+ = P_n U_{n-1}^-,$$

where $U_0^- = u_0$. In this method we L_2-project at each discrete time level t_{n-1} into a space V_n of continuous piecewise linears of mesh size $h_n(x)$, and solve the conservation law exactly on each slab S_n, which corresponds to a method of the form projection-transport. For simplicity we will assume that $\hat{\epsilon} = \epsilon$, a condition which may be included in the stopping criterion if ϵ is not too small, as indicated above. We use the notation $Q = \mathbb{R} \times (0,T)$ and write $\|v\|_Q = \|v\|_{L_2(Q)}$.

Let now $T = t_N > 0$ be a given final time, and let us seek an *a posteriori* error estimate for the error $\|u - U\|_Q$. To this end, we start representing the error in terms of the solution φ of the following linearized dual problem:

$$-\dot{\varphi} - a\varphi' - \epsilon\varphi'' = e, \ x \in R, \ 0 < t < T,$$
$$\varphi(x,t) \to 0, \ x \to \pm\infty, \ 0 < t < T, \qquad (3.80)$$
$$\varphi(x,T) = 0, \ x \in \mathbb{R},$$

where $a = (u + U)/2$. Multiplying (3.80) by e, we obtain by integration by parts over each S_n, $n = 1, \ldots, N$, since $(ae)' = \frac{1}{2}(u^2 - U^2)' = uu' - UU'$,

$$\|e\|_Q^2 = \sum_{n=1}^N \int_{S_n} e\left(-\dot{\varphi} - a\varphi' - \epsilon\varphi''\right) \, dxdt$$

$$\sum_{n=1}^N \int_{S_n} \left((\dot{u} + uu' - \epsilon u'') - (\dot{U} + UU' - \epsilon U'')\right) \varphi \, dxdt$$

$$-\sum_{n=1}^N \int_{\mathbb{R}} (I - P_n)U_{n-1}^- \varphi_{n-1} dx,$$

where we used that $U_{n-1}^+ = P_n U_{n-1}^-$. Thus, by (2.39), (3.79) and the orthogonality of the L_2 projection P_n, we obtain the following representation of the error:

$$\|e\|_Q^2 = -\sum_{n=1}^N \int_{\mathbb{R}} (I - P_n)U_{n-1}^-(I - P_n)\varphi_{n-1} \, dx.$$

Recalling now the basic estimate for the L_2 projection error $(I - P_n)\varphi_{n-1}$:

$$\|(I - P_n)\varphi_{n-1}\| \leq C_i\|h_n^2\varphi_{n-1}''\|,$$

we obtain by Cauchy's inequality

$$\|e\|_Q^2 \leq C_i \left(\sum_{n=1}^{N} \| \frac{h_n^2}{\epsilon k_n}(I - P_n)U_{n-1}^- \|^2 k_n \right)^{1/2} \left(\sum_{n=1}^{N} \left\| \epsilon\varphi_{n-1}'' \right\|^2 k_n \right)^{1/2}.$$

Defining the strong stability factor S_c by

$$S_c = \frac{\left(\sum_{n=1}^{N} \left\| \epsilon\varphi_{n-1}'' \right\|^2 k_n \right)^{1/2}}{\|e\|_Q}, \tag{3.81}$$

and writing

$$\| \frac{h^2}{\epsilon k}(I - P)U \|_{L_2(Q)} = \left(\sum_{n=1}^{N} \| \frac{h_n^2}{\epsilon k_n}(I - P_n)U_{n-1}^- \|^2 k_n \right)^{\frac{1}{2}},$$

we then obtain the following a posteriori error estimate:

Theorem 3.1 *The solution U of (3.79) satisfies*

$$\|u - U\|_Q \leq S_c C_i \| \frac{h^2}{\epsilon k}(I - P)U \|_Q, \tag{3.82}$$

where S_c is defined by (3.81).

This gives an estimate of the computational error $u - U$ in terms of the projection or interpolation error $(I - P)U$, modulo the strong stability constant S_c, and the factor $\frac{h^2}{\epsilon k}$. Below we show that S_c is bounded by a moderate constant in the basic cases of a (i) smooth exact solution, (ii) shock, and (iii) rarefaction, where in the last case the $L_2(Q)$-norm is modified to a weighted variant. Assuming that the factor $\frac{h^2}{\epsilon k}$ is less than one, which may be natural, (3.82) simplifies to

$$\|u - U\|_Q \leq S_c C_i \|(I - P)U\|_Q. \tag{3.83}$$

3.7 A priori error estimate

The above a posteriori error estimate can be turned into an a priori error estimate if we can a priori demonstrate some regularity of U_{n-1}^-, from which we may deduce an estimate for $\|(I-P)U_{n-1}^-\|$, and thus for $\|(I - P)U\|$. The basic energy estimate guarantees that the quantity

$$\sum_{n}(\hat{\epsilon}\nabla U_{n-1}^+, \nabla U_{n-1}^+)$$

is bounded. A similar bound holds with U_{n-1}^+ replaced by U_n^-, if $k_n\|U\|_\infty \leq h$ guaranteeing that the mesh distortion over one slab S_n is controled.

3.8 Stability estimates for smooth solutions, shocks and rarefactions

We now investigate the stability properties of the linearized dual Burgers' equation in some basic cases including smooth solutions, shocks and rarefactions. Recalling (3.80), the dual problem takes the following form if we now linearize at an exact solution $u(x,t)$ (instead of the mean value $(u + U)/2$):

$$-\dot{\varphi} - u\varphi' - \epsilon\varphi'' = e, \ x \in \mathbb{R}, \ 0 < t < T,$$
$$\varphi(x, T) = 0, \ x \in \mathbb{R}. \tag{3.84}$$

The stability properties are largely determined by the sign of u', which reflects the change of the direction u of the characteristics. If $u' \leq 0$, then the characteristics converge with increasing t, which typically occurs in the case of a shock. If $u' \geq 0$, then the characteristics diverge, which typically occurs in the case of a rarefaction. If u' is bounded below by a moderate constant, e.g. $u' \geq 0$, then we may estimate $\|\varphi\|_{L_\infty(L_2(\mathbb{R}))}$ and $\|\sqrt{\epsilon}\varphi'\|_Q$ in terms of a moderate constant times $\|e\|_Q$, which we refer to as *weak stability*. If u' is bounded above by a moderate constant, e.g. $u' \leq 0$, then we may estimate $\|\epsilon\varphi''\|_Q$ in terms of a moderate constant times $\|e\|_Q$, which we refer to as *strong stability*, because we estimate second derivatives of φ, cf. (3.81). These estimates are proved by multiplying by φ and $-\epsilon\varphi''$, respectively, bringing in the positive stabilizing terms $\frac{1}{2}u'\varphi^2$ and $-\frac{1}{2}\epsilon u'(\varphi')^2$, respectively.

Shocks We now prove a strong stability estimate showing that the continuous analog $\|\epsilon\varphi''\|_Q/\|e\|_Q$ of the stability factor S_c defined by (3.81), is of moderate size if u' is bounded above. This assumption is satisfied if u is piecewise smooth with shocks, recalling that u is decreasing in a shock, so that there u' is large negative and in particular $u' \leq 0$. We plot the characteristics of the dual problem in this case below. We thus give evidence that the stability factor S_c of the a posteriori error estimate (3.82) is of moderate size if u is piecewise smooth with shocks, which indicates that such a solution is computable with a computational cost comparable to that of direct interpolation. Thus, a piecewise smooth solution with shocks is not so costly to compute.

Lemma 3.1 *Let $A = \max_Q u'$. The solution φ of (3.84) satisfies*

$$\|\epsilon\varphi''\|_Q + \|\dot{\varphi} + u\varphi'\|_Q + \sup_{0<t<T} \|\epsilon^{1/2}\varphi'(\cdot,t)\|_R \leq 3\exp(AT)\|e\|_Q. \tag{3.85}$$

Proof 3.1 Multiplying the first equation in (3.84) by $-\epsilon\varphi''$, integrating by parts with respect to x, and integrating in time over (τ, T), we get with

$$Q_\tau = R \times (\tau, T)$$

$$\frac{1}{2} \int_R \epsilon \left(\varphi'(\cdot, \tau)\right)^2 dx + \int_{Q_\tau} \left(\epsilon \varphi''\right)^2 dx dt + \int_{Q_\tau} \frac{1}{2} \left(u \epsilon \left(\varphi'\right)^2\right)' dx dt$$

$$\leq \frac{1}{2} \int_{Q_\tau} \left(\epsilon u' \left(\varphi'\right)^2 + e^2 + \left(\hat{\epsilon} \varphi''\right)^2\right) dx dt,$$

(3.86)

which proves the lemma using a Grönwall's inequality, since the last term on the left hand side integrates to zero.

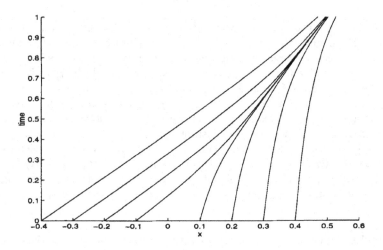

Let us now attempt to derive a weak stability estimate for (3.84) in the case u is a shock. Multiplication by φ and integration over Q_τ gives

$$\frac{1}{2} \int_R \varphi^2(x, \tau) \, dx + \epsilon \int_{Q_\tau} \left(\varphi'(x, t)\right)^2 dx dt$$

$$= -\frac{1}{2} \int_{Q_\tau} u' \varphi^2(x, t) \, dx + \int_R e(x, t) \varphi(x, t) \, dx.$$

Since u' is large negative in the case of a shock, we have large positive term of the right hand side, and using Grönwall's lemma, results in a very large stability factor (of size $\exp(-u')$). The situation is more favorable concerning weak stability estimates for a rarefaction wave solution with $u' \geq 0$, as we demonstrate below.

Summing up, we see that for a shock, the linearized dual problem satisfies a strong stability estimate with a stability factor of moderat size, while a corresponding weak stability estimate appears to have a very large stability factor. These stability features may be understood in a qualitative sense, by pondering the directionality of the characteristics and the nature of the L_2-norm.

Rarefaction wave We now consider the linearized dual Burgers' equation (3.84), linearized at the exact solution $u(x,t) = x/t$, corresponding to a rarefaction wave. Multiplying now (3.84) by $-\epsilon t\varphi''$, and using standard manipulations, we obtain the following weighted norm strong stability estimate for $0 < \tau < T$,

$$\|\tau^{1/2}\epsilon^{1/2}\varphi'\| + \|\omega\epsilon\varphi''\|_{Q_\tau} \leq \|\omega e\|_{Q_\tau}, \qquad (3.87)$$

where $\omega(t) = t^{1/2}$ acts as a weight, and $Q_\tau = \mathbb{R} \times (\tau, T)$.

A weighted norm analog of the a posteriori error estimate (3.83) takes the form

$$\|\omega^{-1}e\|_{L_2(Q_\tau)} \leq S_c\|\omega^{-1}(I - P)U\|_{L_2(Q_\tau)} \qquad (3.88)$$

with S_c defined by the direct weighted norm analog of (3.81). The estimate (3.87) then shows that S_c is of moderate size, and thus a rarefaction wave solution is computable in the weighted norm with computational work corresponding to interpolation. Note that the presence of the weight $t^{-1/2}$ will force more stringent demands on the mesh for t close to zero, which will force an accurate resolution of the initial phase of the rarefaction. This is intuitively reasonable and corresponds to the fact that an initial error in the computation of a rarefaction will get amplified as time goes, because characteristics diverge forward in time. On the other hand, in the case of a shock, an initial error may be eliminated at later times, because of converging characteristics. Thus, a rarefaction is more delicate to compute than a shock.

3.9 Dual-weighted a posteriori error estimates

One may also derive a posteriori error estimates with derivatives of the dual solution occuring as weights in the error representation integral. Starting from (3.82), such a dual-weighted a posteriori error estimate takes the form

$$\|e\|_Q^2 \leq C_i \sum_{n=1}^{N} \int_{\mathbb{R}} \frac{h_n^2}{\epsilon k_n}|(I - P_n)U_{n-1}^-|\omega_{n-1}\ dx\ k_n, \qquad (3.89)$$

where $\omega_{n-1} = \epsilon|\varphi_{n-1}''|$, and φ is the solution of the dual problem (3.80). This estimate may give a more precise error representation, and thus lead to a more optimally adapted mesh size $h(x)$, than the estimate (3.82) with the stability factor S_c as a scalar factor outside the norm.

3.10 A posteriori error estimates in a negative norm

In principle, we may seek to control the error in any given norm by computing the corresponding stability factor by solving a dual problem with suitable data. The norm $L_\infty(H^{-1}(\mathbb{R}))$ offers certain advantages from analysis point of view, and connects to the following data for a dual linearized Burgers' equation, linearized at the solution $u(x, t)$:

$$-\dot{\varphi} - u\varphi' - \epsilon\varphi'' = 0, \quad x \in R, \ 0 < t < T,$$
$$\varphi(x, t) \to 0, \quad x \to \pm\infty, \ 0 < t < T, \quad (3.90)$$
$$\varphi(x, T) = E, \quad x \in R,$$

where $-E'' = e(T)$, $E(x, t) \to 0$ as $x \to \pm\infty$, $0 < t < T$. In this problem, we may multiply by φ'', and derive in the shock wave case with $u' \leq 0$, a stability estimate of the form

$$\max_{0 < \tau < T} (\|\varphi'(\tau)\|^2 + \|\sqrt{\epsilon}\varphi''\|_{Q_\tau}^2) \leq \|\varphi'(T)\|^2 = \|e\|_{H^{-1}}^2. \quad (3.91)$$

An analog of the a posteriori error estimate (3.82) in the norm $L_\infty(H^{-1}(\mathbb{R}))$, reads

$$\|(u - U)(T)\|_{H^{-1}(\mathbb{R})} \leq S_c C_i \|\frac{h^2}{\sqrt{\epsilon k}}(I - P)U\|_{L_\infty(L_2(\mathbb{R}))}, \quad (3.92)$$

with the stability factor is defined by

$$S_c = \frac{\int_0^T \|\sqrt{\epsilon}\varphi''\| dt}{\|\varphi'(T)\|}. \quad (3.93)$$

We notice a gain of a factor $\sqrt{\epsilon}$ as compared to (3.82) due to the asymmetry of norms in (3.92), with the weaker $H^{-1}(\mathbb{R})$-norm on the left hand side.

We now present computed approximations of S_c defined by (3.93) for a typical solution of Burgers' equation, together with plots of dual solutions. Below we present similar results for the compressible Euler equations in 1d, and further results in 2d including two-phase flow are presented in [3]. In all these cases, S_c turns out to be of moderate size, indicating computability of the corresponding solutions (none of which is turbulent). The stability factor S_c related to the norm $L_2(L_2(\mathbb{R}))$ defined by (3.81), is of a similar size in these examples.

3.11 Dual solution and stability factors for Burgers' equation

We consider a case with the exact solution consisting of a rarefaction wave and a shock and $10^{-4} \leq \epsilon \leq 10^{-5}$. We use the cG(1)dG(0) Sd-method on a uniform space mesh with $h = 10^{-3}$ and time step 10^{-4}. We plot the computed solution at t=0, t=0.3 , t=0.8, t=1.

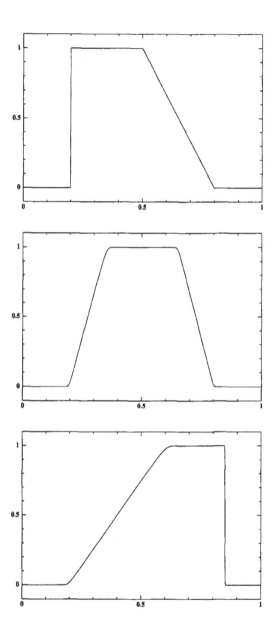

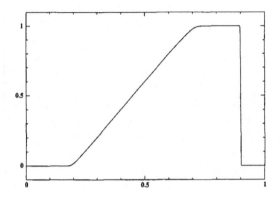

We have solved the dual problem using the following approximations of the coefficient $a = (u + U)/2$ and the error e, with $\bar{u}$ the analytical, inviscid solution, and $U(h)$ the finite element solution on a mesh of size h: (i) $a = (\bar{u} + U(h))$ and $e = \bar{u} - U(h)$, (ii) $a = U(h)$ and $e = \bar{u} - U(h)$, (iii) $a = (U(h/4) + U(h))/2$ and $e = U(h/4) - U(h)$. We plot the corresponding dual solution φ at the same time levels as above, but in reverse order (t=1, t=0.8 , t=0.3 , t=0):

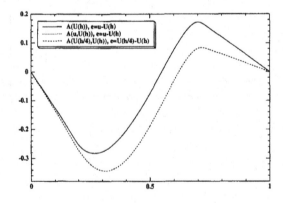

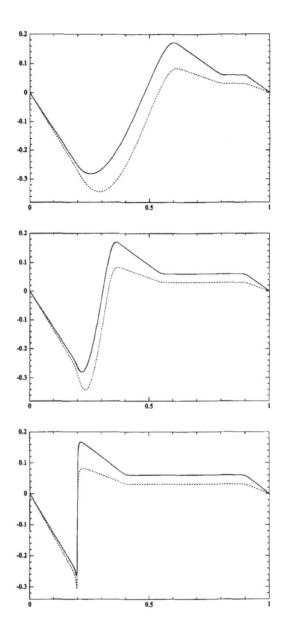

In the following figure we plot the corresponding second derivatives φ'' including a zoom at $t = 0$ around $x = 0.2$:

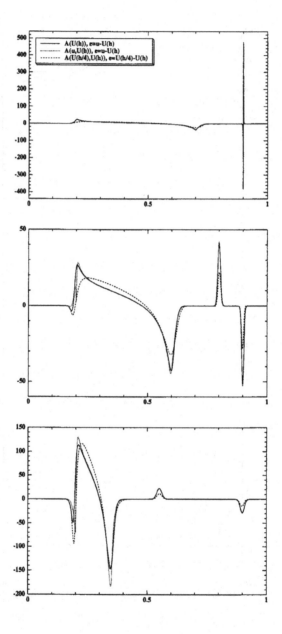

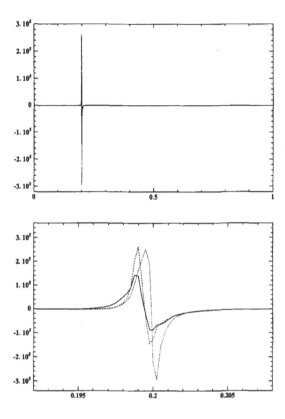

We note the change of the weight $|\varphi''|$ from being large close to the shock at final time towards being large close to the initial discontinuity at $(x, t) = (0.2, 0)$. Note that it is the data from the rarefaction at final time which generates the large values of φ'' at $t = 0$, and not those from the shock.

We next plot the stability factor S_c defined by (3.93) for the three different approaches using three different values on the viscosity ($\epsilon = 0.0001$, $\epsilon = 0.00005$, $\epsilon = 0.00001$):

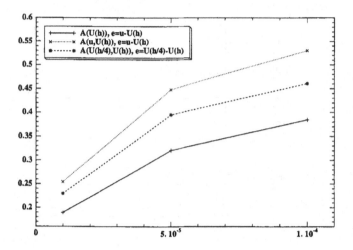

The stability factor defined by (3.81) is of similar size. The results indicate that the Burgers' solution is computable in $L_\infty(H^{-1}(\mathbb{R}))$ and $L_2(L_2(\mathbb{R}))$ with work comparable to interpolation.

3.12 Scalar conservation laws in several dimensions

The material presented above directly generalizes to scalar conservation laws in $\mathbb{R}^d$, $d \geq 1$, of the form

$$\dot{u} + \sum_{i=1}^d f_i(u)_{,i} - \epsilon \Delta u = 0, \quad x \in \mathbb{R}^d, \ t > 0,$$
$$u(x, 0) = u_0(x), \quad x \in \mathbb{R}, \tag{3.94}$$

where $f_i(u)$ is the flux in the x_i direction, and $v_{,i} = \frac{\partial v}{\partial x_i}$. Again, any convex function $\eta(u)$ may be chosen as an entropy, and the corresponding entropy inequality takes the strong form

$$\frac{\partial}{\partial t}\eta(u) + \sum_i \frac{\partial}{\partial x_i} q_i(u) \leq 0, \tag{3.95}$$

where the $q_i(u)$ are the entropy fluxes satisfying $q_i' = f_i'\eta'$.

3.13 The dG-method for scalar conservation laws

We now further generalize to a Galerkin method based on discontinuous piece-wise polynomial approximation of degree q in both space and time, for the inviscid scalar conservation law (3.94) with $\epsilon = 0$. We refer to this method as the dG(q)-method. The dG(0) method coincides with the basic *finite volume method* based on piecewise constant approximation, which attracted much attention in the early development of numerical methods for conservation laws. The dG(q) with $q > 0$ generalizes the basic finite volume method, and gives a solution to the problem of designing high order finite volume methods including the problem of defining fluxes at discontinuities (so called numerical fluxes) and the design of the artificial viscosities, which has been the subject of extensive work in the literature on numerical methods for conservation laws. For a more detailed presentation of the dG method for scalar conservation laws, we refer to [33].

To discretize (3.94) with the dG-method, let as usual $0 = t_0 < t_1 < \ldots < t_n < \ldots$, be a sequence of discrete time levels, let $T_n = \{K\}$ be a (space-time) finite element triangulation of the slab $S_n = \mathbb{R}^2 \times I_n$, where $I_n = (t_{n-1}, t_n)$, into finite elements K of diameter h_K. Typically, the elements K may be tetrahedra or prisms $\kappa \times I_n$, where κ is a triangle, or space-time oriented prisms. Define for a given $q \geq 0$,

$$W_n = \{v \in L_2(S_n) : v|_K \in P_q(K), \quad \forall K \in T_n\},$$

where $P_q(K)$ is a space of polynomials of degree at most q associated with $K \in T_n$, and write $W = \prod_{n \geq 0} W_n$. As above we assume that the functions $v(x,t)$ in W_n vanish for $|x|$ large. With piecewise constant approximation corresponding to $q = 0$, we obtain the dG(0)-method. Choosing piecewise linear approximation on tetrahedra or piecewise bilinear approximation on prisms, we obtain different versions of dG(1). We may also obtain hybrid variants on prisms with piecewise linear approximation in space and piecewise constant approximation in time, and oriented versions thereof.

To formulate the dG(q) method we first rewrite (3.94) in the compact form div $f(u) = 0$, where (assuming for definiteness that $d = 2$), $f(u) = (f_i(u)) \equiv (f_0(u), f_1(u), f_2(u))$ with $f_0(u) = u$ and div now denotes the divergence with respect to $x \equiv (x_0, x_1, x_2)$ with $x_0 = t$, i.e.,

$$\text{div } f(u) = \sum_{i=0}^{2} f_i(u)_{,i} = u_t + \sum_{i=1}^{2} f_i(u)_{,i}. \tag{3.96}$$

The dG(q)-method for (3.94) can now be formulated as follows: find $U \in W$ such that for $n = 0, 1, \ldots,$

$$
\begin{aligned}
&\sum_{K \in T_n} \{ \int_K \text{div} f(U)(v + \delta \sum_{i=0}^{2} f_i'(U) v_{x_i}) dx + \int_K \hat{\epsilon} \nabla U \cdot \nabla v dx \\
&+ \int_{\partial K} (F_K(U) - f(U_K) \cdot n_K) v_K ds \} = 0, \qquad \forall v \in W_n,
\end{aligned} \tag{3.97}
$$

where ∇ is the gradient w.r.t. $x = (x_0, x_1, x_2)$, $U_K = U|_K$, $n_K = n_K(x)$ is the outward unit normal to K at $x \in \partial K$, $F_K(U)$ is the numerical flux (referred to as the Lax-Friedrichs' flux) given by

$$F_K(U)(x) = \frac{1}{2}(f(U_K(x)) + f(U_{K'}(x))) \cdot n_K(x) + C_K(x)(U_K - u_{K'})(x), \tag{3.98}$$

for $x \in \partial K \cap \partial K'$, where K' (for a given $x \in \partial K$) denotes the neighboring element to K, $(K' \in T_m$ with $m = n - 1, n$ or $n + 1)$, and for $x \in \partial K \cap \partial K'$, $f \cdot n_K = \sum_{i=0}^{2} f_i n_{K,i}$, $n_K = (n_{K,i})$,

$$C_K(x) = \tfrac{1}{2} \quad \text{if } n_K(x) = \pm(1,0,0),$$
$$C_k(x) = C_{K'}(x) = C_0 \quad \text{otherwise,}$$

where C_0 is a positive constant satisfying

$$C_0 > \tfrac{1}{2}\|f'\|_\infty = \max_{r \in \mathbb{R}} |f'(r)|, \quad |f'| = (\textstyle\sum_{i=0}^{2}(f_i')^2)^{\frac{1}{2}},$$
$$U_K = U|_K, \quad U_K(x) = \lim_{\substack{y \to x \\ y \in K}} U(y), \quad U_{K'}(x) = \lim_{\substack{y \to x \\ y \in K'}} U(y),$$
$$U_{K'}(\bar{x}, 0) = u^0,$$
$$\delta(U) = \tfrac{1}{2}(k_n^{-2} + h_n^{-2}|f'|)^{-1/2},$$
$$\hat{\epsilon}(U) = \max(\gamma_1 h^2 R(U), \gamma_2 h^{3/2}),$$
$$R(U)|_K = \max_K |\mathrm{div}\ f(U)| + \max_{\partial K} |f(U_K) - f(U_{K'})|/h_K.$$

The dG(0) method is an implicit variant of the classical finite volume method based on piecewise constant approximation and the Lax-Friedrichs' flux, and the dG(q) method is a generalization of order of accurcy at least $q + \tfrac{1}{2}$.

4 The Euler equations for an inviscid perfect gas

In this section we extend the material presented above to the Euler equations for an inviscid perfect gas, including details on the formulation of the G^2-method, and computational results on dual solutions and stability factors. The results on entropy consistency and a posteriori error estimation generalize in a direct way. We also discuss the use of entropy variables.

The Euler equations for a compressible inviscid perfect gas in $\mathbb{R}^3$ read on conservation form: find $u(x, t)$ such that

$$\dot{u} + \sum_{i=1}^{3} f_i(u)_{,i} = 0 \quad x \in \mathbb{R}^3,\ t > 0,$$
$$u(x, 0) = u_0(x) \quad x \in \mathbb{R}^3, \tag{4.99}$$

where u_0 is given initial data,

$$u = \rho \begin{bmatrix} 1 \\ w_1 \\ w_2 \\ w_3 \\ e \end{bmatrix}, \quad f_i = w_i u + p \begin{bmatrix} 0 \\ \delta_{1i} \\ \delta_{2i} \\ \delta_{3i} \\ w_i \end{bmatrix},$$

ρ is the density, $w = (w_1, w_2, w_3)$ is the particle velocity, e is the total energy density, $p = (\gamma - 1)(\rho e - \rho|w|^2)/2)$ is the pressure, δ_{ij} the Kronecker delta,

$\gamma > 1$ is a constant, and $v_{,i} = \partial v/\partial x_i$. These equations generalize the one-dimensional equations (2.46) and express conservation of mass, momentum and energy. For smooth solutions the conservation law (4.99) can be written as a generalized convection problem of the form

$$\dot{u} + \sum_{i=1}^{3} A_i(u)u_{,i} = 0, \qquad (4.100)$$

where the $A_i = \frac{\partial f_i}{\partial u}$ are the Jacobians of the $f_i(u)$.

The function $\eta(u) = \rho\log(p\rho^{-\gamma})$ is a mathematical entropy for (4.99) corresponding to the negative of the physical entropy, which up to trivial modifications is the only known entropy for (4.99). The function $\eta(u)$ is a convex function of u, with symmetric positive definite Hessian η''. Smooth solutions of (4.99) or equivalently (4.100), satisfy the equation

$$\frac{\partial}{\partial t}\eta(u) + \sum_{i=1}^{3} \frac{\partial}{\partial x_i}(q_i(u)) = 0.$$

where $q(u) = (q_i(u)) = \eta w = (\eta w_i)$ is the entropy flux. This equation follows by multiplying (4.100) by the gradient $\eta'(u)$ of $\eta(u)$ and using the compatibility relation

$$\eta'(u)^* A_i(u) = q'_i(u)^*, \qquad (4.101)$$

with $*$ denoting the transpose. The compatibility condition (4.101) is equivalent to the relation

$$\eta'' A_i = A_i^* \eta'', \quad i = 1, 2, 3, \qquad (4.102)$$

stating that the Hessian η'' simultaneously symmetrizes all the A_i. The entropy inequality characterizing physical weak solutions reads in strong form

$$\frac{\partial}{\partial t}\eta(u) + \sum_i \frac{\partial}{\partial x_i}(q_i(u)) \leq 0. \qquad (4.103)$$

4.1 The G²-method

Let $\{S_n\}$ be a sequence of space-time slabs $S_n = \mathbb{R}^3 \times I_n$, where $I_n = (t_{n-1}, t_n)$, with associated space-time piecewise polynomials $V_n \subset [\mathcal{C}(S_n)]^5$, and let $V = \prod_{n\geq 1} V_n$. The G²-method for (4.99) can be formulated as follows (extending the notation used above): find $U \in V$ such that for $n = 1, 2, ...,$

$$
\begin{aligned}
&(\dot{U} + \sum_i A_i(U)U_{,i}, v + \delta(\dot{v} + \sum_i A_i^*(U)v_{,i}))_n \\
&+(\hat{\epsilon}\nabla U, \nabla v)_n + ([U_{n-1}], v_{n-1}^+) = 0 \quad \forall v \in V_n,
\end{aligned}
\qquad (4.104)
$$

where $*$ denotes transpose, and

$$\delta^* = \delta(U)^* = \frac{1}{2}(k_n^{-2}I + h_n^{-2}\sum_{i=1}^{2} A_i(U)^2)^{-\frac{1}{2}} \quad \text{on } S_n, \qquad (4.105)$$

$$\hat{\epsilon} = \max(\gamma_1 h^2 |R(U)|, \gamma_2 h^{3/2}),$$

$$R(U) = |\dot{U} + \sum_i A_i(U)U_{,x_i}| + |[U^n]|/k_n \quad \text{on } S_n.$$

Depending on the space-time orientation of the finite-elements we get different versions of the G²-method, including the Eulerian Sd-method, the chG-method orienting along particle paths, and various orG-methods with orientation according to solution features. The simplest instance of the method is the Sd(1,0) method obtained with a $\mathcal{P}_1 \times \mathcal{P}_{0-}$ approximation in space-time with the basis functions continuous piecewise linear in x and discontinuous piecewise constant in time. Writing $U^n \equiv U|_{S_n}$, the G²-method (4.104) can in this case be formulated as follows: Find $U^n \in W_n$ such that

$$\begin{aligned}(U^n - U^{n-1}, v) + (\sum_i A_i(U^n)U_{,i}^n, v + \delta(\sum_i A_i(U^n)^* v_{,i}))k_n \\ + (\hat{\epsilon}\nabla U^n, \nabla v)k_n = 0 \quad \forall v \in W_n,\end{aligned} \qquad (4.106)$$

where $W_n \subset [H^1(\mathbb{R}^3)]^5$ consists of continuous piecewise linears. To compute U^n from (4.106) approximately, we may iterate in various ways: the simplest possible variant is obtained by lumping the mass matrix related to the inner product $(\cdot, \cdot)$ in the first term on the left, and changing the index n to $n-1$ in the other terms, which gives a fully explicit version of Sd(1,0) of the form

$$\begin{aligned}U_j^n = U_j^{n-1} - k_n(\sum_i f_i(U^{n-1}), \varphi_j) \\ - k_n(\sum_i A_i(U^{n-1})U_{x_i}, \delta \sum_i A_i(U^{n-1})^* \varphi_{j,i}) - k_n(\hat{\epsilon}\nabla U^{n-1}, \nabla \varphi_j),\end{aligned} \qquad (4.107)$$

where $U^n = \sum_j U_j^n \varphi_j(x)$ with $\{\varphi_j(x)\}$ the standard $\mathcal{P}_1$-basis functions of W_n.

To see that square root in (4.105) is well defined, we note that (4.102) implies that $A_0 = (\eta'')^{-1}$ symmetrizes the A_i, so that $\bar{A}_i \equiv A_i A_0$ is symmetric, $i = 1, 2, 3$. Therefore

$$A_0^{-\frac{1}{2}} A_i A_0^{\frac{1}{2}} = A_0^{-\frac{1}{2}} \bar{A}_i A_0^{-\frac{1}{2}}$$

is symmetric, and thus the similarity transform induced by $A_0^{\frac{1}{2}}$ transforms the matrix $M \equiv (k_n^{-2}I + h^{-2}\sum_i A_i^2)$ to an obviously positive definite symmetric matrix. It follows that M has positive eigenvalues and a full set of eigenvectors which shows that $M^{-\frac{1}{2}}$ can be computed. In [21] explicit formulas for the eigenvalues and eigenvectors of M are given.

4.2 Entropy variables

The change of variables $\bar{u} = \eta'(u)$, which is one-to-one since η'' is positive definite, transforms (4.100) into

$$\bar{A}_0 \dot{\bar{u}} + \sum \bar{A}_i \bar{u}_{,i} = 0 \qquad (4.108)$$

where $\bar{A}_0(\bar{u}) = (\eta'')^{-1}(\bar{u})$ is positive definite symmetric, and $\bar{A}_i = A_i \bar{A}_0$ are symmetric because of (4.102). We refer to $\bar{u} = \eta'(u)$ as the *entropy variables*.

In the entropy variables $\bar{u} = \eta'(u)$, the conservation law (4.100) takes the form of the symmetric hyperbolic system (4.108) with the $\bar{A}_i$ symmetric and $\bar{A}_0$ positive definite. The entropy inequality is obtained multiplying a viscous variant of (4.108) by $\bar{u}\varphi$ with φ a non-negative test function, since $\bar{u} = \eta'(u)$

We may apply the G^2 method to (4.108) seeking a piecewise polynomial approximation $\bar{U}$ of $\bar{u}$ in some finite element space $\bar{V}$. In this case we may choose as a discrete test function $\bar{v} = \bar{U} \in \bar{V}$, from which the global form of the entropy inequality follows. Choosing further $\bar{v} = \pi_h(\bar{U}\varphi)$, and using the interpolation estimate (referred to as "super-approximation" because no derivative of U appears)

$$\|\bar{U}\varphi - \pi_h(\bar{U}\varphi)\| \le C_\varphi \|h\bar{U}\|, \qquad (4.109)$$

we obtain entropy consistency of the G^2-method, this time without using the artificial viscosity term. We conclude that in the G^2 method in entropy variables, the entropy condition is more directly built in because it results from multiplication with the entropy variable $\bar{U}$ itself.

4.3 Dual solution and stability factors in 1d

We apply the simplest $Sd(1,0)$ method in explicit form erefexplversion1 to the standard Riemann problem suggested by Sod, consisting of a rarefaction, contact discontinuity, and a shock, We obtain the following approximate solution on a uniform mesh with $h = 10^{-3}$ and time step 10^{-4} (at time T=0.2055)

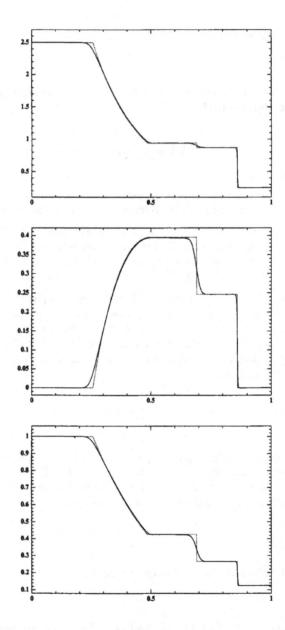

We notice the sharp resolution of the shock, and the smearing of the contact discontinuity (and the rarefaction) caused by the first order artificial viscosity in this variant. The smearing is much smaller in the chG(1,0) and orG(1,0) variants.

The dual solutions take the following form at $t = T$, $t = 0.9T$, $t = 0.5T$, $t = 0$:

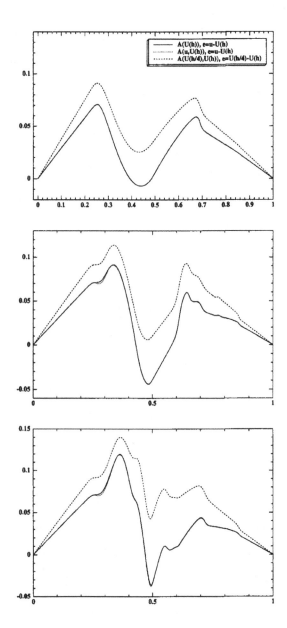

316

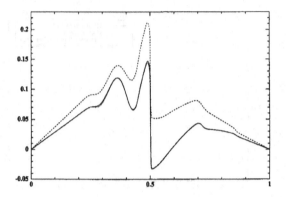

We next plot the corresponding second derivatives :

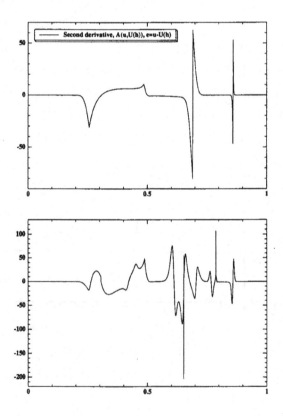

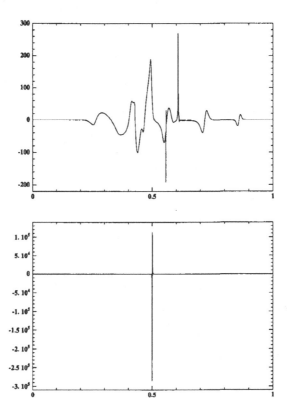

We compare the curves the different approaches at time t=0.9T and t=0.5T:

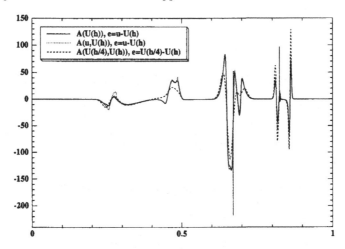

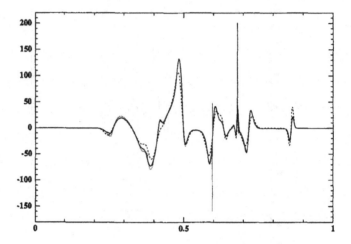

We finally plot the stability factors defined by (3.93) for different ϵ:

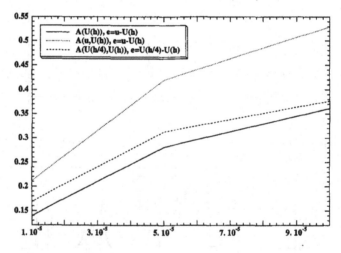

5 The Navier-Stokes equations for viscid compressible flow

The Navier-Stokes equations for a compressible viscous (Newtonian) fluid is the following generalization of the Euler equations (4.99):

$$
\begin{aligned}
\dot{u} + \sum_{i=1}^3 f_i(u)_{,i} + \sum_{i=1}^3 f_i^v(u)_{,i} = 0 \quad & x \in \mathbb{R}^3,\, t > 0, \\
u(x,0) = u_0(x) \quad & x \in \mathbb{R}^3,
\end{aligned}
\tag{5.110}
$$

where the f_i^v are diffusive fluxes given by

$$f_i^v = \begin{bmatrix} 0 \\ \tau_{1i} \\ \tau_{2i} \\ \tau_{3i} \\ \sum_i \tau_{ij} w_j \end{bmatrix},$$

where the τ_{ij} are viscous stresses given by

$$\tau_{ij} = 2\mu\epsilon_{ij}(w) + \lambda\nabla \cdot w\delta_{ij} \tag{5.111}$$

with μ and λ parameters.

6 Reactive compressible flow

We consider a fluid mixture of N species, which may undergo M chemical reactions. The Euler equations (4.99) generalize to

$$\dot{u} + \sum_{i=1}^3 f_i(u)_{,i} = S(u) \quad x \in \mathbb{R}^3, t > 0, \\ u(x,0) = u_0(x) \quad x \in \mathbb{R}^3, \tag{6.112}$$

where

$$u = \rho \begin{bmatrix} y_1 \\ \cdot \\ \cdot \\ y_N \\ w_1 \\ w_2 \\ w_3 \\ e \end{bmatrix}, \quad f_i = w_i u + p \begin{bmatrix} 0 \\ \cdot \\ \cdot \\ 0 \\ \delta_{1i} \\ \delta_{2i} \\ \delta_{3i} \\ w_i \end{bmatrix} - \begin{bmatrix} q_i^{D,1} \\ \cdot \\ \cdot \\ q_i^{D,N} \\ 0 \\ \cdot \\ \cdot \\ 0 \\ q_i \end{bmatrix},$$

where ρ is the mixture density, y_i is the mass-fraction for species i, $q = (q_j)$ is the mixture heat flux and $q^{D,i}$ is the diffusive flux for species i, given by

$$q = -\kappa\nabla(\rho c_v T), \qquad q^{D,i} = -D_i\nabla\rho_i$$

where κ and the D_i are diffusion coefficients, and c_v a heat capacity, and finally $S(u) = [s_1 \cdots s_N \, 0 \cdots 0]^*$ is a reaction term, typically given by

$$s_i = W_i \sum_{k=1}^M (\nu_{i,k}'' - \nu_{i,k}') B_k T^{\alpha_k} e^{-\frac{E_{\alpha,k}}{RT}} \prod_{j=1}^N (\frac{X_j p}{RT})^{\nu_{j,k}'}$$

where W_i is the molecular weight of species i, $\nu'_{i,k}$, and $\nu''_{i,k}$ are stoichiometric coefficients, $E_{\alpha,k}$ represents an activation energy, and $B_k T^{\alpha_k}$ is the frequency factor for reaction step k,

$$X_i = \frac{\frac{y_i}{W_i}}{\sum_j \frac{y_j}{W_j}}$$

is the mole fraction of species i, and finally

$$p = \rho R T \sum_i \frac{y_i}{W_i}.$$

The Navier-Stokes equations (5.110) generalize similary.

The G^2-method and a posteriori estimation The G^2-method for (4.99) directly generalizes to (5.110) and (6.112) including viscous variants, see [37], [2], [3]. As an example we present the Sd-method. Let $\{S_n\}$ be a sequence of space-time slabs $S_n = \mathbb{R}^3 \times I_n$, where $I_n = (t_{n-1}, t_n)$, with associated space-time piecewise polynomials $V_n \subset [\mathcal{C}(S_n)]^{N+4}$, and let $V = \prod_{n \geq 1} V_n$. The Sd-method for (6.112) reads: find $U \in V$ such that for $n = 1, 2, ...,$

$$(\dot{U} + \sum_i A_i(U)U_{,i}, v + \delta(\dot{v} + \sum_i A_i^*(U)v_{,i}))_n$$
$$+ (\hat{\epsilon} \nabla U, \nabla v)_n + ([U_{n-1}], v_{n-1}^+)$$
$$= (S(U), v + \delta(\dot{v} + \sum_i A_i^*(U)v_{,i}))_n \quad \forall v \in V_n,$$

where

$$\delta^* = \delta(U)^* = \frac{1}{2}(k_n^{-2}I + h_n^{-2} \sum_{i=1}^{3} A_i(U)^2)^{-\frac{1}{2}} \quad \text{on } S_n,$$

$$\hat{\epsilon} = \max(\gamma_1 h^2 |R(U)|, \gamma_2 h^{3/2}),$$

$$R(U) = |\dot{U} + \sum_i A_i(U)U_{x_i}| + \|[U^n]\|/k_n \quad \text{on} \quad S_n.$$

For explicit formulas for the computation of δ, we refer to [37]. The a posteriori error estimatation technique generalizes directly. Again, the main issue is the size of the stability factors S_c. As indicated, in three dimensions these factors may be expected to increase with decreasing viscosity as the complexity of the solutions increase and eventually turbulence develops. Computational results in two dimensions are presented in [37] and in [3]. For computational results on stability factors for incompressible flow, we refer to [34].

7 Conclusion

We have given evidence that a general framework for the numerical solution of conservation laws including the basic concepts of reliability and efficiency, may be developed on the basis of generalized Galerkin methods. A fundamental role is played by a posteriori error estimates involving certain computable stability factors or weights. The computational work required to reach a certain error tolerance increases with increasing stability factors. We have given theoretical and numerical evidence that for problems in one dimension, stability factors typically are of moderate size even for small viscosities, and accordingly solutions in general are computable with work compared to that of direct interpolation. In three dimensions stability factors are expected to increase with decreasing viscosity. In [30], we show that stability factors are inversely proportional to the viscosity in the basic case of parallel flow in three dimensions. As the viscosity decreases, ultimately solutions in three dimensions in general become turbulent and corresponding stability factors are very large. To obtain models with computable solutions, turbulence modeling is required, where on the one hand the effective viscosity is larger and stability factors smaller, and on the other hand the modeling error (in suitable weak norms) is controled quantitatively. The design of turbulence models is a major open problem, where computational techniques today open new possibilities. In particular, the techniques presented in this note of a posteriori error estimation, opens the possibility of separating and controling computational and modeling errors.

References

1. S. Bertoluzza, F. Brezzi, C. Johnson and A. Russo, *Dynamic computational subgrid modeling*, Laboratorio Analisi Numerica, Univ of Pavia/Chalmers Finite Element Center. To appear 1997

2. E. Burman, *Adaptive finite element methods for two-phase flow*, Licentiate Thesis, Department of Mathematics, Chalmers University of Technology, Göteborg, 1996.

3. E. Burman, *Adaptive finite element methods for two-phase flow*, Ph d Thesis, Department of Mathematics, Chalmers University of Technology, Göteborg, to appear 1996.

4. K. Eriksson, D. Estep, P. Hansbo and C. Johnson, *Computational Differential Equations*, Cambridge University Press/Studentlitteratur, 1996.

5. ———, *Introduction to Computational Mathematical Modeling*, to appear 1998.

6. ———, *Introduction to Engineering Mathematics*, to appear 1999.

7. ———, *Advanced Computational Differential Equations*, to appear 1998.

8. ———, *Adaptive methods for differential equations*, Acta Numerica, 1995.

9. K. Eriksson and C. Johnson, *Adaptive finite element methods for parabolic problems I: A linear model problem*, SIAM J. Numer. Anal., 28 (1991), pp. 43–77.

10. ———, *Adaptive finite element methods for parabolic problems II: Optimal error estimates in $L_\infty L_2$ and $L_\infty L_\infty$*, SIAM J. Numer. Anal., 32 (1995), pp. 706–740.

322

11. ———, *Adaptive finite element methods for parabolic problems III: Time steps variable in space.* To appear.

12. ———, *Adaptive finite element methods for parabolic problems IV: Non-linear problems*, SIAM J. Numer. Anal., 32 (1995), pp. 1729–1749.

13. ———, *Adaptive finite element methods for parabolic problems V: Long-time integration*, SIAM J. Numer. Anal., 32 (1995), pp. 1750–1763.

14. ———, *Adaptive finite element methods for parabolic problems VI: Analytic semigroups* , Preprint #1996-32, Department of Mathematics, Chalmers University of Technology, to appear in Siam J. Numer. Anal.

15. ———, *Adaptive streamline diffusion finite element methods for stationary convection-diffusion problems*, Math. Comp., 60 (1993), pp. 167–188.

16. D. ESTEP, *A posteriori error bounds and global error control for approximations of ordinary differential equations*, Siam J.Numer. Anal. 32 (1995), 1-48.

17. C. JOHNSON AND D. ESTEP, *On the computability of the Lorenz system*, to appear in M3AS.

18. D. ESTEP, M. LARSON AND R. WILLIAMS, *Estimating the error of numerical solutions of nonlinear reaction-diffusion equations*, 1997.

19. P. HANSBO, *The characteristic streamline diffusion method for convection-diffusion problems*, Comput. Methods Appl. Mech. Engrg., 96 (1992), pp. 239–253.

20. ———, *The characteristic streamline diffusion method for the time-dependent incompressible Navier-Stokes equations*, Comput. Methods Appl. Mech. Engrg., 99 (1992), pp. 171–186.

21. ———, *Explicit streamline diffusion finite element methods for the compressible Euler equations in conservation variables*, J. Comput. Phys., (1993). In press.

22. ———, *Space-time oriented streamline diffusion methods for nonlinear conservation laws in one dimension*, Commun. Numer. Methods Engrg., (1993). In press.

23. P. HANSBO AND C. JOHNSON, *Adaptive streamline diffusion finite element methods for compressible flow using conservation variables*, Comput. Methods Appl. Mech. Engrg., 87 (1991), pp. 267–280.

24. P. HANSBO AND A. SZEPESSY, *A velocity-pressure streamline diffusion finite element method for the incompressible Navier-Stokes equations*, Comput. Methods Appl. Mech. Engrg., 84 (1990), pp. 175–192.

25. ———, *A new approach to algorithms for convection problems based on exact transport + projection*, Comput. Methods Appl. Mech. Engrg., 100 (1992), pp. 45–62.

26. ———, *Discontinuous Galerkin finite element methods for second order hyperbolic problems*, Comput. Methods Appl. Mech. Engrg., 107 (1993), pp 117-129.

27. C. JOHNSON AND P. HANSBO, *Adaptive finite element methods for small strain elasto-plasticity*, in Finite Inelastic Deformations - Theory and Applications, D. Besdo and E. Stein, eds., Springer, Berlin, 1992, pp. 273–288.

28. ———, *Adaptive finite element methods in computational mechanics*, Comput. Methods Appl. Mech. Engrg., 101 (1992), pp. 143–181.

29. C. JOHNSON AND A. LOGG, *Mechlab*, to appear.

30. C. JOHNSON, R. RANNACHER, AND M. BOMAN, *Numerics and hydrodynamic stability: Towards error control in CFD*, SIAM J. Numer. Anal. 32 (1995), pp. 1058-1079.

31. C. JOHNSON, R. RANNACHER, AND R. BECKER, *Adaptive multigrid methods.* 1995.

32. C. JOHNSON AND A. SZEPESSY, *Adaptive finite element methods for conservation laws based on a posteriori error estimates*, Comm Appl Pure Math. XLVIII (1995], 199-234.

33. ——, *Convergence of the Discontinuous Galerkin method for scalar conservation laws*, M3AS, 1996.

34. M. LARSON, *Analysis of adaptive finite element methods*, Ph D Thesis, Department of Mathematics, Chalmers University of Technology, Göteborg, 1997.

35. M. LEVENSTAM, *Adaptive finite element simulation of welding*, Department of Mathematics, Chalmers University of Technology, Göteborg.

36. P. MÖLLER AND P. HANSBO, *On advancing front mesh generation in three dimensions*, Preprint #1993-33, Department of Mathematics, Chalmers University of Technology, 1993.

37. R. SANDBOGE, *Adaptive finite element methods for reactive flow*, Ph D Thesis, Department of Mathematics, Chalmers University of Technology, Göteborg, 1997.

Chapter 4

Essentially Non-oscillatory and Weighted Essentially Non-oscillatory Schemes for Hyperbolic Conservation Laws

Chi-Wang Shu

Division of Applied Mathematics
Brown University
Providence, RI 02912, USA
E-mail: shu@cfm.brown.edu

ABSTRACT

In these lecture notes we describe the construction, analysis, and application of ENO (Essentially Non-Oscillatory) and WENO (Weighted Essentially Non-Oscillatory) schemes for hyperbolic conservation laws and related Hamilton-Jacobi equations. ENO and WENO schemes are high order accurate finite difference schemes designed for problems with piecewise smooth solutions containing discontinuities. The key idea lies at the approximation level, where a nonlinear adaptive procedure is used to automatically choose the locally smoothest stencil, hence avoiding crossing discontinuities in the interpolation procedure as much as possible. ENO and WENO schemes have been quite successful in applications, especially for problems containing both shocks and complicated smooth solution structures, such as compressible turbulence simulations and aeroacoustics.

These lecture notes are basically self-contained. It is our hope that with these notes and with the help of the quoted references, the readers can understand the algorithms and code them up for applications. Sample codes are also available from the author.

Contents

1 Introduction

ENO (Essentially Non-Oscillatory) schemes were started with the classic paper of Harten, Engquist, Osher and Chakravarthy in 1987 [38]. This paper has been cited at least 144 times by early 1997, according to the ISI database. The Journal of Computational Physics decided to republish this classic paper as part of the celebration of the journal's 30th birthday [68].

Finite difference and related finite volume schemes are based on interpolations of discrete data using polynomials or other simple functions. In the approximation theory, it is well known that the wider the stencil, the higher the order of accuracy of the interpolation, *provided the function being interpolated is smooth inside the stencil*. Traditional finite difference methods are based on fixed stencil interpolations. For example, to obtain an interpolation for cell i to third order accuracy, the information of the three cells $i-1$, i and $i+1$ can be used, to build a second order interpolation polynomial. In other words, one always looks one cell to the left, one cell to the right, plus the center cell itself, regardless of where in the domain one is situated. This works well for globally smooth problems. The resulting scheme is linear for linear PDEs, hence stability can be easily analyzed by Fourier transforms (for the uniform grid case). However, fixed stencil interpolation of second or higher order accuracy is necessarily *oscillatory* near a discontinuity, see Fig.2.1, left, in Sect.2.2. Such oscillations, which are called the Gibbs phenomena in spectral methods, do not decay in magnitude when the mesh is refined. It is a nuisance to say the least for practical calculations, and often leads to numerical instabilities in nonlinear problems containing discontinuities.

Before 1987, there were mainly two common ways to eliminate or reduce such spurious oscillations near discontinuities. One way was to add an artificial viscosity. This could be tuned so that it was large enough near the discontinuity to suppress, or at least to reduce the oscillations, but was small elsewhere to maintain high-order accuracy. One disadvantage of this approach is that fine tuning of the parameter controlling the size of the artificial viscosity is problem dependent. Another way was to apply limiters to eliminate the oscillations. In effect, one reduced the order of accuracy of the interpolation near the discontinuity (e.g. reducing the slope of a linear interpolant, or using a linear rather than a quadratic interpolant near the shock). By carefully designing such limiters, the TVD (total variation diminishing) property could be achieved for nonlinear scalar one dimensional problems. One disadvantage of this approach is that accuracy necessarily degenerates to first order near *smooth* extrema. We will not discuss the method of adding explicit artificial viscosity or the TVD method in these lecture notes. We refer to the books by Sod [75] and by LeVeque [52], and the references listed therein, for details.

The ENO idea proposed in [38] seems to be the first successful attempt to obtain a self similar (i.e. no mesh size dependent parameter), uniformly high order accurate, yet essentially non-oscillatory interpolation (i.e. the magnitude of the oscillations decays as $O(\Delta x^k)$ where k is the order of accuracy) for piecewise smooth functions. The generic solution for hyperbolic conservation laws is in the class of piecewise smooth functions. The reconstruction in [38] is a natural extension of an earlier second order version of Harten and Osher [37]. In [38], Harten, Engquist, Osher and Chakravarthy investigated different ways of measuring local smoothness to determine the local stencil, and developed a hierarchy that begins with one or two cells, then adds one cell at a time to the stencil from the two candidates on the left and right, based on the size of the two relevant Newton divided differences. Although there are other reasonable strategies to choose the stencil based on local smoothness, such as comparing the magnitudes of the highest degree divided differences among all candidate stencils and picking the one with the least absolute value, experience seems to show that the hierarchy proposed in [38] is the most robust for a wide range of grid sizes, Δx, both *before* and inside the asymptotic regime.

As one can see from the numerical examples in [38] and in later papers, many of which being mentioned in these lecture notes or in the references listed, ENO schemes are indeed uniformly high order accurate and resolve shocks with sharp and monotone (to the eye) transitions. ENO schemes are especially suitable for problems containing both shocks and complicated smooth flow structures, such as those occurring in shock interactions with a turbulent flow and shock interaction with vortices.

Since the publication of the original paper of Harten, Engquist, Osher and Chakravarthy [38], the original authors and many other researchers have followed the pioneer work, improving the methodology and expanding the area of its applications. ENO schemes based on point values and TVD Runge-Kutta time discretizations, which can save computational costs significantly for multi space dimensions, were developed in [69] and [70]. Later biasing in the stencil choosing process to enhance stability and accuracy were developed in [28] and [67]. Weighted ENO (WENO) schemes were developed, using a convex combination of all candidate stencils instead of just one as in the original ENO, [53], [43]. ENO schemes based on other than polynomial building blocks were constructed in [40], [16]. Sub-cell resolution and artificial compression to sharpen contact discontinuities were studied in [35], [83], [70] and [43]. Multidimensional ENO schemes based on general triangulation were developed in [1]. ENO and WENO schemes for Hamilton-Jacobi type equations were designed and applied in [59], [60], [50] and [45]. ENO schemes using one-sided Jocobians for field by field decomposition, which improves the robustness for calculations of systems, were discussed in [25]. Combination of ENO with multiresolution ideas was pursued in [7]. Combination of ENO with spectral method using a domain decomposition approach was carried out in [8]. On the application side, ENO and WENO have been successfully used to simulate shock turbulence interactions [70], [71], [2]; to the direct simulation of compressible turbulence [71], [80], [49]; to relativistic hydrodynamics equations [24]; to shock vortex interactions and other gas dynamics problems [12], [27], [43]; to incompressible flow problems [26], [31]; to viscoelasticity equations with fading memory [72]; to semi-conductor device simulation [28], [41], [42]; to image processing [59], [64], [73]; etc. This list is definitely incomplete and may be biased by the author's own research experience, but one can already see that ENO and WENO have been applied quite extensively in many different fields. Most of the problems solved by ENO and WENO schemes are of the type in which solutions contain both strong shocks and rich smooth region structures. Lower order methods usually have difficulties for such problems and it is thus attractive and efficient to use high order stable methods such as ENO and WENO to handle them.

Today the study and application of ENO and WENO schemes are still very active. We expect the schemes and the basic methodology to be developed further and to become even more successful in the future.

In these lecture notes we describe the construction, analysis, and application of ENO and WENO schemes for hyperbolic conservation laws and related Hamilton-Jacobi equations. They are basically self-contained. Our hope is that with these notes and with the help of the quoted references, the readers can understand the algorithms and code them up for applications. Sample codes are also available from the author.

2 One Space Dimension

2.1 Reconstruction and Approximation in 1D

In this section we concentrate on the problems of interpolation and approximation in one space dimension.

Given a grid

$$a = x_{\frac{1}{2}} < x_{\frac{3}{2}} < ... < x_{N-\frac{1}{2}} < x_{N+\frac{1}{2}} = b. \tag{2.1}$$

We define cells, cell centers, and cell sizes by

$$I_i \equiv \left[x_{i-\frac{1}{2}}, x_{i+\frac{1}{2}} \right], \ x_i \equiv \tfrac{1}{2} \left(x_{i-\frac{1}{2}} + x_{i+\frac{1}{2}} \right),$$
$$\Delta x_i \equiv x_{i+\frac{1}{2}} - x_{i-\frac{1}{2}}, \qquad i = 1, 2, ..., N. \tag{2.2}$$

We denote the maximum cell size by

$$\Delta x \equiv \max_{1 \leq i \leq N} \Delta x_i. \tag{2.3}$$

Reconstruction from cell averages. The first approximation problem we will face, in solving hyperbolic conservation laws using cell averages (finite volume schemes, see Sect.2.3), is the following *reconstruction* problem [38].

Problem 2.1. One dimensional reconstruction.

Given the cell averages of a function $v(x)$:

$$\bar{v}_i \equiv \frac{1}{\Delta x_i} \int_{x_{i-\frac{1}{2}}}^{x_{i+\frac{1}{2}}} v(\xi) \, d\xi, \qquad i = 1, 2, ..., N, \tag{2.4}$$

find a polynomial $p_i(x)$, of degree at most $k-1$, for each cell I_i, such that it is a k-th order accurate approximation to the function $v(x)$ inside I_i:

$$p_i(x) = v(x) + O(\Delta x^k), \qquad x \in I_i, \ i = 1, ..., N. \tag{2.5}$$

In particular, this gives approximations to the function $v(x)$ at the cell boundaries

$$v_{i+\frac{1}{2}}^- = p_i(x_{i+\frac{1}{2}}), \qquad v_{i-\frac{1}{2}}^+ = p_i(x_{i-\frac{1}{2}}), \qquad i = 1, ..., N, \tag{2.6}$$

which are k-th order accurate:

$$v_{i+\frac{1}{2}}^- = v(x_{i+\frac{1}{2}}) + O(\Delta x^k), \quad v_{i-\frac{1}{2}}^+ = v(x_{i-\frac{1}{2}}) + O(\Delta x^k), \quad i = 1, ..., N. \tag{2.7}$$

$\square$

The polynomial $p_i(x)$ in Problem 2.1 can be replaced by other simple functions, such as trigonometric polynomials. See Sect.4.1.

We will not discuss boundary conditions in this section. We thus assume that $\bar{v}_i$ is also available for $i \leq 0$ and $i > N$ if needed.

In the following we describe a procedure to solve Problem 2.1.

Given the location I_i and the order of accuracy k, we first choose a "stencil", based on r cells to the left, s cells to the right, and I_i itself if $r, s \geq 0$, with $r + s + 1 = k$:

$$S(i) \equiv \{I_{i-r}, ..., I_{i+s}\}. \tag{2.8}$$

There is a unique polynomial of degree at most $k - 1 = r + s$, denoted by $p(x)$ (we will drop the subscript i when it does not cause confusion), whose cell average in each of the cells in $S(i)$ agrees with that of $v(x)$:

$$\frac{1}{\Delta x_j} \int_{x_{j-\frac{1}{2}}}^{x_{j+\frac{1}{2}}} p(\xi)\, d\xi = \bar{v}_j, \qquad j = i - r, ..., i + s. \tag{2.9}$$

This polynomial $p(x)$ is the k-th order approximation we are looking for, as it is easy to prove (2.5), see the discussion below, as long as the function $v(x)$ is smooth in the region covered by the stencil $S(i)$.

For solving Problem 2.1, we also need the approximations to the values of $v(x)$ at the cell boundaries, (2.6). Since the mappings from the given cell averages $\bar{v}_j$ in the stencil $S(i)$ to the values $v_{i+\frac{1}{2}}^-$ and $v_{i-\frac{1}{2}}^+$ in (2.6) are linear, there exist constants c_{rj} and $\tilde{c}_{rj}$, which depend on the left shift of the stencil r of the stencil $S(i)$ in (2.8), on the order of accuracy k, and on the cell sizes Δx_j in the stencil S_i, but *not* on the function v itself, such that

$$v_{i+\frac{1}{2}}^- = \sum_{j=0}^{k-1} c_{rj} \bar{v}_{i-r+j}, \qquad v_{i-\frac{1}{2}}^+ = \sum_{j=0}^{k-1} \tilde{c}_{rj} \bar{v}_{i-r+j}. \tag{2.10}$$

We note that the difference between the values with superscripts $\pm$ at the same location $x_{i+\frac{1}{2}}$ is due to the possibility of different stencils for cell I_i and for cell I_{i+1}. If we identify the left shift r not with the cell I_i but with the point of reconstruction $x_{i+\frac{1}{2}}$, i.e. using the stencil (2.8) to approximate $x_{i+\frac{1}{2}}$, then we can drop the superscripts $\pm$ and also eliminate the need to consider $\tilde{c}_{rj}$ in (2.10), as it is clear that

$$\tilde{c}_{rj} = c_{r-1,j}.$$

We summarize this as follows: given the k cell averages

$$\bar{v}_{i-r}, \quad ..., \quad \bar{v}_{i-r+k-1},$$

there are constants c_{rj} such that the reconstructed value at the cell boundary $x_{i+\frac{1}{2}}$:

$$v_{i+\frac{1}{2}} = \sum_{j=0}^{k-1} c_{rj} \bar{v}_{i-r+j}, \qquad (2.11)$$

is k-th order accurate:

$$v_{i+\frac{1}{2}} = v(x_{i+\frac{1}{2}}) + O(\Delta x^k). \qquad (2.12)$$

To understand how the constants $\{c_{rj}\}$ are obtained, as well as how the accuracy property (2.5) is proven, we look at the primitive function of $v(x)$:

$$V(x) \equiv \int_{-\infty}^{x} v(\xi) \, d\xi, \qquad (2.13)$$

where the lower limit $-\infty$ is not important and can be replaced by any fixed number. Clearly, $V(x_{i+\frac{1}{2}})$ can be expressed by the cell averages of $v(x)$ using (2.4):

$$V(x_{i+\frac{1}{2}}) = \sum_{j=-\infty}^{i} \int_{x_{j-\frac{1}{2}}}^{x_{j+\frac{1}{2}}} v(\xi) \, d\xi = \sum_{j=-\infty}^{i} \bar{v}_j \Delta x_j, \qquad (2.14)$$

thus with the knowledge of the cell averages $\{\bar{v}_j\}$ we also know the primitive function $V(x)$ at the cell boundaries exactly. If we denote the unique polynomial of degree at most k, which interpolates $V(x_{j+\frac{1}{2}})$ at the following $k+1$ points:

$$x_{i-r-\frac{1}{2}}, \quad \cdots, \quad x_{i+s+\frac{1}{2}}, \qquad (2.15)$$

by $P(x)$, and denote its derivative by $p(x)$:

$$p(x) \equiv P'(x), \qquad (2.16)$$

then it is easy to verify (2.9):

$$\frac{1}{\Delta x_j} \int_{x_{j-\frac{1}{2}}}^{x_{j+\frac{1}{2}}} p(\xi) \, d\xi = \frac{1}{\Delta x_j} \int_{x_{j-\frac{1}{2}}}^{x_{j+\frac{1}{2}}} P'(\xi) \, d\xi$$

$$= \frac{1}{\Delta x_j} \left(P(x_{j+\frac{1}{2}}) - P(x_{j-\frac{1}{2}}) \right)$$

$$= \frac{1}{\Delta x_j} \left(V(x_{j+\frac{1}{2}}) - V(x_{j-\frac{1}{2}}) \right)$$

$$= \frac{1}{\Delta x_j} \left(\int_{-\infty}^{x_{j+\frac{1}{2}}} v(\xi) \, d\xi - \int_{-\infty}^{x_{j-\frac{1}{2}}} v(\xi) \, d\xi \right)$$

$$= \frac{1}{\Delta x_j} \int_{x_{j-\frac{1}{2}}}^{x_{j+\frac{1}{2}}} v(\xi)\, d\xi = \bar{v}_j, \quad j = i-r, ..., i+s,$$

where the third equality holds because $P(x)$ interpolates $V(x)$ at the points $x_{j-\frac{1}{2}}$ and $x_{j+\frac{1}{2}}$ whenever $j = i-r, ..., i+s$. This implies that $p(x)$ is the polynomial we are looking for. Standard approximation theory (see an elementary numerical analysis book) tells us that

$$P'(x) = V'(x) + O(\Delta x^k), \qquad x \in I_i.$$

This is the accuracy requirement (2.5).

Now let us look at the practical issue of how to obtain the constants $\{c_{rj}\}$ in (2.11). For this we could use the Lagrange form of the interpolation polynomial:

$$P(x) = \sum_{m=0}^{k} V(x_{i-r+m-\frac{1}{2}}) \prod_{\substack{l=0 \\ l \neq m}}^{k} \frac{x - x_{i-r+l-\frac{1}{2}}}{x_{i-r+m-\frac{1}{2}} - x_{i-r+l-\frac{1}{2}}}. \qquad (2.17)$$

For easier manipulation we subtract a constant $V(x_{i-r-\frac{1}{2}})$ from (2.17), and use the fact that

$$\sum_{m=0}^{k} \prod_{\substack{l=0 \\ l \neq m}}^{k} \frac{x - x_{i-r+l-\frac{1}{2}}}{x_{i-r+m-\frac{1}{2}} - x_{i-r+l-\frac{1}{2}}} = 1,$$

to obtain:

$$P(x) - V(x_{i-r-\frac{1}{2}}) \qquad (2.18)$$

$$= \sum_{m=0}^{k} \left(V(x_{i-r+m-\frac{1}{2}}) - V(x_{i-r-\frac{1}{2}}) \right) \prod_{\substack{l=0 \\ l \neq m}}^{k} \frac{x - x_{i-r+l-\frac{1}{2}}}{x_{i-r+m-\frac{1}{2}} - x_{i-r+l-\frac{1}{2}}}.$$

Taking derivative on both sides of (2.18), and noticing that

$$V(x_{i-r+m-\frac{1}{2}}) - V(x_{i-r-\frac{1}{2}}) = \sum_{j=0}^{m-1} \bar{v}_{i-r+j} \Delta x_{i-r+j}$$

because of (2.14), we obtain

$$p(x) = \sum_{m=0}^{k} \sum_{j=0}^{m-1} \bar{v}_{i-r+j} \Delta x_{i-r+j} \left(\frac{\sum_{\substack{l=0 \\ l \neq m}}^{k} \prod_{\substack{q=0 \\ q \neq m,l}}^{k} \left(x - x_{i-r+q-\frac{1}{2}} \right)}{\prod_{\substack{l=0 \\ l \neq m}}^{k} \left(x_{i-r+m-\frac{1}{2}} - x_{i-r+l-\frac{1}{2}} \right)} \right). \qquad (2.19)$$

Evaluating the expression (2.19) at $x = x_{i+\frac{1}{2}}$, we finally obtain

$$
v_{i+\frac{1}{2}} = p(x_{i+\frac{1}{2}}) =
$$

$$
\sum_{j=0}^{k-1} \left(\sum_{m=j+1}^{k} \frac{\displaystyle\sum_{\substack{l=0 \\ l\neq m}}^{k} \prod_{\substack{q=0 \\ q\neq m,l}}^{k} \left(x_{i+\frac{1}{2}} - x_{i-r+q-\frac{1}{2}}\right)}{\displaystyle\prod_{\substack{l=0 \\ l\neq m}}^{k} \left(x_{i-r+m-\frac{1}{2}} - x_{i-r+l-\frac{1}{2}}\right)} \right) \Delta x_{i-r+j}\, \overline{v}_{i-r+j} ,
$$

i.e. the constants c_{rj} in (2.11) are given by

$$
c_{rj} = \left(\sum_{m=j+1}^{k} \frac{\displaystyle\sum_{\substack{l=0 \\ l\neq m}}^{k} \prod_{\substack{q=0 \\ q\neq m,l}}^{k} \left(x_{i+\frac{1}{2}} - x_{i-r+q-\frac{1}{2}}\right)}{\displaystyle\prod_{\substack{l=0 \\ l\neq m}}^{k} \left(x_{i-r+m-\frac{1}{2}} - x_{i-r+l-\frac{1}{2}}\right)} \right) \Delta x_{i-r+j} . \qquad (2.20)
$$

Although there are many zero terms in the inner sum of (2.20) when $x_{i+\frac{1}{2}}$ is a node in the interpolation, we will keep this general form so that it applies also to the case where $x_{i+\frac{1}{2}}$ is not an interpolation point.

For a nonuniform grid, one would want to pre-compute the constants $\{c_{rj}\}$ as in (2.20), for $0 \leq i \leq N$, $-1 \leq r \leq k - 1$, and $0 \leq j \leq k - 1$, and store them before solving the PDE.

For a uniform grid, $\Delta x_i = \Delta x$, the expression for c_{rj} does not depend on i or Δx any more:

$$
c_{rj} = \sum_{m=j+1}^{k} \frac{\displaystyle\sum_{\substack{l=0 \\ l\neq m}}^{k} \prod_{\substack{q=0 \\ q\neq m,l}}^{k} (r - q + 1)}{\displaystyle\prod_{\substack{l=0 \\ l\neq m}}^{k} (m - l)} . \qquad (2.21)
$$

We list in Table 2.1 the constants c_{rj} in this uniform grid case (2.21), for order of accuracy between $k = 1$ and $k = 7$.

From Table 2.1, we would know, for example, that

$$
v_{i+\frac{1}{2}} = -\frac{1}{6}\overline{v}_{i-1} + \frac{5}{6}\overline{v}_i + \frac{1}{3}\overline{v}_{i+1} + O(\Delta x^3) .
$$

338

Table 2.1. The constants c_{rj} in (2.21).

k	r	j=0	j=1	j=2	j=3	j=4	j=5	j=6
1	-1	1						
	0	1						
2	-1	3/2	-1/2					
	0	1/2	1/2					
	1	-1/2	3/2					
3	-1	11/6	-7/6	1/3				
	0	1/3	5/6	-1/6				
	1	-1/6	5/6	1/3				
	2	1/3	-7/6	11/6				
4	-1	25/12	-23/12	13/12	-1/4			
	0	1/4	13/12	-5/12	1/12			
	1	-1/12	7/12	7/12	-1/12			
	2	1/12	-5/12	13/12	1/4			
	3	-1/4	13/12	-23/12	25/12			
5	-1	137/60	-163/60	137/60	-21/20	1/5		
	0	1/5	77/60	-43/60	17/60	-1/20		
	1	-1/20	9/20	47/60	-13/60	1/30		
	2	1/30	-13/60	47/60	9/20	-1/20		
	3	-1/20	17/60	-43/60	77/60	1/5		
	4	1/5	-21/20	137/60	-163/60	137/60		
6	-1	49/20	-71/20	79/20	-163/60	31/30	-1/6	
	0	1/6	29/20	-21/20	37/60	-13/60	1/30	
	1	-1/30	11/30	19/20	-23/60	7/60	-1/60	
	2	1/60	-2/15	37/60	37/60	-2/15	1/60	
	3	-1/60	7/60	-23/60	19/20	11/30	-1/30	
	4	1/30	-13/60	37/60	-21/20	29/20	1/6	
	5	-1/6	31/30	-163/60	79/20	-71/20	49/20	
7	-1	363/140	-617/140	853/140	-2341/420	667/210	-43/42	1/7
	0	1/7	223/140	-197/140	153/140	-241/420	37/210	-1/42
	1	-1/42	13/42	153/140	-241/420	109/420	-31/420	1/105
	2	1/105	-19/210	107/210	319/420	-101/420	5/84	-1/140
	3	-1/140	5/84	-101/420	319/420	107/210	-19/210	1/105
	4	1/105	-31/420	109/420	-241/420	153/140	13/42	-1/42
	5	-1/42	37/210	-241/420	153/140	-197/140	223/140	1/7
	6	1/7	-43/42	667/210	-2341/420	853/140	-617/140	363/140

Conservative approximation to the derivative from point values.
The second approximation problem we will face, in solving hyperbolic conservation laws using point values (finite difference schemes, see Sect.2.3), is the following problem in obtaining high order *conservative* approximation to the derivative from point values [69,70].

Problem 2.2. One dimensional conservative approximation.
Given the point values of a function $v(x)$:

$$v_i \equiv v(x_i), \qquad i = 1, 2, ..., N, \tag{2.22}$$

find a numerical flux function

$$\hat{v}_{i+\frac{1}{2}} \equiv \hat{v}(v_{i-r}, ..., v_{i+s}), \qquad i = 0, 1, ..., N, \tag{2.23}$$

such that the flux difference approximates the derivative $v'(x)$ to k-th order accuracy:

$$\frac{1}{\Delta x_i} \left(\hat{v}_{i+\frac{1}{2}} - \hat{v}_{i-\frac{1}{2}} \right) = v'(x_i) + O(\Delta x^k), \qquad i = 0, 1, ..., N. \tag{2.24}$$

□

We again ignore the boundary conditions here and assume that v_i is available for $i \leq 0$ and $i > N$ if needed.

The solution of this problem is essential for the high order conservative schemes based on point values (finite difference) rather than on cell averages (finite volume).

This problem looks quite different from Problem 2.1. However, we will see that there is a close relationship between these two. *We assume that the grid is uniform,* $\Delta x_i = \Delta x$. This assumption is, unfortunately, essential in the following development.

If we can find a function $h(x)$, which may depend on the grid size Δx, such that

$$v(x) = \frac{1}{\Delta x} \int_{x-\frac{\Delta x}{2}}^{x+\frac{\Delta x}{2}} h(\xi) d\xi, \tag{2.25}$$

then clearly

$$v'(x) = \frac{1}{\Delta x} \left[h\left(x + \frac{\Delta x}{2}\right) - h\left(x - \frac{\Delta x}{2}\right) \right],$$

hence all we need to do is to use

$$\hat{v}_{i+\frac{1}{2}} = h(x_{i+\frac{1}{2}}) + O(\Delta x^k) \tag{2.26}$$

to achieve (2.24). We note here that it would look like an $O(\Delta x^{k+1})$ term in (2.26) is needed in order to get (2.24), due to the Δx term in the denominator.

However, in practice, the $O(\Delta x^k)$ term in (2.26) is usually smooth, hence the difference in (2.24) would give an extra $O(\Delta x)$, just to cancel the one in the denominator.

It is not easy to approximate $h(x)$ via (2.25), as it is only implicitly defined there. However, we notice that the known function $v(x)$ is the cell average of the unknown function $h(x)$, so to find $h(x)$ we just need to use the *reconstruction* procedure described in Sect.2.1. If we take the primitive of $h(x)$:

$$H(x) = \int_{-\infty}^{x} h(\xi)d\xi, \qquad (2.27)$$

then (2.25) clearly implies

$$H(x_{i+\frac{1}{2}}) = \sum_{j=-\infty}^{i} \int_{x_{j-\frac{1}{2}}}^{x_{j+\frac{1}{2}}} h(\xi)d\xi = \Delta x \sum_{j=-\infty}^{i} v_j. \qquad (2.28)$$

Thus, given the point values $\{v_j\}$, we "identify" them as cell averages of another function $h(x)$ in (2.25), then the primitive function $H(x)$ is exactly known at the cell interfaces $x = x_{i+\frac{1}{2}}$. We thus use the same reconstruction procedure described in Sect.2.1, to get a k-th order approximation to $h(x_{i+\frac{1}{2}})$, which is then taken as the numerical flux $\hat{v}_{i+\frac{1}{2}}$ in (2.23).

In other words, if the "stencil" for the flux $\hat{v}_{i+\frac{1}{2}}$ in (2.23) is the following k points:

$$x_{i-r}, \ ..., \ x_{i+s}, \qquad (2.29)$$

where $r + s = k - 1$, then the flux $\hat{v}_{i+\frac{1}{2}}$ is expressed as

$$\hat{v}_{i+\frac{1}{2}} = \sum_{j=0}^{k-1} c_{rj} v_{i-r+j}, \qquad (2.30)$$

where the constants $\{c_{rj}\}$ are given by (2.21) and Table 2.1.

From Table 2.1 we would know, for example, that if

$$\hat{v}_{i+\frac{1}{2}} = -\frac{1}{6}v_{i-1} + \frac{5}{6}v_i + \frac{1}{3}v_{i+1},$$

then

$$\frac{1}{\Delta x}\left(\hat{v}_{i+\frac{1}{2}} - \hat{v}_{i-\frac{1}{2}}\right) = v'(x_i) + O(\Delta x^3).$$

We emphasize again that, unlike in the reconstruction procedure in Sect.2.1, here the grid *must* be uniform: $\Delta x_j = \Delta x$. Otherwise, it can be proven that no choice of constants c_{rj} in (2.30) (which may depend on the local grid sizes but not on the function $v(x)$) could make the conservative approximation to the derivative (2.24) higher than second order accurate ($k > 2$). The proof is a simple exercise of Taylor expansions. Thus, the high order finite difference (third order and higher) discussed in these lecture notes can apply only to uniform or smoothly varying grids.

Because of this equivalence of obtaining a conservative approximation to the derivative (2.23)-(2.24) and the reconstruction problem discussed in Sect.2.1, we will only need to consider the reconstruction problem in the following sections.

Fixed stencil approximation. By fixed stencil, we mean that the left shift r in (2.8) or (2.29) is *the same for all locations* i. Usually, for a globally smooth function $v(x)$, the best approximation is obtained either by a central approximation $r = s - 1$ for even k (here central is relative to the location $x_{i+\frac{1}{2}}$), or by a one point upwind biased approximation $r = s$ or $r = s - 2$ for odd k. For example, if the grid is uniform $\Delta x_i = \Delta x$, then a central 4th order reconstruction for $v_{i+\frac{1}{2}}$, in (2.11), is given by

$$v_{i+\frac{1}{2}} = -\frac{1}{12}\overline{v}_{i-1} + \frac{7}{12}\overline{v}_i + \frac{7}{12}\overline{v}_{i+1} - \frac{1}{12}\overline{v}_{i+2} + O(\Delta x^4),$$

and the two one point upwind biased 3rd order reconstructions for $v_{i+\frac{1}{2}}$ in (2.11), are given by

$$v_{i+\frac{1}{2}} = -\frac{1}{6}\overline{v}_{i-1} + \frac{5}{6}\overline{v}_i + \frac{1}{3}\overline{v}_{i+1} + O(\Delta x^3)$$

$$or \quad v_{i+\frac{1}{2}} = \frac{1}{3}\overline{v}_i + \frac{5}{6}\overline{v}_{i+1} - \frac{1}{6}\overline{v}_{i+2} + O(\Delta x^3).$$

Similarly, a central 4th order flux (2.30) is

$$\hat{v}_{i+\frac{1}{2}} = -\frac{1}{12}v_{i-1} + \frac{7}{12}v_i + \frac{7}{12}v_{i+1} - \frac{1}{12}v_{i+2},$$

which gives

$$\frac{1}{\Delta x}\left(\hat{v}_{i+\frac{1}{2}} - \hat{v}_{i-\frac{1}{2}}\right) = v'(x_i) + O(\Delta x^4),$$

and the two one point upwind biased 3rd order fluxes (2.30) are given by

$$\hat{v}_{i+\frac{1}{2}} = -\frac{1}{6}v_{i-1} + \frac{5}{6}v_i + \frac{1}{3}v_{i+1}$$

$$or \quad \hat{v}_{i+\frac{1}{2}} = \frac{1}{3}v_i + \frac{5}{6}v_{i+1} - \frac{1}{6}v_{i+2},$$

which gives

$$\frac{1}{\Delta x}\left(\hat{v}_{i+\frac{1}{2}} - \hat{v}_{i-\frac{1}{2}}\right) = v'(x_i) + O(\Delta x^3).$$

Traditional central and upwind schemes, either finite volume or finite difference, can be derived by these fixed stencil reconstructions or flux differenced approximations to the derivatives.

2.2 ENO and WENO Approximations in 1D

For solving hyperbolic conservation laws, we are interested in the class of piecewise smooth functions. A piecewise smooth function $v(x)$ is smooth (i.e. it has as many derivatives as the scheme calls for) except for at finitely many isolated points. At these points, $v(x)$ and its derivatives are assumed to have finite left and right limits. Such functions are "generic" for solutions to hyperbolic conservation laws.

For such piecewise smooth functions, the order of accuracy we refer to in these lecture notes are *formal*, that is, it is defined as whatever accuracy determined by the local truncation error in the *smooth regions* of the function.

If the function $v(x)$ is only piecewise smooth, a fixed stencil approximation described in Sect.2.1 may not be adequate near discontinuities. Fig.2.1 (left) gives the 4-th order (piecewise cubic) interpolation with a central stencil for the step function, i.e. the polynomial approximation inside the interval $[x_{i-\frac{1}{2}}, x_{i+\frac{1}{2}}]$ interpolates the step function at the four points $x_{i-\frac{3}{2}}, x_{i-\frac{1}{2}}, x_{i+\frac{1}{2}}, x_{i+\frac{3}{2}}$. Notice the obvious over/undershoots for the cells near the discontinuity.

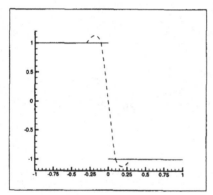 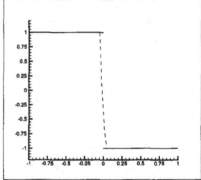

Fig. 2.1: Fixed central stencil cubic interpolation (left) and ENO cubic interpolation (right) for the step function. Solid: exact function; Dashed: interpolant piecewise cubic polynomials.

These oscillations (termed *the Gibbs Phenomena* in spectral methods) happen because the stencils, as defined by (2.15), actually contain the discontinuous cell for x_i close enough to the discontinuity. As a result, the approximation property (2.5) is no longer valid in such stencils.

ENO approximation. A closer look at Fig.2.1 (left) motivates the idea of "adaptive stencil", namely, the left shift r changes with the location x_i. The basic idea is to avoid including the discontinuous cell in the stencil, if possible.

To achieve this effect, we need to look at the Newton formulation of the interpolation polynomial.

We first review the definition of the Newton divided differences. The 0-th degree divided differences of the function $V(x)$ in (2.13)-(2.14) are defined by:

$$V[x_{i-\frac{1}{2}}] \equiv V(x_{i-\frac{1}{2}});$$
(2.31)

and in general the j-th degree divided differences, for $j \geq 1$, are defined inductively by

$$V[x_{i-\frac{1}{2}}, ..., x_{i+j-\frac{1}{2}}] \equiv \frac{V[x_{i+\frac{1}{2}}, ..., x_{i+j-\frac{1}{2}}] - V[x_{i-\frac{1}{2}}, ..., x_{i+j-\frac{3}{2}}]}{x_{i+j-\frac{1}{2}} - x_{i-\frac{1}{2}}}.$$
(2.32)

Similarly, the divided differences of the cell averages $\bar{v}$ in (2.4) are defined by

$$\bar{v}[x_i] \equiv \bar{v}_i;$$
(2.33)

and in general

$$\bar{v}[x_i, ..., x_{i+j}] \equiv \frac{\bar{v}[x_{i+1}, ..., x_{i+j}] - \bar{v}[x_i, ..., x_{i+j-1}]}{x_{i+j} - x_i}.$$
(2.34)

We note that, by (2.14),

$$V[x_{i-\frac{1}{2}}, x_{i+\frac{1}{2}}] = \frac{V(x_{i+\frac{1}{2}}) - V(x_{i-\frac{1}{2}})}{x_{i+\frac{1}{2}} - x_{i-\frac{1}{2}}} = \bar{v}_i,$$
(2.35)

i.e. the 0-th degree divided differences of $\bar{v}$ are the first degree divided differences of $V(x)$. We can thus write the divided differences of $V(x)$ of first degree and higher by those of $\bar{v}$ of 0-th degree and higher, using (2.35) and (2.32).

The Newton form of the k-th degree interpolation polynomial $P(x)$, which interpolates $V(x)$ at the $k+1$ points (2.15), can be expressed using the divided differences (2.31)-(2.32) by

$$P(x) = \sum_{j=0}^{k} V[x_{i-r-\frac{1}{2}}, ..., x_{i-r+j-\frac{1}{2}}] \prod_{m=0}^{j-1} \left(x - x_{i-r+m-\frac{1}{2}}\right).$$
(2.36)

We can take the derivative of (2.36) to get $p(x)$ in (2.16):

$$p(x) = \sum_{j=1}^{k} V[x_{i-r-\frac{1}{2}}, ..., x_{i-r+j-\frac{1}{2}}] \sum_{m=0}^{j-1} \prod_{\substack{l=0 \\ l \neq m}}^{j-1} \left(x - x_{i-r+l-\frac{1}{2}}\right).$$
(2.37)

Notice that only first and higher degree divided differences of $V(x)$ appear in (2.37). Hence by (2.35), we can express $p(x)$ completely by the divided differences of $\bar{v}$, without any need to reference $V(x)$.

Let us now recall an important property of divided differences:

$$V[x_{i-\frac{1}{2}}, ..., x_{i+j-\frac{1}{2}}] = \frac{V^{(j)}(\xi)}{j!}, \tag{2.38}$$

for some ξ inside the stencil: $x_{i-\frac{1}{2}} < \xi < x_{i+j-\frac{1}{2}}$, as long as the function $V(x)$ is smooth in this stencil. If $V(x)$ is discontinuous at some point inside the stencil, then it is easy to verify that

$$V[x_{i-\frac{1}{2}}, ..., x_{i+j-\frac{1}{2}}] = O\left(\frac{1}{\Delta x^j}\right). \tag{2.39}$$

Thus the divided difference is a measurement of the smoothness of the function inside the stencil.

We now describe the ENO idea by using (2.36). Suppose our job is to find a stencil of $k+1$ consecutive points, which must include $x_{i-\frac{1}{2}}$ and $x_{i+\frac{1}{2}}$, such that $V(x)$ is "the smoothest" in this stencil comparing with other possible stencils. We perform this job by breaking it into steps, in each step we only add one point to the stencil. We thus start with the two point stencil

$$\tilde{S}_2(i) = \{x_{i-\frac{1}{2}}, x_{i+\frac{1}{2}}\}, \tag{2.40}$$

where we have used $\tilde{S}$ to denote a stencil for the primitive function V. Notice that the stencil $\tilde{S}$ for V has a corresponding stencil S for $\bar{v}$ through (2.35), for example (2.40) corresponds to a single cell stencil

$$S(i) = \{I_i\}$$

for $\bar{v}$. The linear interpolation on the stencil $\tilde{S}_2(i)$ in (2.40) can be written in the Newton form as

$$P^1(x) = V[x_{i-\frac{1}{2}}] + V[x_{i-\frac{1}{2}}, x_{i+\frac{1}{2}}]\left(x - x_{i-\frac{1}{2}}\right).$$

At the next step, we have only two choices to expand the stencil by adding one point: we can either add the left neighbor $x_{i-\frac{3}{2}}$, resulting in the following quadratic interpolation

$$R(x) = P^1(x) + V[x_{i-\frac{3}{2}}, x_{i-\frac{1}{2}}, x_{i+\frac{1}{2}}]\left(x - x_{i-\frac{1}{2}}\right)\left(x - x_{i+\frac{1}{2}}\right), \tag{2.41}$$

or add the right neighbor $x_{i+\frac{3}{2}}$, resulting in the following quadratic interpolation

$$S(x) = P^1(x) + V[x_{i-\frac{1}{2}}, x_{i+\frac{1}{2}}, x_{i+\frac{3}{2}}]\left(x - x_{i-\frac{1}{2}}\right)\left(x - x_{i+\frac{1}{2}}\right). \tag{2.42}$$

We note that the deviations from $P^1(x)$ in (2.41) and (2.42), are the same function

$$\left(x - x_{i-\frac{1}{2}}\right)\left(x - x_{i+\frac{1}{2}}\right)$$

multiplied by two different constants

$$V[x_{i-\frac{3}{2}}, x_{i-\frac{1}{2}}, x_{i+\frac{1}{2}}], \quad and \quad V[x_{i-\frac{1}{2}}, x_{i+\frac{1}{2}}, x_{i+\frac{3}{2}}]. \qquad (2.43)$$

These two constants are the two second degree divided differences of $V(x)$ in two different stencils. We have already noticed before, in (2.38) and (2.39), that a smaller divided difference implies the function is "smoother" in that stencil. We thus decide upon which point to add to the stencil, by comparing the two relevant divided differences (2.43), and picking the one with a smaller absolute value. Thus, if

$$\left| V[x_{i-\frac{3}{2}}, x_{i-\frac{1}{2}}, x_{i+\frac{1}{2}}] \right| < \left| V[x_{i-\frac{1}{2}}, x_{i+\frac{1}{2}}, x_{i+\frac{3}{2}}] \right|, \qquad (2.44)$$

we will take the 3 point stencil as

$$\tilde{S}_3(i) = \{x_{i-\frac{3}{2}}, x_{i-\frac{1}{2}}, x_{i+\frac{1}{2}}\};$$

otherwise, we will take

$$\tilde{S}_3(i) = \{x_{i-\frac{1}{2}}, x_{i+\frac{1}{2}}, x_{i+\frac{3}{2}}\}.$$

This procedure can be continued, with one point added to the stencil at each step, according to the smaller of the absolute values of the two relevant divided differences, until the desired number of points in the stencil is reached.

We note that, for the uniform grid case $\Delta x_i = \Delta x$, there is no need to compute the divided differences as in (2.32). We should use undivided differences instead:

$$V < x_{i-\frac{1}{2}}, x_{i+\frac{1}{2}} >= V[x_{i-\frac{1}{2}}, x_{i+\frac{1}{2}}] = \bar{v}_i \qquad (2.45)$$

(see (2.35)), and

$$V < x_{i-\frac{1}{2}}, ..., x_{i+j+\frac{1}{2}} > \equiv V < x_{i+\frac{1}{2}}, ..., x_{i+j+\frac{1}{2}} > - $$
$$V < x_{i-\frac{1}{2}}, ..., x_{i+j-\frac{1}{2}} >, \ j \geq 1 \qquad (2.46)$$

The Newton interpolation formulae (2.36)-(2.37) should also be adjusted accordingly. This both saves computational time and reduces round-off effects.

The FORTRAN program for this ENO choosing process is very simple:

```
* assuming the m-th degree divided (or undivided) differen-
* ces of V(x), with x_i as the left-most point in the argu-
* ments, are stored in V(i,m), also assuming that "is" is
* the left-most point in the stencil for cell i for a k-th
* degree polynomial

      is=i
      do m=2,k
      if(abs(V(is-1,m)).lt.abs(V(is,m))) is=is-1
      enddo
```

Once the stencil $\tilde{S}(i)$, hence $S(i)$, in (2.8) is found, one could use (2.11), with the prestored values of the constants c_{rj}, (2.20) or (2.21), to compute the reconstructed values at the cell boundary. Or, one could use (2.30) to compute the fluxes. An alternative way is to compute the values or fluxes using the Newton form (2.37) directly. The computational cost is about the same.

We summarize the ENO reconstruction procedure in the following

Procedure 2.1. 1D ENO reconstruction.

Given the cell averages $\{\bar{v}_i\}$ of a function $v(x)$, we obtain a piecewise polynomial reconstruction, of degree at most $k - 1$, using ENO, in the following way:

1. Compute the divided differences of the primitive function $V(x)$, for degrees 1 to k, using $\bar{v}$, (2.35) and (2.32).
   If the grid is uniform $\Delta x_i = \Delta x$, at this stage, undivided differences (2.45)-(2.46) should be computed instead.
2. In cell I_i, start with a two point stencil

$$\tilde{S}_2(i) = \{x_{i-\frac{1}{2}}, x_{i+\frac{1}{2}}\}$$

for $V(x)$, which is equivalent to a one point stencil,

$$S_1(i) = \{I_i\}$$

for $\bar{v}$.

3. For $l = 2, ..., k$, assuming

$$\tilde{S}_l(i) = \{x_{j+\frac{1}{2}}, ..., x_{j+l-\frac{1}{2}}\}$$

is known, add one of the two neighboring points, $x_{j-\frac{1}{2}}$ or $x_{j+l+\frac{1}{2}}$, to the stencil, following the ENO procedure:

   – If

$$\left|V[x_{j-\frac{1}{2}}, ..., x_{j+l-\frac{1}{2}}]\right| < \left|V[x_{j+\frac{1}{2}}, ..., x_{j+l+\frac{1}{2}}]\right|, \tag{2.47}$$

   add $x_{j-\frac{1}{2}}$ to the stencil $\tilde{S}_l(i)$ to obtain

$$\tilde{S}_{l+1}(i) = \{x_{j-\frac{1}{2}}, ..., x_{j+l-\frac{1}{2}}\}.$$

   – Otherwise, add $x_{j+l+\frac{1}{2}}$ to the stencil $\tilde{S}_l(i)$ to obtain

$$\tilde{S}_{l+1}(i) = \{x_{j+\frac{1}{2}}, ..., x_{j+l+\frac{1}{2}}\}.$$

4. Use the Lagrange form (2.19) or the Newton form (2.37) to obtain $p_i(x)$,
   which is a polynomial of degree at most $k-1$ in I_i, satisfying the accuracy
   condition (2.5), *as long as $v(x)$ is smooth in I_i.*
   We could use $p_i(x)$ to get the approximations at the cell boundaries:

$$v_{i+\frac{1}{2}}^- = p_i(x_{i+\frac{1}{2}}), \qquad v_{i-\frac{1}{2}}^+ = p_i(x_{i-\frac{1}{2}}).$$

However, it is usually more convenient, when the stencil is known, to use
(2.10), with c_{rj} defined by (2.20) for a nonuniform grid, or by (2.21) and
Table 2.1 for a uniform grid, to compute an approximation to $v(x)$ at the
cell boundaries.

$\square$

For the same piecewise cubic interpolation to the step function, but this
time using the ENO procedure with a two point stencil $\tilde{S}_2(i) = \{x_{i-\frac{1}{2}}, x_{i+\frac{1}{2}}\}$
in the Step 2 of Procedure 2.1, we obtain a non-oscillatory interpolation, in
Fig.2.1 (right).

For a piecewise smooth function $V(x)$, ENO interpolation starting with
a two point stencil $\tilde{S}_2(i) = \{x_{i-\frac{1}{2}}, x_{i+\frac{1}{2}}\}$ in the Step 2 of Procedure 2.1, as
was shown in Fig.2.1 (right), has the following properties [39]:

1. The accuracy condition

$$P_i(x) = V(x) + O(\Delta x^{k+1}), \qquad x \in I_i$$

is valid for any cell I_i which does not contain a discontinuity.
This implies that the ENO interpolation procedure can recover the full
high order accuracy right up to the discontinuity.
2. $P_i(x)$ is monotone in any cell I_i which *does* contain a discontinuity of
$V(x)$.
3. The reconstruction is TVB (total variation bounded). That is, there exists
a function $z(x)$, satisfying

$$z(x) = P_i(x) + O(\Delta x^{k+1}), \qquad x \in I_i$$

for any cell I_i, including those cells which contain discontinuities, such
that

$$TV(z) \leq TV(V).$$

Property 3 is clearly a consequence of Properties 1 and 2 (just take $z(x)$ to
be $V(x)$ in the smooth cells and take $z(x)$ to be $P_i(x)$ in the cells containing
discontinuities). It is quite interesting that Property 2 holds. One would have
expected trouble in those "shocked cells", i.e. cells I_i which contain discontinuities, for ENO would not help for such cases as the stencil starts with two
points already containing a discontinuity. We will give a proof of Property 2
for a simple but illustrative case, i.e. when $V(x)$ is a step function

$$V(x) = \begin{cases} 0, x \leq 0; \\ 1, x > 0 \end{cases}$$

and the k-th degree polynomial $P(x)$ interpolates $V(x)$ at $k+1$ points

$$x_{\frac{1}{2}} < x_{\frac{3}{2}} < \ldots < x_{k+\frac{1}{2}}$$

containing the discontinuity

$$x_{j_0-\frac{1}{2}} < 0 < x_{j_0+\frac{1}{2}}$$

for some j_0 between 1 and k. For any interval which does not contain the discontinuity 0:

$$[x_{j-\frac{1}{2}}, x_{j+\frac{1}{2}}], \qquad j \neq j_0, \tag{2.48}$$

we have

$$P(x_{j-\frac{1}{2}}) = V(x_{j-\frac{1}{2}}) = V(x_{j+\frac{1}{2}}) = P(x_{j+\frac{1}{2}}),$$

hence there is at least one point ξ_j in between, $x_{j-\frac{1}{2}} < \xi_j < x_{j+\frac{1}{2}}$, such that $P'(\xi_j) = 0$. This way we can find $k-1$ distinct zeroes for $P'(x)$, as there are $k-1$ intervals (2.48) which do not contain the discontinuity 0. However, $P'(x)$ is a non-zero polynomial of degree at most $k-1$, hence can have at most $k-1$ distinct zeroes. This implies that $P'(x)$ *does not have any zero inside the shocked interval* $[x_{j_0-\frac{1}{2}}, x_{j_0+\frac{1}{2}}]$, i.e. $P(x)$ is *monotone* in this shocked interval. This proof can be generalized to a proof for Property 2 [39].

WENO approximation. In this subsection we describe the recently developed WENO (weighted ENO) reconstruction procedure [53,43]. WENO is based on ENO, of course. For simplicity of presentation, in this subsection we assume the grid is uniform, i.e. $\Delta x_i = \Delta x$.

As we can see from Sect.2.2, ENO reconstruction is uniformly high order accurate right up to the discontinuity. It achieves this effect by adaptively choosing the stencil based on the absolute values of divided differences. However, one could make the following remarks about ENO reconstruction, indicating rooms for improvements:

1. The stencil might change even by a round-off error perturbation near zeroes of the solution and its derivatives. That is, when both sides of (2.47) are near 0, a small change at the round off level would change the direction of the inequality and hence the stencil. In smooth regions, this "free adaptation" of stencils is clearly not necessary. Moreover, this may cause loss of accuracy when applied to a hyperbolic PDE [35,67].
2. The resulting numerical flux (2.23) is not smooth, as the stencil pattern may change at neighboring points.
3. In the stencil choosing process, k candidate stencils are considered, covering $2k-1$ cells, but only one of the stencils is actually used in forming the reconstruction (2.10) or the flux (2.30), resulting in k-th order accuracy. If all the $2k-1$ cells in the potential stencils are used, one could get $(2k-1)$-th order accuracy in smooth regions.

4. ENO stencil choosing procedure involves many logical "if" structures, or equivalent mathematical formulae, which are not very efficient on certain vector computers such as CRAYs (however they are friendly to parallel computers).

There have been attempts in the literature to remedy the first problem, the "free adaptation" of stencils. In [28] and [67], the following "biasing" strategy was proposed. One first identity a "preferred" stencil

$$\tilde{S}_{pref}(i) = \{x_{i-r+\frac{1}{2}}, ..., x_{i-r+k+\frac{1}{2}}\}, \tag{2.49}$$

which might be central or one-point upwind. One then replaces (2.47) by

$$\left| V[x_{j-\frac{1}{2}}, ..., x_{j+l-\frac{1}{2}}] \right| < b \left| V[x_{j+\frac{1}{2}}, ..., x_{j+l+\frac{1}{2}}] \right|,$$

if

$$x_{j+\frac{1}{2}} > x_{i-r+\frac{1}{2}},$$

i.e. if the left-most point $x_{j+\frac{1}{2}}$ in the current stencil $\tilde{S}_l(i)$ has not reached the left-most point $x_{i-r+\frac{1}{2}}$ of the preferred stencil $S_{pref}(i)$ in (2.49) yet; otherwise, if

$$x_{j+\frac{1}{2}} \leq x_{i-r+\frac{1}{2}},$$

one replaces (2.47) by

$$b \left| V[x_{j-\frac{1}{2}}, ..., x_{j+l-\frac{1}{2}}] \right| < \left| V[x_{j+\frac{1}{2}}, ..., x_{j+l+\frac{1}{2}}] \right|.$$

Here, $b > 1$ is the so-called biasing parameter. Analysis in [67] indicates a good choice of the parameter $b = 2$. The philosophy is to stay as close as possible to the preferred stencil, unless the alternative candidate is, roughly speaking, a factor $b > 1$ better in smoothness.

WENO is a more recent attempt to improve upon ENO in these four points. The basic idea is the following: instead of using only one of the candidate stencils to form the reconstruction, one uses a convex combination of all of them. To be more precise, suppose the k candidate stencils

$$S_r(i) = \{x_{i-r}, ..., x_{i-r+k-1}\}, \qquad r = 0, ..., k-1 \tag{2.50}$$

produce k different reconstructions to the value $v_{i+\frac{1}{2}}$, according to (2.11),

$$v_{i+\frac{1}{2}}^{(r)} = \sum_{j=0}^{k-1} c_{rj} \bar{v}_{i-r+j}, \qquad r = 0, ..., k-1, \tag{2.51}$$

WENO reconstruction would take a convex combination of all $v_{i+\frac{1}{2}}^{(r)}$ defined in (2.51) as a new approximation to the cell boundary value $v(x_{i+\frac{1}{2}})$:

$$v_{i+\frac{1}{2}} = \sum_{r=0}^{k-1} \omega_r v_{i+\frac{1}{2}}^{(r)}. \tag{2.52}$$

Apparently, the key to the success of WENO would be the choice of the weights ω_r. We require

$$\omega_r \geq 0, \qquad \sum_{r=0}^{k-1} \omega_r = 1 \tag{2.53}$$

for stability and consistency.

If the function $v(x)$ is smooth in all of the candidate stencils (2.50), there are constants d_r such that

$$v_{i+\frac{1}{2}} = \sum_{r=0}^{k-1} d_r v_{i+\frac{1}{2}}^{(r)} = v(x_{i+\frac{1}{2}}) + O(\Delta x^{2k-1}). \tag{2.54}$$

For example, d_r for $1 \leq k \leq 3$ are given by

$$
\begin{aligned}
d_0 &= 1, \qquad k = 1; \\
d_0 &= \frac{2}{3}, \quad d_1 = \frac{1}{3}, \qquad k = 2; \\
d_0 &= \frac{3}{10}, \quad d_1 = \frac{3}{5}, \quad d_2 = \frac{1}{10}, \qquad k = 3.
\end{aligned}
$$

We can see that d_r is always positive and, due to consistency,

$$\sum_{r=0}^{k-1} d_r = 1. \tag{2.55}$$

In this smooth case, we would like to have

$$\omega_r = d_r + O(\Delta x^{k-1}), \qquad r = 0, ..., k-1, \tag{2.56}$$

which would imply $(2k-1)$-th order accuracy:

$$v_{i+\frac{1}{2}} = \sum_{r=0}^{k-1} \omega_r v_{i+\frac{1}{2}}^{(r)} = v(x_{i+\frac{1}{2}}) + O(\Delta x^{2k-1}) \tag{2.57}$$

because

$$
\begin{aligned}
\sum_{r=0}^{k-1} \omega_r v_{i+\frac{1}{2}}^{(r)} - \sum_{r=0}^{k-1} d_r v_{i+\frac{1}{2}}^{(r)} &= \sum_{r=0}^{k-1} (\omega_r - d_r) \left(v_{i+\frac{1}{2}}^{(r)} - v(x_{i+\frac{1}{2}}) \right) \\
&= \sum_{r=0}^{k-1} O(\Delta x^{k-1}) O(\Delta x^k) = O(\Delta x^{2k-1})
\end{aligned}
$$

where in the first equality we used (2.53) and (2.55).

When the function $v(x)$ has a discontinuity in one or more of the stencils (2.50), we would hope the corresponding weight(s) ω_r to be essentially 0, to emulate the successful ENO idea.

Another consideration is that the weights should be smooth functions of the cell averages involved. In fact, the weights designed in [43] and described below are C^∞.

Finally, we would like to have weights which are computationally efficient. Thus, polynomials or rational functions are preferred over exponential type functions.

All these considerations lead to the following form of weights:

$$\omega_r = \frac{\alpha_r}{\sum_{s=0}^{k-1} \alpha_s}, \qquad r = 0, ..., k-1 \tag{2.58}$$

with

$$\alpha_r = \frac{d_r}{(\epsilon + \beta_r)^2} . \tag{2.59}$$

Here $\epsilon > 0$ is introduced to avoid the denominator to become 0. We take $\epsilon = 10^{-6}$ in all our numerical tests [43]. β_r are the so-called "smooth indicators" of the stencil $S_r(i)$: if the function $v(x)$ is smooth in the stencil $S_r(i)$, then

$$\beta_r = O(\Delta x^2) ,$$

but if $v(x)$ has a discontinuity inside the stencil $S_r(i)$, then

$$\beta_r = O(1).$$

Translating into the weights ω_r in (2.58), we will have

$$\omega_r = O(1)$$

when the function $v(x)$ is smooth in the stencil $S_r(i)$, and

$$\omega_r = O(\Delta x^4)$$

if $v(x)$ has a discontinuity inside the stencil $S_r(i)$. Emulation of ENO near a discontinuity is thus achieved.

One also has to worry about the accuracy requirement (2.56), which must be checked when the specific form of the smooth indicator β_r is given. For any smooth indicator β_r, it is easy to see that the weights defined by (2.58) satisfies (2.53). To satisfy (2.56), it suffices to have, through a Taylor expansion analysis:

$$\beta_r = D\left(1 + O(\Delta x^{k-1})\right), \qquad r = 0, ..., k-1, \tag{2.60}$$

where D is a nonzero quantity independent of r (but may depend on Δx).

As we have seen in Sect.2.2, the ENO reconstruction procedure chooses the "smoothest" stencil by comparing a hierarchy of divided or undivided differences. This is because these differences can be used to measure the smoothness of the function on a stencil, (2.38)-(2.39). In [43], after extensive experiments, a robust (for third and fifth order at least) choice of smooth indicators β_r is given. As we know, on each stencil $S_r(i)$, we can construct a $(k-1)$-th degree reconstruction polynomial, which if evaluated at $x = x_{i+\frac{1}{2}}$, renders the approximation to the value $v(x_{i+\frac{1}{2}})$ in (2.51). Since the total variation is a good measurement for smoothness, it would be desirable to minimize the total variation for this reconstruction polynomial inside I_i. Consideration for a smooth flux and for the role of higher order variations leads us to the following measurement for smoothness: let the reconstruction polynomial on the stencil $S_r(i)$ be denoted by $p_r(x)$, we define

$$\beta_r = \sum_{l=1}^{k-1} \int_{x_{i-\frac{1}{2}}}^{x_{i+\frac{1}{2}}} \Delta x^{2l-1} \left(\frac{\partial^l p_r(x)}{\partial^l x} \right)^2 dx. \tag{2.61}$$

The right hand side of (2.61) is just a sum of the squares of scaled L^2 norms for all the derivatives of the interpolation polynomial $p_r(x)$ over the interval $(x_{i-\frac{1}{2}}, x_{i+\frac{1}{2}})$. The factor Δx^{2l-1} is introduced to remove any Δx dependency in the derivatives, in order to preserve self-similarity when used to hyperbolic PDEs (Sect.2.3).

We remark that (2.61) is similar to but smoother than the total variation measurement based on the L^1 norm. It also renders a more accurate WENO scheme for the case $k=2$ and 3.

When $k=2$, (2.61) gives the following smoothness measurement [53,43]:

$$\beta_0 = (\bar{v}_{i+1} - \bar{v}_i)^2,$$
$$\beta_1 = (\bar{v}_i - \bar{v}_{i-1})^2. \tag{2.62}$$

For $k=3$, (2.61) gives [43]:

$$\beta_0 = \frac{13}{12}(\bar{v}_i - 2\bar{v}_{i+1} + \bar{v}_{i+2})^2 + \frac{1}{4}(3\bar{v}_i - 4\bar{v}_{i+1} + \bar{v}_{i+2})^2,$$
$$\beta_1 = \frac{13}{12}(\bar{v}_{i-1} - 2\bar{v}_i + \bar{v}_{i+1})^2 + \frac{1}{4}(\bar{v}_{i-1} - \bar{v}_{i+1})^2, \tag{2.63}$$
$$\beta_2 = \frac{13}{12}(\bar{v}_{i-2} - 2\bar{v}_{i-1} + \bar{v}_i)^2 + \frac{1}{4}(\bar{v}_{i-2} - 4\bar{v}_{i-1} + 3\bar{v}_i)^2.$$

We can easily verify that the accuracy condition (2.60) is satisfied, even near smooth extrema [43]. This indicates that (2.62) gives a third order WENO scheme, and (2.63) gives a fifth order one.

Notice that the discussion here has a one point upwind bias in the optimal linear stencil, suitable for a problem with wind blowing from left to right. If the wind blows the other way, the procedure should be modified symmetrically with respect to $x_{i+\frac{1}{2}}$.

In summary, we have the following WENO reconstruction procedure:

Procedure 2.2. 1D WENO reconstruction.

Given the cell averages $\{\bar{v}_i\}$ of a function $v(x)$, for each cell I_i, we obtain upwind biased $(2k-1)$-th order approximations to the function $v(x)$ at the cell boundaries, denoted by $v^+_{i-\frac{1}{2}}$ and $v^-_{i+\frac{1}{2}}$, in the following way:

1. Obtain the k reconstructed values $v^{(r)}_{i+\frac{1}{2}}$, of k-th order accuracy, in (2.51), based on the stencils (2.50), for $r = 0, ..., k-1$;
   Also obtain the k reconstructed values $v^{(r)}_{i-\frac{1}{2}}$, of k-th order accuracy, using (2.10), again based on the stencils (2.50), for $r = 0, ..., k-1$;
2. Find the constants d_r and $\tilde{d}_r$, such that (2.54) and

$$v_{i-\frac{1}{2}} = \sum_{r=0}^{k-1} \tilde{d}_r v^{(r)}_{i-\frac{1}{2}} = v(x_{i-\frac{1}{2}}) + O(\Delta x^{2k-1})$$

are valid. By symmetry,

$$\tilde{d}_r = d_{k-1-r}.$$

3. Find the smooth indicators β_r in (2.61), for all $r = 0, ..., k-1$. Explicit formulae for $k = 2$ and $k = 3$ are given in (2.62) and (2.63) respectively.
4. Form the weights ω_r and $\tilde{\omega}_r$ using (2.58)-(2.59) and

$$\tilde{\omega}_r = \frac{\tilde{\alpha}_r}{\sum_{s=0}^{k-1} \tilde{\alpha}_s}, \qquad \tilde{\alpha}_r = \frac{\tilde{d}_r}{(\epsilon + \beta_r)^2}, \qquad r = 0, ..., k-1.$$

5. Find the $(2k-1)$-th order reconstruction

$$v^-_{i+\frac{1}{2}} = \sum_{r=0}^{k-1} \omega_r v^{(r)}_{i+\frac{1}{2}}, \qquad v^+_{i-\frac{1}{2}} = \sum_{r=0}^{k-1} \tilde{\omega}_r v^{(r)}_{i-\frac{1}{2}} . \tag{2.64}$$

□

We can obtain weights for higher orders of k (corresponding to seventh and higher order WENO schemes) using the same recipe. However, these schemes of seventh and higher order have not been extensively tested yet.

2.3 ENO and WENO Schemes for 1D Conservation Laws

In this section we describe the ENO and WENO schemes for 1D conservation laws:

$$u_t(x,t) + f_x(u(x,t)) = 0 \tag{2.65}$$

equipped with suitable initial and boundary conditions.

We will concentrate on the discussion of spatial discretization, and will leave the time variable t continuous (the method-of-lines approach). Time discretizations will be discussed in Sect.4.2.

Our computational domain is $a \leq x \leq b$. We have a grid defined by (2.1), with the notations (2.2)-(2.3). Except for in Sect.2.3, we do not consider boundary conditions. We thus assume that the values of the numerical solution are also available outside the computational domain whenever they are needed. This would be the case for periodic or compactly supported problems.

Finite volume formulation in the scalar case. For finite volume schemes, or schemes based on cell averages, we do not solve (2.65) directly, but its integrated version. We integrate (2.65) over the interval I_i to obtain

$$\frac{d\bar{u}(x_i, t)}{dt} = -\frac{1}{\Delta x_i} \left(f(u(x_{i+\frac{1}{2}}, t)) - f(u(x_{i-\frac{1}{2}}, t)) \right), \tag{2.66}$$

where

$$\bar{u}(x_i, t) \equiv \frac{1}{\Delta x_i} \int_{x_{i-\frac{1}{2}}}^{x_{i+\frac{1}{2}}} u(\xi, t)\, d\xi \tag{2.67}$$

is the cell average. We approximate (2.66) by the following conservative scheme

$$\frac{d\bar{u}_i(t)}{dt} = -\frac{1}{\Delta x_i} \left(\hat{f}_{i+\frac{1}{2}} - \hat{f}_{i-\frac{1}{2}} \right), \tag{2.68}$$

where $\bar{u}_i(t)$ is the numerical approximation to the cell average $\bar{u}(x_i, t)$, and the numerical flux $\hat{f}_{i+\frac{1}{2}}$ is defined by

$$\hat{f}_{i+\frac{1}{2}} = h\left(u^-_{i+\frac{1}{2}}, u^+_{i+\frac{1}{2}} \right) \tag{2.69}$$

with the values $u^{\pm}_{i+\frac{1}{2}}$ obtained by the ENO reconstruction Procedure 2.1, or by the WENO reconstruction Procedure 2.2.

The two argument function h in (2.69) is a monotone flux. It satisfies:

- $h(a, b)$ is a Lipschitz continuous function in both arguments;
- $h(a, b)$ is a nondecreasing function in a and a nonincreasing function in b. Symbolically $h(\uparrow, \downarrow)$;
- $h(a, b)$ is consistent with the physical flux f, that is, $h(a, a) = f(a)$.

Examples of monotone fluxes include:

1. Godunov flux:

$$h(a, b) = \begin{cases} \min_{a \leq u \leq b} f(u) & \text{if } a \leq b \\ \max_{b \leq u \leq a} f(u) & \text{if } a > b \end{cases}. \tag{2.70}$$

2. Engquist-Osher flux:

$$h(a, b) = \int_0^a \max(f'(u), 0)du + \int_0^b \min(f'(u), 0)du + f(0). \tag{2.71}$$

3. Lax-Friedrichs flux:

$$h(a, b) = \frac{1}{2}\left[f(a) + f(b) - \alpha(b - a) \right] \tag{2.72}$$

where $\alpha = \max_u |f'(u)|$ is a constant. The maximum is taken over the relevant range of u.

We have listed the monotone fluxes from the least dissipative (less smearing of discontinuities) to the most. For lower order methods (order of reconstruction is 1 or 2), there is a big difference between results obtained by different monotone fluxes. However, this difference becomes much smaller for higher order reconstructions. In Fig. 2.2, we plot the results of a right moving shock for the Burgers' equation ($f(u) = \frac{u^2}{2}$ in (2.65)), with first order reconstruction using Godunov and Lax-Friedrichs monotone fluxes (top), and with fourth order ENO reconstruction using Godunov and Lax-Friedrichs monotone fluxes (bottom). We can clearly see that, while the Godunov flux behaves much better for the first order scheme, the two fourth order ENO schemes behave similarly. We thus use the simple and inexpensive Lax-Friedrichs flux in most of our high order calculations.

We remark that, by the classic Lax-Wendroff theorem [51], the solution to the conservative scheme (2.68), *if converges*, will converge to a weak solution of (2.65).

In summary, to build a finite volume ENO scheme (2.68), given the cell averages $\{\overline{u}_i\}$ (we will often drop the explicit reference to the time variable t), we proceed as follows:

Procedure 2.3. Finite volume 1D scalar ENO and WENO.

1. Follow the Procedure 2.1 in Sect.2.2 for ENO, or the Procedure 2.2 in Sect.2.2 for WENO, to obtain the k-th order reconstructed values $u^-_{i+\frac{1}{2}}$ and $u^+_{i+\frac{1}{2}}$ for all i;
2. Choose a monotone flux (e.g., one of (2.70) to (2.72)), and use (2.69) to compute the flux $\hat{f}_{i+\frac{1}{2}}$ for all i;
3. Form the scheme (2.68).

$\square$

Notice that the finite volume scheme can be applied to arbitrary nonuniform grids.

Finite difference formulation in the scalar case. We first assume the grid is uniform and solve (2.65) directly using a conservative approximation to the spatial derivative:

$$\frac{du_i(t)}{dt} = -\frac{1}{\Delta x}\left(\hat{f}_{i+\frac{1}{2}} - \hat{f}_{i-\frac{1}{2}}\right) \tag{2.73}$$

where $u_i(t)$ is the numerical approximation to the point value $u(x_i, t)$, and the numerical flux

$$\hat{f}_{i+\frac{1}{2}} = \hat{f}(u_{i-r}, ..., u_{i+s})$$

satisfies the following conditions:

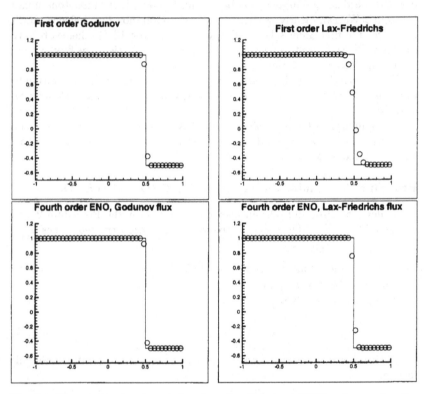

Fig. 2.2: First order (top) and fourth order (bottom) ENO schemes for the Burgers equation, with the Godunov flux (left) and the Lax-Friedrichs flux (right). Solid lines: exact solution; Circles: the computed solution at $t = 4$.

– $\hat{f}$ is a Lipschitz continuous function in all the arguments;
– $\hat{f}$ is consistent with the physical flux f, that is, $\hat{f}(u, ..., u) = f(u)$.

Again the Lax-Wendroff theorem [51] applies. The solution to the conservative scheme (2.73), *if converges*, will converge to a weak solution of (2.65).

The numerical flux $\hat{f}_{i+\frac{1}{2}}$ is obtained by the ENO or WENO reconstruction procedures, Procedure 2.1 or 2.2, *with* $\overline{v}(x) = f(u(x,t))$. For stability, it is important that upwinding is used in constructing the flux. The easiest and the most inexpensive way to achieve upwinding is the following: compute the Roe speed

$$\overline{a}_{i+\frac{1}{2}} \equiv \frac{f(u_{i+1}) - f(u_i)}{u_{i+1} - u_i}, \tag{2.74}$$

and

– if $\overline{a}_{i+\frac{1}{2}} \geq 0$, then the the wind blows from the left to the right. We would use $v^-_{i+\frac{1}{2}}$ for the numerical flux $\hat{f}_{i+\frac{1}{2}}$;
– if $\overline{a}_{i+\frac{1}{2}} < 0$, then the wind blows from the right to the left. We would use $v^+_{i+\frac{1}{2}}$ for the numerical flux $\hat{f}_{i+\frac{1}{2}}$.

This produces the Roe scheme [62] at the first order level. For this reason, the ENO scheme based on this approach was termed "ENO-Roe" in [70].

In summary, to build a finite difference ENO scheme (2.73) *using the ENO-Roe approach*, given the point values $\{u_i\}$ (we again drop the explicit reference to the time variable t), we proceed as follows:

Procedure 2.4. Finite difference 1D scalar ENO-and WENO-Roe.

1. Compute the Roe speed $\overline{a}_{i+\frac{1}{2}}$ for all i using (2.74);
2. Identify $\overline{v}_i = f(u_i)$ and use the ENO reconstruction Procedure 2.1 or the WENO reconstruction Procedure 2.2, to obtain the cell boundary values $v^-_{i+\frac{1}{2}}$ if $\overline{a}_{i+\frac{1}{2}} \geq 0$, or $v^+_{i+\frac{1}{2}}$ if $\overline{a}_{i+\frac{1}{2}} < 0$;
3. If the Roe speed at $x_{i+\frac{1}{2}}$ is positive

$$\overline{a}_{i+\frac{1}{2}} \geq 0,$$

then take the numerical flux as:

$$\hat{f}_{i+\frac{1}{2}} = v^-_{i+\frac{1}{2}};$$

otherwise, take the the numerical flux as:

$$\hat{f}_{i+\frac{1}{2}} = v^+_{i+\frac{1}{2}};$$

4. Form the scheme (2.73).

□

One disadvantage of the ENO-Roe approach is that entropy violating solutions may be obtained, just like in the first order Roe scheme case. For example, if ENO-Roe is applied to the Burgers equation

$$u_t + \left(\frac{u^2}{2}\right)_x = 0$$

with the following initial condition

$$u(x,0) = \begin{cases} -1, \text{ if } x < 0, \\ 1, \quad \text{ if } x \geq 0, \end{cases}$$

it will converge to the entropy violating expansion shock:

$$u(x,t) = \begin{cases} -1, \text{ if } x < 0, \\ 1, \quad \text{ if } x \geq 0. \end{cases}$$

Local entropy correction could be used to remedy this [70]. However, it is usually more robust to use a global "flux splitting":

$$f(u) = f^+(u) + f^-(u) \tag{2.75}$$

where

$$\frac{df^+(u)}{du} \geq 0, \qquad \frac{df^-(u)}{du} \leq 0. \tag{2.76}$$

We would need the positive and negative fluxes $f^\pm(u)$ to have as many derivatives as the order of the scheme. This unfortunately rules out many popular fluxes (such as those of van Leer [79] and Osher [58]) for high order methods in this framework.

The simplest smooth splitting is the Lax-Friedrichs splitting:

$$f^\pm(u) = \frac{1}{2}(f(u) \pm \alpha u) \tag{2.77}$$

where α is again taken as $\alpha = \max_u |f'(u)|$ over the relevant range of u.

We note that there is a close relationship between a flux splitting (2.75) and a monotone flux (2.69). In fact, for any flux splitting (2.75) satisfying (2.76),

$$h(a,b) = f^+(a) + f^-(b) \tag{2.78}$$

is clearly a monotone flux. However, not every monotone flux can be written in the flux split form (2.75). For example, the Godunov flux (2.70) cannot.

With the flux splitting (2.75), we apply the the ENO or WENO reconstruction procedures, Procedure 2.1 or 2.2, with $\overline{v}(x) = f^+(u(x,t))$ and

$\bar{v}(x) = f^-(u(x,t))$ separately, to obtain two numerical fluxes $\hat{f}^+_{i+\frac{1}{2}}$ and $\hat{f}^-_{i+\frac{1}{2}}$, and then sum them to get the numerical flux $\hat{f}_{i+\frac{1}{2}}$.

In summary, to build a finite difference (FD) ENO or WENO scheme (2.73) *using the flux splitting approach*, given the point values $\{u_i\}$, we proceed as follows:

Procedure 2.5. FD 1D scalar flux splitting ENO and WENO.

1. Find a smooth flux splitting (2.75), satisfying (2.76);
2. Identify $\bar{v}_i = f^+(u_i)$ and use the ENO or WENO reconstruction procedure, Procedure 2.1 or 2.2, to obtain the cell boundary values $v^-_{i+\frac{1}{2}}$ for all i;
3. Take the positive numerical flux as

$$\hat{f}^+_{i+\frac{1}{2}} = v^-_{i+\frac{1}{2}};$$

4. Identify $\bar{v}_i = f^-(u_i)$ and use the ENO or WENO reconstruction procedures, Procedure 2.1 or 2.2, to obtain the cell boundary values $v^+_{i+\frac{1}{2}}$ for all i;
5. Take the negative numerical flux as

$$\hat{f}^-_{i+\frac{1}{2}} = v^+_{i+\frac{1}{2}};$$

6. Form the numerical flux as

$$\hat{f}_{i+\frac{1}{2}} = \hat{f}^+_{i+\frac{1}{2}} + \hat{f}^-_{i+\frac{1}{2}};$$

7. Form the scheme (2.73).

□

We remark that the finite difference scheme in this section and the finite volume scheme in Sect.2.3 are equivalent for one dimensional, linear PDE with constant coefficients: the only difference is in the initial conditions (one uses point values and the other uses cell averages of the exact initial condition). Notice that the schemes are still nonlinear in this case. However, this equivalence does not hold for a nonlinear PDE. Moreover, we will see later that there are significant differences in efficiency of the two approaches for multidimensional problems.

In the following we test the accuracy of the fifth order finite difference WENO schemes on the linear equation:

$$u_t + u_x = 0, \qquad -1 \leq x \leq 1$$
$$u(x,0) = u_0(x) \qquad \text{periodic}.$$

In Table 2.2, we show the errors of the fifth order WENO scheme given by the weights (2.58)-(2.59) with the smooth indicator (2.63), at time $t = 1$ for the initial condition $u_0(x) = \sin(\pi x)$, and compare them with the errors of the linear 5th order upstream central scheme (referred to as CENTRAL-5 in the following tables). We can see that fifth order WENO gives the expected order of accuracy starting at about 40 grid points.

Table 2.2. Accuracy on $u_t + u_x = 0$ with $u_0(x) = \sin(\pi x)$.

Method	N	L_∞ error	L_∞ order	L_1 error	L_1 order
WENO-5	10	2.98e-2	-	1.60e-2	-
	20	1.45e-3	4.36	7.41e-4	4.43
	40	4.58e-5	4.99	2.22e-5	5.06
	80	1.48e-6	4.95	6.91e-7	5.01
	160	4.41e-8	5.07	2.17e-8	4.99
	320	1.35e-9	5.03	6.79e-10	5.00
CENTRAL-5	10	4.98e-3	-	3.07e-3	-
	20	1.60e-4	4.96	9.92e-5	4.95
	40	5.03e-6	4.99	3.14e-6	4.98
	80	1.57e-7	5.00	9.90e-8	4.99
	160	4.91e-9	5.00	3.11e-9	4.99
	320	1.53e-10	5.00	9.73e-11	5.00

In Table 2.3, we show errors for the initial condition $u_0(x) = \sin^4(\pi x)$. The order of accuracy for the fifth order WENO settles down later than in the previous example. Notice that this is the example for which ENO schemes lose their accuracy [63], [67].

Table 2.3. Accuracy on $u_t + u_x = 0$ with $u_0(x) = \sin^4(\pi x)$.

Method	N	L_∞ error	L_∞ order	L_1 error	L_1 order
WENO-5	20	1.08e-1	-	4.91e-2	-
	40	8.90e-3	3.60	3.64e-3	3.75
	80	1.80e-3	2.31	5.00e-4	2.86
	160	1.22e-4	3.88	2.17e-5	4.53
	320	4.37e-6	4.80	6.17e-7	5.14
	640	9.79e-8	5.48	1.57e-8	5.30
CENTRAL-5	20	5.23e-2	-	3.35e-2	-
	40	2.47e-3	4.40	1.52e-3	4.46
	80	8.32e-5	4.89	5.09e-5	4.90
	160	2.65e-6	4.97	1.60e-6	4.99
	320	8.31e-8	5.00	4.99e-8	5.00
	640	2.60e-9	5.00	1.56e-9	5.00

We emphasize again that the high order conservative finite difference ENO and WENO schemes of third or higher order accuracy can only be applied to a uniform grid or a smoothly varying grid, i.e. a grid such that a smooth transformation

$$\xi = \xi(x)$$

will result in a uniform grid in the new variable ξ. Here ξ must contain as many derivatives as the order of accuracy of scheme calls for. If this is the case, then (2.65) is transformed to

$$u_t + \xi_x f(u)_\xi = 0$$

and the conservative ENO or WENO derivative approximation is then applied to $f(u)_\xi$. It is proven in [58] that this way the scheme is still conservative, i.e. Lax-Wendroff theorem [51] still applies.

Boundary conditions. For periodic boundary conditions, or problems with compact support for the entire computation (not just the initial data), there is no difficulty in implementing boundary conditions: one simply set as many ghost points as needed using either the periodicity condition or the compactness of the solution.

Other types of boundary conditions should be handled according to their type: for reflective or symmetry boundary conditions, one would set as many ghost points as needed, then use the symmetry/antisymmetry properties to prescribe solution values at those ghost points. For inflow or partially inflow (e.g. a subsonic outflow where one of the characteristic waves flows in) boundary conditions, one would usually use the physical inflow boundary condition at the exact boundary (for example, if $x_{\frac{1}{2}}$ is the left boundary and a finite volume scheme is used, one would use the given boundary value u_b as $u_{\frac{1}{2}}^-$ in the monotone flux at $x_{\frac{1}{2}}$; if x_0 is the left boundary and a finite difference scheme is used, one would use the given boundary value u_b as u_0). Apart from that, the most natural way of treating boundary conditions for the ENO scheme is to *use only the available values inside the computational domain when choosing the stencil.* In other words, only stencils completely contained inside the computational domain is used in the ENO stencil choosing process described in Procedure 2.1. In practical implementation, in order to avoid logical structures to distinguish whether a given stencil is completely inside the computational domain, one could set all the ghost values outside the computational domain to be very large with large variations (e.g. setting $u_{-j} = (10j)^{10}$ if x_{-j}, for $j = 1, 2, ...$, are ghost points). This way the ENO stencil choosing procedure will automatically avoid choosing any stencil containing ghost points. Another way of treating boundary conditions is to use extrapolation of suitable order to set the values of the solution in all necessary ghost points. For scalar problems this is actually equivalent to the approach of using only the stencils inside the computational domain in the ENO procedure. WENO can be handled in a similar fashion.

Stability analysis (GKS analysis [30], [76]) can be used to study the linear stability when the boundary treatment described above is applied to a fixed stencil upwind biased scheme. For most practical situations the schemes are linearly stable [3].

Provable properties in the scalar case. Second order ENO schemes are also TVD (total variation diminishing), hence have at least subsequences which converge to weak solutions. There is no known convergence result for ENO schemes of degree higher than 2, even for smooth solutions.

WENO schemes have better convergence results, mainly because their numerical fluxes are smoother. It is proven [43] that WENO schemes converge for smooth solutions. Also, Jiang and Yu [44] have obtained an existence proof for traveling waves for WENO schemes. This is an important first step towards the proof of convergence for shocked cases.

Even though there are very little theoretical results about ENO or WENO schemes, in practice these schemes are very robust and stable. We caution against any attempts to modify the schemes solely for the purpose of stability or convergence proofs. In [69] we gave a remark about a modification of ENO schemes, which keeps the formal uniform high order accuracy and makes them stable and convergent for general multi dimensional scalar equations. However it was pointed out there that the modification is not computationally useful, hence the convergence result has little value.

The remark in [69] is illustrative hence we reproduce it here. We start with a flux splitting (2.75) satisfying (2.76), and notice that the first order monotone scheme

$$\frac{du_i}{dt} = -\frac{1}{\Delta x_i} \left(f^+(u_i) - f^+(u_{i-1}) + f^-(u_{i+1}) - f^-(u_i) \right) \equiv R_1(u)_i \quad (2.79)$$

is convergent (also for multi space dimensions). We now construct a high order ENO approximation in the following way: starting from the two point stencil $\{x_{i-1}, x_i\}$, we expand it into a $k+1$ point stencil in an ENO fashion using the divided differences of $f^+(u(x))$. We then build the k-th degree polynomial $P^+(x)$ which interpolates $f^+(u(x))$ in this stencil. $P^-(x)$ is constructed in a similar way, starting from the two point stencil $\{x_i, x_{i+1}\}$. The scheme is finally defined as

$$\frac{du_i}{dt} = -\frac{d}{dx} \left(P^+(x) + P^-(x) \right)\big|_{x=x_i} \equiv R_k(u)_i \quad (2.80)$$

This scheme is clearly k-th order accurate but is not conservative. We now denote the difference between the high order scheme (2.80) and the first order monotone scheme (2.79) by

$$D(u)_i \equiv R_k(u)_i - R_1(u)_i, \quad (2.81)$$

and limit it by

$$\tilde{D}(u)_i = \overline{m}(D(u)_i, M\Delta x^\alpha), \tag{2.82}$$

where $M > 0$ and $0 < \alpha \leq 1$ are constants, and the capping function $\overline{m}$ is defined by

$$\overline{m}(a, b) = \begin{cases} a, & if \ |a| \leq b; \\ b, & if \ a > b; \\ -b, & if \ a < -b. \end{cases}$$

The modified ENO scheme is then defined by

$$\frac{du_i}{dt} = \tilde{R}_k(u)_i \equiv R_1(u)_i + \tilde{D}(u)_i. \tag{2.83}$$

We notice that, in smooth regions, the difference between the first order and high order residues, $D(u)_i$, as defined in (2.81), is of the size $O(\Delta x)$, hence the capping (2.82) does not take effect in such regions, if $\alpha < 1$ or if $\alpha = 1$ and M is large enough, when Δx is sufficiently small. This implies that the scheme (2.83) is uniformly accurate. Moreover, since

$$\left| \tilde{R}_k(u)_i - R_1(u)_i \right| \leq M\Delta x^\alpha$$

by (2.82), the high order scheme (2.83) shares every good property of the first order monotone scheme (2.79), such as total variation boundedness, entropy conditions, and convergence. From a theoretical point of view, this is the strongest result one could possibly hope for a high order scheme. However, the mesh size dependent limiting (2.82) renders the scheme highly impractical: the quality of the numerical solution will depend strongly on the choice of the parameters M and α, as well as on the mesh size Δx.

Systems. We only consider hyperbolic $m \times m$ systems, i.e. the Jocobian $f'(u)$ has m real eigenvalues

$$\lambda_1(u) \leq ... \leq \lambda_m(u) \tag{2.84}$$

and a complete set of independent eigenvectors

$$r_1(u), ..., r_m(u). \tag{2.85}$$

We denote the matrix whose columns are eigenvectors (2.85) by

$$R(u) = (r_1(u), ..., r_m(u)) \tag{2.86}$$

Then clearly

$$R^{-1}(u) \, f'(u) \, R(u) \, = \, \Lambda(u) \tag{2.87}$$

where $\Lambda(u)$ is the diagonal matrix with $\lambda_1(u), ..., \lambda_m(u)$ on the diagonal. Notice that the rows of $R^{-1}(u)$, denoted by $l_1(u), ..., l_m(u)$ (row vectors), are left eigenvectors of $f'(u)$:

$$l_i(u)f'(u) = \lambda_i(u)l_i(u), \qquad i = 1, ..., m. \tag{2.88}$$

There are several ways to generalize scalar ENO or WENO schemes to systems.

The easiest way is to apply the ENO or WENO schemes in a component by component fashion. For the finite volume (FV) formulation, this means that we make the reconstruction using ENO or WENO for each of the components of u separately. This produces the left and right values $u_{i+\frac{1}{2}}^{\pm}$ at the cell interface $x_{i+\frac{1}{2}}$. An exact or approximate Riemann solver, $h(u_{i+\frac{1}{2}}^-, u_{i+\frac{1}{2}}^+)$, is then used to build the scheme (2.68)-(2.69). The exact Riemann solver is given by the exact solution of (2.65) with the following step function as initial condition

$$u(x,0) = \begin{cases} u_{i+\frac{1}{2}}^-, & x \leq 0; \\ u_{i+\frac{1}{2}}^+, & x > 0, \end{cases} \tag{2.89}$$

evaluated at the center $x = 0$. Notice that the solution to (2.65) with the initial condition (2.89) is self-similar, that is, it is a function of the variable $\xi = \frac{x}{t}$, hence is constant along $x = 0$. If we denote this solution by $u_{i+\frac{1}{2}}$, then the flux is taken as

$$h(u_{i+\frac{1}{2}}^-, u_{i+\frac{1}{2}}^+) = f(u_{i+\frac{1}{2}}).$$

In the scalar case, the exact Riemann solver gives the Godunov flux (2.70). Exact Riemann solver can be obtained for many systems including the Euler equations of compressible gas, which is used very often in practice. However, it is usually very costly to get this solution (for Euler equations of compressible gas, an iterative procedure is needed to obtain this solution, see [74]). In practice, approximate Riemann solvers are usually good enough. As in the scalar case, the quality of the solution is usually very sensitive to the choice of approximate Riemann solvers for *lower order* schemes (first or second order), but this sensitivity decreases with an increasing order of accuracy. The simplest approximate Riemann solver (albeit the most dissipative) is again the Lax-Friedrichs solver (2.72), except that now the constant α is taken as

$$\alpha = \max_u \max_{1 \leq j \leq m} |\lambda_j(u)| \tag{2.90}$$

where $\lambda_j(u)$ are the eigenvalues of the Jacobian $f'(u)$, (2.84). The maximum is again taken over the relevant range of u.

We summarize the procedure in the following

Procedure 2.6. Component-wise FV 1D system ENO and WENO.

1. For each component of the solution $\bar{u}$, apply the scalar ENO Procedure 2.1 or WENO Procedure 2.2 to reconstruct the corresponding component of the solution at the cell interfaces, $u^{\pm}_{i+\frac{1}{2}}$ for all i;
2. Apply an exact or approximate Riemann solver to compute the flux $\hat{f}_{i+\frac{1}{2}}$ for all i in (2.69);
3. Form the scheme (2.68).

□

For the finite difference formulation, a smooth flux splitting (2.75) is again needed. The condition (2.76) now becomes that the two Jacobians

$$\frac{\partial f^+(u)}{\partial u}, \qquad \frac{\partial f^-(u)}{\partial u} \tag{2.91}$$

are still diagonalizable (preferably by the same eigenvectors $R(u)$ as for $f'(u)$), and have only non-negative / non-positive eigenvalues, respectively. We again recommend the Lax-Friedrichs flux splitting (2.77), with α given by (2.90), because of its simplicity and smoothness. A somewhat more complicated Lax-Friedrichs type flux splitting is:

$$f^{\pm}(u) = \frac{1}{2}(f(u) \pm R(u)\,\overline{\Lambda}\,R^{-1}(u)\,u)\,,$$

where $R(u)$ and $R^{-1}(u)$ are defined in (2.86), and

$$\overline{\Lambda} = diag(\overline{\lambda}_1, ..., \overline{\lambda}_m)$$

where $\overline{\lambda}_j = max_u|\lambda_j(u)|$, and the maximum is again taken over the relevant range of u. This way the dissipation is added in each field according to the maximum size of eigenvalues in that field, not globally. One could also use other flux splittings, such as the van Leer splitting for gas dynamics [79]. However, for higher order schemes, the flux splitting must be sufficiently smooth in order to retain the order of accuracy.

With these flux splittings, we can again use the scalar recipes to form the finite difference scheme: just compute the positive and negative fluxes $\hat{f}^+_{i+\frac{1}{2}}$ and $\hat{f}^-_{i+\frac{1}{2}}$ component by component.

We summarize the procedure in the following

Procedure 2.7. Component-wise FD 1D system ENO and WENO.

1. Find a flux splitting (2.75). The simplest example is the Lax-Friedrichs flux splitting (2.77), with α given by (2.90);
2. For each component of the solution u, apply the scalar Procedure 2.5 to reconstruct the corresponding component of the numerical flux $\hat{f}_{i+\frac{1}{2}}$;
3. Form the scheme (2.73).

□

These component by component versions of ENO and WENO schemes are simple and cost effective. They work reasonably well for many problems, especially when the order of accuracy is not high (second or sometimes third order). However, for more demanding test problems, or when the order of accuracy is high, we would need the following more costly, but much more robust characteristic decompositions.

To explain the characteristic decomposition, we start with a simple example where $f(u) = Au$ in (2.65) is linear and A is a constant matrix. In this situation, the eigenvalues (2.84), the eigenvectors (2.85), and the related matrices R, R^{-1} and Λ (2.86)-(2.87), are all constant matrices. If we define a change of variable

$$v = R^{-1} u, \qquad (2.92)$$

then the PDE (2.65) becomes diagonal:

$$v_t + \Lambda v_x = 0 \qquad (2.93)$$

that is, the m equations in (2.93) are decoupled and each one is a scalar linear convection equation of the form

$$w_t + \lambda_j w_x = 0. \qquad (2.94)$$

We can thus use the reconstruction or flux evaluation techniques for the scalar equations, discussed in Sections 2.3 and 2.3, to handle each of the equations in (2.94). After we obtain the results, we can "come back" to the physical space u by using the inverse of (2.92):

$$u = R v. \qquad (2.95)$$

For example, if the reconstructed polynomial for each component j in (2.93) is denoted by $q_j(x)$, then we form

$$q(x) = \begin{pmatrix} q_1(x) \\ \cdot \\ \cdot \\ \cdot \\ q_m(x) \end{pmatrix} \qquad (2.96)$$

and obtain the reconstruction in the physical space by using (2.95):

$$p(x) = R q(x). \qquad (2.97)$$

The flux evaluations for the finite difference schemes can be handled similarly.

We now come to the situation where $f'(u)$ is not constant. The trouble is that now all the matrices $R(u)$, $R^{-1}(u)$ $\Lambda(u)$ are dependent upon u. We must "freeze" them locally in order to carry out a similar procedure as in the constant coefficient case. Thus, to compute the flux at the cell boundary $x_{i+\frac{1}{2}}$, we would need an approximation to the Jocobian at the middle value $u_{i+\frac{1}{2}}$. This can be simply taken as the arithmetic mean

$$u_{i+\frac{1}{2}} = \frac{1}{2} \left(u_i + u_{i+1} \right), \tag{2.98}$$

or as a more elaborate average satisfying some nice properties, e.g. the mean value theorem

$$f(u_{i+1}) - f(u_i) = f'(u_{i+\frac{1}{2}})(u_{i+1} - u_i). \tag{2.99}$$

Roe average [62] is such an example for the compressible Euler equations of gas dynamics and some other physical systems. It is also possible to use two different one-sided Jacobians at a higher computational cost [25].

Once we have this $u_{i+\frac{1}{2}}$, we will use $R(u_{i+\frac{1}{2}})$, $R^{-1}(u_{i+\frac{1}{2}})$ and $\Lambda(u_{i+\frac{1}{2}})$ to help evaluating the numerical flux at $x_{i+\frac{1}{2}}$. We thus omit the notation $i + \frac{1}{2}$ and still denote these matrices by R, R^{-1} and Λ, etc. We then repeat the procedure described above for linear systems. The difference here being, the matrices R, R^{-1} and Λ are different at different locations $x_{i+\frac{1}{2}}$, hence the cost of the operation is greatly increased.

In summary, we have the following procedures:

Procedure 2.8. Characteristic-wise FV 1D ENO and WENO.

1. Compute the divided or undivided differences of the cell averages $\bar{u}$, for all i;
2. At each fixed $x_{i+\frac{1}{2}}$, do the following:
   (a) Compute an average state $u_{i+\frac{1}{2}}$, using either the simple mean (2.98) or a Roe average satisfying (2.99);
   (b) Compute the right eigenvectors, the left eigenvectors, and the eigenvalues of the Jacobian $f'(u_{i+\frac{1}{2}})$, (2.84)-(2.87), and denote them by

$$R = R(u_{i+\frac{1}{2}}), \qquad R^{-1} = R^{-1}(u_{i+\frac{1}{2}}), \qquad \Lambda = \Lambda(u_{i+\frac{1}{2}});$$

   (c) Transform all those differences computed in Step 1, which are in the potential stencil of the ENO and WENO reconstructions for obtaining $u_{i+\frac{1}{2}}^{\pm}$, to the local characteristic fields by using (2.92). For example,

$$\bar{v}_j = R^{-1}\bar{u}_j, \qquad j \text{ in a neighborhood of i;}$$

   (d) Perform the scalar ENO or WENO reconstruction Procedure 2.3, for each component of the characteristic variables $\bar{v}$, to obtain the corresponding component of the reconstruction $v_{i+\frac{1}{2}}^{\pm}$;

(e) Transform back into physical space by using (2.95):

$$u^{\pm}_{i+\frac{1}{2}} = R\, v^{\pm}_{i+\frac{1}{2}}$$

3. Apply an exact or approximate Riemann solver to compute the flux $\hat{f}_{i+\frac{1}{2}}$ for all i in (2.69); then form the scheme (2.68).

$\square$

Similarly, the procedure to obtain a finite difference ENO-Roe type scheme using the local characteristic variables is:

Procedure 2.9. Characteristic-wise FD 1D system, Roe-type.

1. Compute the divided or undivided differences of the flux $f(u)$ for all i;
2. At each fixed $x_{i+\frac{1}{2}}$, do the following:
   (a) Compute an average state $u_{i+\frac{1}{2}}$, using either the simple mean (2.98) or a Roe average satisfying (2.99);
   (b) Compute the right eigenvectors, the left eigenvectors, and the eigenvalues of the Jacobian $f'(u_{i+\frac{1}{2}})$, (2.84)-(2.87), and denote them by

   $$R = R(u_{i+\frac{1}{2}}), \qquad R^{-1} = R^{-1}(u_{i+\frac{1}{2}}), \qquad \Lambda = \Lambda(u_{i+\frac{1}{2}});$$

   (c) Transform all those differences computed in Step 1, which are in the potential stencil of the ENO and WENO reconstructions for obtaining the flux $\hat{f}_{i+\frac{1}{2}}$, to the local characteristic fields by using (2.92). For example,

   $$v_j = R^{-1} f(u_j), \qquad \text{j in a neighborhood of i};$$

   (d) Perform the scalar ENO or WENO Roe-type Procedure 2.4, for each component of the characteristic variables v, to obtain the corresponding component of the flux $\hat{v}_{i+\frac{1}{2}}$. The Roe speed $\bar{a}_{i+\frac{1}{2}}$ is replaced by the eigenvalue $\lambda_l(u_{i+\frac{1}{2}})$ for the l-th component of the characteristic variables v;
   (e) Transform back into physical space by using (2.95):

   $$\hat{f}_{i+\frac{1}{2}} = R\, \hat{v}_{i+\frac{1}{2}}$$

3. Form the scheme (2.73).

$\square$

Finally, the procedure to obtain a finite difference flux splitting ENO or WENO scheme using the local characteristic variables is:

Procedure 2.10. Characteristic-wise FD 1D system, flux splitting.

1. Compute the divided or undivided differences of the flux $f(u)$ and the solution u for all i;
2. At each fixed $x_{i+\frac{1}{2}}$, do the following:
   (a) Compute an average state $u_{i+\frac{1}{2}}$, using either the simple mean (2.98) or a Roe average satisfying (2.99);
   (b) Compute the right eigenvectors, the left eigenvectors, and the eigenvalues of the Jacobian $f'(u_{i+\frac{1}{2}})$, (2.84)-(2.87), and denote them by

$$R = R(u_{i+\frac{1}{2}}), \qquad R^{-1} = R^{-1}(u_{i+\frac{1}{2}}), \qquad \Lambda = \Lambda(u_{i+\frac{1}{2}});$$

   (c) Transform all those differences computed in Step 1, which are in the potential stencil of the ENO and WENO reconstructions for obtaining the flux $\hat{f}_{i+\frac{1}{2}}$, to the local characteristic fields by using (2.92). For example,

$$v_j = R^{-1} u_j, \quad g_j = R^{-1} f(u_j), \quad j \text{ in a neighborhood of } i;$$

   (d) Perform the scalar flux splitting ENO or WENO Procedure 2.5, for each component of the characteristic variables, to obtain the corresponding component of the flux $\hat{g}^{\pm}_{i+\frac{1}{2}}$. For the most commonly used Lax-Friedrichs flux splitting, we can use, for the l-th component of the characteristic variables, the viscosity coefficient

$$\alpha = \max_{1 \leq j \leq N} |\lambda_l(u_j)|;$$

   (e) Transform back into physical space by using (2.95):

$$\hat{f}^{\pm}_{i+\frac{1}{2}} = R \hat{g}^{\pm}_{i+\frac{1}{2}}$$

3. Form the flux by taking

$$\hat{f}_{i+\frac{1}{2}} = \hat{f}^{+}_{i+\frac{1}{2}} + \hat{f}^{-}_{i+\frac{1}{2}}$$

and then form the scheme (2.73).

There are attempts recently to simplify this characteristic decomposition. For example, for the compressible Euler equations of gas dynamics, Jiang and Shu [43] used smooth indicators based on density and pressure to perform the so-called pseudo characteristic decompositions. There are also second and sometimes third order component ENO type schemes [56], [54], with limited success for higher order methods.

3 Multi Space Dimensions

3.1 Reconstruction and Approximation in Multi Dimensions

In this section we describe how the ideas of reconstruction and approximation in Sect.2.1 are generalized to multi space dimensions. We will concentrate our discussion in 2D, although things carry over to higher dimensions as well.

In most part of this section we will consider Cartesian grids, that is, the domain is a rectangle

$$[a, b] \times [c, d] \tag{3.1}$$

covered by cells

$$I_{ij} \equiv [x_{i-\frac{1}{2}}, x_{i+\frac{1}{2}}] \times [y_{j-\frac{1}{2}}, y_{j+\frac{1}{2}}], \qquad 1 \le i \le N_x, \ 1 \le j \le N_y \tag{3.2}$$

where

$$a = x_{\frac{1}{2}} < x_{\frac{3}{2}} < ... < x_{N_x - \frac{1}{2}} < x_{N_x + \frac{1}{2}} = b,$$

and

$$c = y_{\frac{1}{2}} < y_{\frac{3}{2}} < ... < y_{N_y - \frac{1}{2}} < y_{N_y + \frac{1}{2}} = d.$$

The centers of the cells are

$$(x_i, y_j), \qquad x_i \equiv \frac{1}{2} \left(x_{i-\frac{1}{2}} + x_{i+\frac{1}{2}} \right), \ y_j \equiv \frac{1}{2} \left(y_{j-\frac{1}{2}} + y_{j+\frac{1}{2}} \right), \tag{3.3}$$

and we still use

$$\Delta x_i \equiv x_{i+\frac{1}{2}} - x_{i-\frac{1}{2}}, \qquad i = 1, 2, ..., N_x \tag{3.4}$$

and

$$\Delta y_j \equiv y_{j+\frac{1}{2}} - y_{j-\frac{1}{2}}, \qquad j = 1, 2, ..., N_y \tag{3.5}$$

to denote the grid sizes. We denote the maximum grid sizes by

$$\Delta x \equiv \max_{1 \le i \le N_x} \Delta x_i, \qquad \Delta y \equiv \max_{1 \le j \le N_y} \Delta y_j, \tag{3.6}$$

and assume that Δx and Δy are of the same magnitude (their ratio is bounded from above and below during refinement). Finally,

$$\Delta \equiv \max(\Delta x, \Delta y). \tag{3.7}$$

Reconstruction from cell averages. The approximation problem we will face, in solving hyperbolic conservation laws using cell averages (finite volume schemes, see Sect.3.3), is still the following *reconstruction* problem.

Problem 3.1. Two dimensional reconstruction.
Given the cell averages of a function $v(x, y)$:

$$\bar{v}_{ij} \equiv \frac{1}{\Delta x_i \Delta y_j} \int_{y_{j-\frac{1}{2}}}^{y_{j+\frac{1}{2}}} \int_{x_{i-\frac{1}{2}}}^{x_{i+\frac{1}{2}}} v(\xi, \eta) \, d\xi \, d\eta,$$

$$i = 1, 2, ..., N_x, \ j = 1, 2, ..., N_y, \tag{3.8}$$

find a polynomial $p_{ij}(x, y)$, preferably of degree at most $k - 1$, for each cell I_{ij}, such that it is a k-th order accurate approximation to the function $v(x, y)$ inside I_{ij}:

$$p_{ij}(x, y) = v(x, y) + O(\Delta^k), \qquad (x, y) \in I_{ij}, \ i = 1, ..., N_x, j = 1, ..., N_y. \tag{3.9}$$

In particular, this gives approximations to the function $v(x, y)$ at the cell boundaries

$$v^-_{i+\frac{1}{2}, y} = p_{ij}(x_{i+\frac{1}{2}}, y), \quad v^+_{i-\frac{1}{2}, y} = p_{ij}(x_{i-\frac{1}{2}}, y),$$

$$i = 1, ..., N_x, \ y_{j-\frac{1}{2}} \leq y \leq y_{j+\frac{1}{2}}$$

$$v^-_{x, j+\frac{1}{2}} = p_{ij}(x, y_{j+\frac{1}{2}}), \quad v^+_{x, j-\frac{1}{2}} = p_{ij}(x, y_{j-\frac{1}{2}}),$$

$$j = 1, ..., N_y, \ x_{i-\frac{1}{2}} \leq x \leq x_{i+\frac{1}{2}}$$

which are k-th order accurate:

$$v^\pm_{i+\frac{1}{2}, y} = v(x_{i+\frac{1}{2}}, y) + O(\Delta^k), \qquad i = 0, 1, ..., N_x, \ y_{j-\frac{1}{2}} \leq y \leq y_{j+\frac{1}{2}} \tag{3.10}$$

and

$$v^\pm_{x, j+\frac{1}{2}} = v(x, y_{j+\frac{1}{2}}) + O(\Delta^k), \qquad j = 0, 1, ..., N_y, \ x_{i-\frac{1}{2}} \leq x \leq x_{i+\frac{1}{2}}. \tag{3.11}$$

□

Again we will not discuss boundary conditions in this section. We thus assume that $\bar{v}_{ij}$ is also available for $i \leq 0, i > N_x$ and for $j \leq 0, j > N_y$ if needed.

In the following we describe the general procedure to solve Problem 3.1.

Given the location I_{ij} and the order of accuracy k, we again first choose a "stencil", based on $\frac{k(k+1)}{2}$ neighboring cells, the collection of these cells still being denoted by $S(i, j)$. We then try to find a polynomial of degree at most

$k - 1$, denoted by $p(x, y)$ (we again drop the subscript ij when it does not cause confusion), whose cell average in each of the cells in $S(i, j)$ agrees with that of $v(x, y)$:

$$\frac{1}{\Delta x_l \Delta y_m} \int_{y_{m-\frac{1}{2}}}^{y_{m+\frac{1}{2}}} \int_{x_{l-\frac{1}{2}}}^{x_{l+\frac{1}{2}}} p(\xi, \eta) \, d\xi \, d\eta = \bar{v}_{lm}, \qquad if \ I_{lm} \in S(i, j). \quad (3.12)$$

We first remark that there are now many more candidate stencils $S(i, j)$ than in the 1D case, More importantly, unlike in the 1D case, here we encounter the following essential difficulties:

– Not all of the candidate stencils can be used to obtain a polynomial $p(x, y)$ of degree at most $k - 1$ satisfying condition (3.12).

For example, it is an easy exercise to show that neither existence nor uniqueness holds, if one wants to reconstruct a first degree polynomial $p(x, y)$ satisfying (3.12) for the three horizontal cells

$$S(i, j) = \{I_{i-1,j}, I_{ij}, I_{i+1,j}\}.$$

To see this, let's assume that

$$I_{i-1,j} = [-2\Delta, -\Delta] \times [0, \Delta], \ I_{i,j} = [-\Delta, 0] \times [0, \Delta],$$
$$I_{i+1,j} = [0, \Delta] \times [0, \Delta],$$

and the first degree polynomial $p(x, y)$ is given by

$$p(x, y) = \alpha + \beta x + \gamma y$$

then condition (3.12) implies

$$\begin{cases} \alpha - \frac{3}{2}\Delta\beta + \frac{1}{2}\Delta\gamma = \bar{v}_{i-1,j} \\ \alpha - \frac{1}{2}\Delta\beta + \frac{1}{2}\Delta\gamma = \bar{v}_{i,j} \\ \alpha + \frac{1}{2}\Delta\beta + \frac{1}{2}\Delta\gamma = \bar{v}_{i+1,j} \end{cases}$$

which is a singular linear system for α, β and γ.

– Even if one obtains such a polynomial $p(x, y)$, there is no guarantee that the accuracy conditions (3.9) will hold. We again use the same simple example. If we pick the function

$$v(x, y) = 0,$$

then one of the polynomials of degree one satisfying the condition (3.12) is

$$p(x, y) = \Delta - 2y$$

clearly the difference

$$v(x, 0) - p(x, 0) = -\Delta$$

is not at the size of $O(\Delta^2)$ in $x_{i-\frac{1}{2}} \leq x \leq x_{i+\frac{1}{2}}$, as is required by (3.9).

This difficulty will be more profound for unstructured meshes such as triangles. See, for example, [1].

For rectangular meshes, if we use the tensor products of 1D polynomials, i.e. use polynomials in Q^{k-1}:

$$p(x,y) = \sum_{m=0}^{k-1} \sum_{l=0}^{k-1} a_{lm} x^l y^m$$

then things can proceed as in 1D. We restrict ourselves in the following tensor product stencils:

$$S_{rs}(i,j) = \{I_{lm} : i - r \le l \le i + k - 1 - r, \ j - s \le m \le j + k - 1 - s\}$$

then we can address Problem 3.1 by introducing the two dimensional primitives:

$$V(x,y) = \int_{-\infty}^{y} \int_{-\infty}^{x} v(\xi, \eta) d\xi d\eta \,.$$

Clearly

$$V(x_{i+\frac{1}{2}}, y_{j+\frac{1}{2}}) = \int_{-\infty}^{y_{j+\frac{1}{2}}} \int_{-\infty}^{x_{i+\frac{1}{2}}} v(\xi, \eta) d\xi d\eta$$

$$= \sum_{m=-\infty}^{j} \sum_{l=-\infty}^{i} \bar{v}_{lm} \Delta x_l \Delta y_m \,,$$

hence as in the 1D case, with the knowledge of the cell averages $\bar{v}$ we know the primitive function V exactly at cell corners.

On a tensor product stencil

$$\tilde{S}_{rs}(i,j) = \{(x_{l+\frac{1}{2}}, y_{m+\frac{1}{2}}) : i - r - 1 \le l \le i + k - 1 - r, \tag{3.13}$$
$$j - s - 1 \le m \le j + k - 1 - s\}$$

there is a unique polynomial $P(x,y)$ in Q^k which interpolates V at every point in $\tilde{S}_{rs}(i,j)$. We take the mixed derivative of the polynomial P to get:

$$p(x,y) = \frac{\partial^2 P(x,y)}{\partial x \partial y}$$

then $p(x,y)$ is in Q^{k-1}, approximates $v(x,y)$, which is the mixed derivative of $V(x,y)$, to k-th order:

$$v(x,y) - p(x,y) = O(\Delta^k)$$

and also satisfies (3.12):

$$\frac{1}{\Delta x_l \Delta y_m} \int_{y_{m-\frac{1}{2}}}^{y_{m+\frac{1}{2}}} \int_{x_{l-\frac{1}{2}}}^{x_{l+\frac{1}{2}}} p(\xi, \eta) \, d\xi \, d\eta$$

$$
= \frac{1}{\Delta x_l \Delta y_m} \int_{y_{m-\frac{1}{2}}}^{y_{m+\frac{1}{2}}} \int_{x_{l-\frac{1}{2}}}^{x_{l+\frac{1}{2}}} \frac{\partial^2 P}{\partial \xi \partial \eta}(\xi, \eta) \, d\xi \, d\eta
$$

$$
= \frac{1}{\Delta x_l \Delta y_m} \left(P(x_{l+\frac{1}{2}}, y_{m+\frac{1}{2}}) - P(x_{l+\frac{1}{2}}, y_{m-\frac{1}{2}}) \right.
$$

$$
\left. - P(x_{l-\frac{1}{2}}, y_{m+\frac{1}{2}}) + P(x_{l-\frac{1}{2}}, y_{m-\frac{1}{2}}) \right)
$$

$$
= \frac{1}{\Delta x_l \Delta y_m} \left(V(x_{l+\frac{1}{2}}, y_{m+\frac{1}{2}}) - V(x_{l+\frac{1}{2}}, y_{m-\frac{1}{2}}) \right.
$$

$$
\left. - V(x_{l-\frac{1}{2}}, y_{m+\frac{1}{2}}) + V(x_{l-\frac{1}{2}}, y_{m-\frac{1}{2}}) \right)
$$

$$
= \frac{1}{\Delta x_l \Delta y_m} \int_{y_{m-\frac{1}{2}}}^{y_{m+\frac{1}{2}}} \int_{x_{l-\frac{1}{2}}}^{x_{l+\frac{1}{2}}} v(\xi, \eta) \, d\xi \, d\eta = \bar{v}_{lm},
$$

$$
i - r \le l \le i + k - 1 - r, \quad j - s \le m \le j + k - 1 - s.
$$

This gives us a practical way to perform the reconstruction in 2D. We first perform a one dimensional reconstruction (Problem 2.1), say in the y direction, obtaining one dimensional cell averages of the function v in the other direction (say in the x direction). We then perform a reconstruction in the other direction.

It should be remarked that the cost to do this 2D reconstruction is very high: For each grid point, if the cost to perform a one dimensional reconstruction is c, then we need $2c$ per grid point to perform this 2D reconstruction. In general n space dimensions, the cost grows to nc.

We also remark that to use polynomials in Q^{k-1} is a waste: to get the correct order of accuracy only polynomials in P^{k-1} is needed. However, there is no natural way of utilizing polynomials in P^{k-1} (see the comments above and also the paper of Abgrall [1]).

The reconstruction problem, Problem 3.1, can also be raised for general, non-Cartesian meshes, such as triangles. However, the solution becomes much more complicated. For discussions, see for example [1].

Conservative approximation to the derivative from point values.
The second approximation problem we will face, in solving hyperbolic conservation laws using point values (finite difference schemes, see Sect.3.2), is again the following problem in obtaining high order conservative approximation to the derivative from point values [69,70]. As in the 1D case, here we also assume that the grid is uniform in each direction. We again ignore the boundary conditions and assume that v_{ij} is available for $i \le 0$ and $i > N_x$, and for $j \le 0$ and $j > N_y$.

Problem 3.2. Two dimensional conservative approximation.
Given the point values of a function $v(x, y)$:

$$
v_{ij} \equiv v(x_i, y_j), \qquad i = 1, 2, ..., N_x, \quad j = 1, 2, ..., N_y, \tag{3.14}
$$

find numerical flux functions

$$\hat{v}_{i+\frac{1}{2},j} \equiv \hat{v}(v_{i-r,j}, ..., v_{i+k-1-r,j}), \qquad i = 0, 1, ..., N_x \qquad (3.15)$$

and

$$\hat{v}_{i,j+\frac{1}{2}} \equiv \hat{v}(v_{i,j-s}, ..., v_{i,j+k-1-s}), \qquad j = 0, 1, ..., N_y \qquad (3.16)$$

such that the flux differences approximate the derivatives $v_x(x, y)$ and $v_y(x, y)$ to k-th order accuracy:

$$\frac{1}{\Delta x} \left(\hat{v}_{i+\frac{1}{2},j} - \hat{v}_{i-\frac{1}{2},j} \right) = v_x(x_i, y_j) + O(\Delta x^k), \ i = 0, 1, ..., N_x, \qquad (3.17)$$

and

$$\frac{1}{\Delta y} \left(\hat{v}_{i,j+\frac{1}{2}} - \hat{v}_{i,j-\frac{1}{2}} \right) = v_y(x_i, y_j) + O(\Delta y^k), \ j = 0, 1, ..., N_y, \qquad (3.18)$$

$\square$

The solution of this problem is essential for the high order conservative schemes based on point values (finite difference) rather than on cell averages (finite volume).

Having seen the complication of reconstructions in the previous subsection for multi space dimensions, it is a good relieve to see that conservative approximation to the derivative from point values is as simple in multi dimensions as in 1D. In fact, for fixed j, if we take

$$w(x) = v(x, y_j)$$

then to obtain $v_x(x_i, y_j) = w'(x_i)$ we only need to perform the one dimensional procedure in Sect.2.1, Problem 2.2, to the one dimensional function $w(x)$. Same thing for $v_y(x, y)$.

As in the 1D case, the conservative approximation to derivatives, of third order accuracy or higher, can only be applied to uniform or smoothly varying meshes (curvilinear coordinates). It cannot be applied to general unstructured meshes such as triangles, unless conservative is given up.

3.2 ENO and WENO Approximations in Multi Dimensions

For solving hyperbolic conservation laws in multi space dimensions, we are again interested in the class of piecewise smooth functions. We define a piecewise smooth function $v(x, y)$ to be such that, for each fixed y, the one dimensional function $w(x) = v(x, y)$ is piecewise smooth in the sense described in Sect.2.2. Likewise, for each fixed x, the one dimensional function $w(y) = v(x, y)$ is also assumed to be piecewise smooth. Such functions are again "generic" for solutions to multi dimensional hyperbolic conservation laws in practice.

In the previous section, we have already discussed the problems of reconstruction and conservative approximations to derivatives in multi space dimensions. At least for the Cartesian type grids, both the reconstruction and the conservative approximation can be obtained from one dimensional procedures.

For the reconstruction, we first use a one dimensional ENO or WENO reconstruction, Procedure 2.1 or 2.2, on the two dimensional cell averages, say in the y direction, to obtain one dimensional cell averages in x only. Then, another one dimensional reconstruction in the remaining direction, say in the x direction, is performed to recover the function itself, again using the one dimensional ENO or WENO methodology, Procedure 2.1 or 2.2.

For the conservative approximation to derivatives, since they are already formulated in a dimension by dimension fashion, one dimensional ENO and WENO procedures can be trivially applied. In effect, the FORTRAN program for the 2D problem is the same as the one for the 1D problem, with an outside "do loop".

What happens to general geometry which cannot be covered by a Cartesian grid?

If the domain is smooth enough, it usually can be mapped *smoothly* to a rectangle (or at least to a union of non-overlapping rectangles). That is, the transformation

$$\xi = \xi(x, y), \qquad \eta = \eta(x, y) \tag{3.19}$$

maps the physical domain Ω where (x, y) belongs, to a rectangular computational domain

$$a \leq \xi \leq b, \qquad c \leq \eta \leq d. \tag{3.20}$$

We require the transformation functions (3.19) to be smooth (i.e. it has as many derivatives as the accuracy of the scheme calls for). Using chain rule, we could write, for example,

$$v_x = \xi_x v_\xi + \eta_x v_\eta \tag{3.21}$$

We can then use our ENO or WENO approximations on v_ξ and v_η, as they are now defined in rectangular domains. The smoothness of ξ_x and η_x will guarantee that this leads to a high order approximation to v_x as well through (3.21).

If the domain is really ugly, or if one wants to use unstructured meshes for other purposes (e.g. for adaptivity), then ENO and WENO approximations for unstructured meshes must be studied. This is a much less matured subject at present. We refer the readers to [1] for some efforts in this direction.

3.3 ENO and WENO Schemes for Multi Dimensional Conservation Laws

In this section we describe the ENO and WENO schemes for 2D conservation laws:

$$u_t(x, y, t) + f_x(u(x, y, t)) + g_y(u(x, y, t)) = 0 \qquad (3.22)$$

again equipped with suitable initial and boundary conditions.

Although we present everything in 2D, most of the discussion is also valid for higher dimensions.

We again concentrate on the discussion of spatial discretizations, and will leave the time variable t continuous (the method-of-lines approach). Time discretizations will be discussed in Sect.4.2.

In most of the discussion in this section, our computational domain is rectangular, given by (3.1). Our grids will thus be Cartesian, given by (3.2) and (3.3). Unstructured meshes will only be mentioned briefly.

We do not discuss boundary conditions in this section. We thus assume that the values of the numerical solution are also available outside the computational domain whenever they are needed. This would be the case for periodic or compactly supported problems. Two dimensional boundary condition treatments are similar to the one dimensional case discussed in Sect.2.3.

Finite volume formulation in the scalar case. For finite volume schemes, or schemes based on cell averages, we do not solve (3.22) directly, but its integrated version. We integrate (3.22) over the interval I_{ij} to obtain

$$\frac{d\bar{u}_{ij}(t)}{dt} = -\frac{1}{\Delta x_i \Delta y_j} \left(\int_{y_{j-\frac{1}{2}}}^{y_{j+\frac{1}{2}}} f(u(x_{i+\frac{1}{2}}, y, t))\, dy + \right.$$
$$- \int_{y_{j-\frac{1}{2}}}^{y_{j+\frac{1}{2}}} f(u(x_{i-\frac{1}{2}}, y, t))\, dy + \int_{x_{i-\frac{1}{2}}}^{x_{i+\frac{1}{2}}} g(u(x, y_{j+\frac{1}{2}}, t))\, dx + \qquad (3.23)$$
$$\left. - \int_{x_{i-\frac{1}{2}}}^{x_{i+\frac{1}{2}}} g(u(x, y_{j-\frac{1}{2}}, t))\, dx \right)$$

where

$$\bar{u}_{ij}(t) \equiv \frac{1}{\Delta x_i \Delta y_j} \int_{y_{j-\frac{1}{2}}}^{y_{j+\frac{1}{2}}} \int_{x_{i-\frac{1}{2}}}^{x_{i+\frac{1}{2}}} u(\xi, \eta, t)\, d\xi\, d\eta \qquad (3.24)$$

is the cell average. We approximate (3.23) by the following conservative scheme

$$\frac{d\bar{u}_{ij}(t)}{dt} = -\frac{1}{\Delta x_i} \left(\hat{f}_{i+\frac{1}{2},j} - \hat{f}_{i-\frac{1}{2},j} \right) - \frac{1}{\Delta y_j} \left(\hat{g}_{i,j+\frac{1}{2}} - \hat{g}_{i,j-\frac{1}{2}} \right), \qquad (3.25)$$

where the numerical flux $\hat{f}_{i+\frac{1}{2},j}$ is defined by

$$\hat{f}_{i+\frac{1}{2},j} = \sum_\alpha w_\alpha h \left(u^-_{i+\frac{1}{2},y_j+\beta_\alpha \Delta y_j}, u^+_{i+\frac{1}{2},y_j+\beta_\alpha \Delta y_j} \right), \tag{3.26}$$

where β_α and w_α are Gaussian quadrature nodes and weights, for approximating the integration in y:

$$\frac{1}{\Delta y_j} \int_{y_{j-\frac{1}{2}}}^{y_{j+\frac{1}{2}}} f(u(x_{i+\frac{1}{2}}, y, t)\, dy$$

inside the integral form of the PDE (3.23), and $u^\pm_{i+\frac{1}{2},y}$ are the k-th order accurate reconstructed values obtained by ENO or WENO reconstruction described in the previous section. As before, the superscripts $\pm$ imply the values are obtained within the cell I_{ij} (for the superscript -) and the cell $I_{i+1,j}$ (for the superscript +), respectively. The flux $\hat{g}_{i,j+\frac{1}{2}}$ is defined similarly by

$$\hat{g}_{i,j+\frac{1}{2}} = \sum_\alpha w_\alpha h \left(u^-_{x_i+\beta_\alpha \Delta x_i, j+\frac{1}{2}}, u^+_{x_i+\beta_\alpha \Delta x_i, j+\frac{1}{2}} \right), \tag{3.27}$$

for approximating the integration in x:

$$\frac{1}{\Delta x_i} \int_{x_{i-\frac{1}{2}}}^{x_{i+\frac{1}{2}}} g(u(x, y_{j+\frac{1}{2}}, t)\, dx$$

inside the integral form of the PDE (3.23). $u^\pm_{x,j+\frac{1}{2}}$ are again the k-th order accurate reconstructed values obtained by ENO or WENO reconstruction described in the previous section. h is again the one dimensional monotone flux, examples being given in (2.70)-(2.72).

We summarize the procedure to build a finite volume ENO or WENO 2D scheme (3.25), given the cell averages $\{\bar{u}_{ij}\}$ (we again drop the explicit reference to the time variable t), and a one dimensional monotone flux h, as follows:

Procedure 3.1. Finite volume 2D scalar ENO and WENO.

1. Follow the procedures described in Sect.3.2, to obtain ENO or WENO reconstructed values at the Gaussian points, $u^\pm_{i+\frac{1}{2},y_j+\beta_\alpha \Delta y_j}$ and $u^\pm_{x_i+\beta_\alpha \Delta x_i, j+\frac{1}{2}}$. Notice that this step involves two one dimensional reconstructions, each one to remove a one dimensional cell average in one of the two directions. Also notice that the optimal weights used in the WENO reconstruction procedure are different for different Gaussian points indexed by α;

2. Compute the flux $\hat{f}_{i+\frac{1}{2},j}$ and $\hat{g}_{i,j+\frac{1}{2}}$ using (3.26) and (3.27);

3. Form the scheme (3.25).

$\square$

We remark that the finite volume scheme in 2D, as described above, is very expensive due to the following reasons:

- A two dimensional reconstruction, at the cost of two one dimensional reconstruction per grid point, is needed. For general n space dimensions, the cost becomes n one dimensional reconstruction per grid point;
- More than one quadrature points are needed in formulating the flux (3.26)-(3.27), for order of accuracy higher than two. Thus, for ENO, although the stencil choosing process needs to be done only once, the reconstruction (2.10) has to be done for each quadrature point used in the flux formulation. For WENO, the optimal weights are also different for each quadrature point. This becomes much more costly for $n > 2$ dimension, as then the fluxes are defined by integrals in $n - 1$ dimension and a $n - 1$ dimensional quadrature rule must be used.

This is why multidimensional finite volume schemes of order of accuracy higher than 2 are rarely used. For 2D, based on [34], Casper [11] has coded up a fourth order finite volume ENO scheme for Cartesian grids, see also [12]. 3D finite volume ENO code of order of accuracy higher than 2 does not exist yet, to the author's knowledge.

At the second order level, the cost is greatly reduced because:

- There is no need to perform a reconstruction, as the cell average $\bar{u}_{ij}$ agrees with the point value at the center $u(x_i, y_j)$ to second order $O(\Delta^2)$;
- The quadrature rule in defining the flux (3.26)-(3.27) needs only one (mid) point.

One advantage of finite volume ENO or WENO schemes is that they can be defined on arbitrary meshes, provided that an ENO or WENO reconstruction on that mesh is available. See, for example, [1].

Finite difference formulation in the scalar case. Here we assume a uniform grid and solve (3.22) directly using a conservative approximation to the spatial derivative:

$$\frac{du_{ij}(t)}{dt} = -\frac{1}{\Delta x}\left(\hat{f}_{i+\frac{1}{2},j} - \hat{f}_{i-\frac{1}{2},j}\right) - \frac{1}{\Delta y}\left(\hat{g}_{i,j+\frac{1}{2}} - \hat{g}_{i,j-\frac{1}{2}}\right) \qquad (3.28)$$

where $u_{ij}(t)$ is the numerical approximation to the point value $u(x_i, y_j, t)$.

The numerical flux $\hat{f}_{i+\frac{1}{2},j}$ is obtained by the one dimensional ENO or **WENO approximation procedure, Procedure 2.4 or 2.5,** with $v(x) = f(u(x, y_j, t))$ and with j fixed. Likewise, the numerical flux $\hat{g}_{i,j+\frac{1}{2}}$ is obtained by the one dimensional ENO or WENO approximation procedure, with $v(y) = f(u(x_i, y, t))$ and with i fixed.

All the one dimensional discussions in Sect.2.3, such as upwinding, ENO-Roe, flux splitting, etc., can be applied here dimension by dimension.

The discussion here is also valid for higher spatial dimension n. In effect, it is the same one dimensional conservative derivative approximation applied to each space dimension.

It is a straightforward exercise [13] to show that, in terms of operation count, the finite difference ENO or WENO schemes are about a factor of 4 less than the finite volume counterpart of the same order. In 3D this factor becomes about 9.

We thus strongly recommend the usage of the finite difference version of ENO and WENO schemes (also called ENO and WENO schemes based on point values), whenever possible.

Provable properties in the scalar case. Second order ENO schemes are also maximum norm non-increasing. Of course, this stability is too weak to imply any convergence. As was mentioned before, there is no known convergence result for ENO schemes of order higher than 2, even for smooth solutions.

WENO schemes have better convergence results also in the current multi-D case, mainly because their numerical fluxes are smoother. It is proven [43] that WENO schemes converge for smooth solutions.

We again emphasize that, even though there are very little theoretical results about ENO or WENO schemes, in practice they are very robust and stable. We once again caution against any attempts to modify the schemes solely for the purpose of stability or convergence proofs. In fact the modification of ENO schemes in [69], presented in Sect.2.3, which keeps the formal uniform high order accuracy, actually produces schemes which are convergent to entropy solutions for general multi dimensional scalar equations. However it was pointed out there that the modification is not computationally useful, hence the convergence result has little value.

Systems. The advice here is that, when the fluxes are computed along a cell boundary, a one dimensional local characteristic decomposition normal to the boundary is performed. Also, the monotone flux is replaced with a one dimensional exact or approximate Riemann solver. Thus, the discussion in Sect.2.3 can be applied here.

There are discussions in the literature about truly multi-dimensional recipes. However, these tend to become extremely complicated for order of accuracy higher than two, so they have not been used extensively in practice. Another reason to suggest against using such complicated truly multidimensional recipes for order of accuracy higher than two is that, while dimension by dimension schemes as advocated in these lecture notes are not rotationally invariant, the direction related non-symmetry actually diminishes with increased order [13].

4 Further Topics

4.1 Further Topics in ENO and WENO Schemes

In this section we discuss some miscellaneous (but not necessarily unimportant!) topics in ENO and WENO schemes.

Subcell resolution. This idea was first raised by Harten [35]. The observation is that, since in interpolating the primitive V, *two* points must be included in the initial stencil (see Procedure 2.1), one cannot avoid having at least one cell for each discontinuity, inside which the reconstructed polynomial is not accurate ($O(1)$ error there). We can clearly see this $O(1)$ error in the ENO interpolation in Figure 2.1. The reconstruction in this shocked cell, although inaccurate, will always be monotone (Property 2 in Sect.2.2), so stability will not be a problem. However, it does cause a smearing of the discontinuity (over one cell, initially).

If we are solving a truly nonlinear shock, then characteristics flow into the shock, thus any error one makes during time evolution tends to be absorbed into the shock (we also say that the shock has a self sharpening mechanism). However, we are less lucky with a linear discontinuity, such as a discontinuity carried by the linear equation $u_t + u_x = 0$. Such linear discontinuities are also called contact discontinuities in gas dynamics. The characteristics for such cases are parallel to the discontinuity, hence any numerical smearing tends to accumulate and the discontinuity becomes progressively more smeared with time (Harten argues that the smearing of the discontinuity is at the rate of $O(\Delta x^{1-\frac{1}{k+1}})$ where k is the order of the scheme. Although higher order schemes have less smearing, when time is large the smearing is still very significant.

Harten [35] makes the following simple observation: in the shocked cell I_i, instead of using the reconstruction polynomial $p_i(x)$, which is highly inaccurate (the only useful information it carries is the cell average in the cell), one could try to find the location of the discontinuity inside the cell I_i, say at x_s, and then use the neighboring reconstructions $p_{i-1}(x)$ extended to x_s from left and $p_{i+1}(x)$ extended to x_s from right. To find the shock location, one could argue that $p_{i-1}(x)$ is a very accurate approximation to $v(x)$ up to the discontinuity x_s from left, and $p_{i+1}(x)$ is a very accurate approximation to $v(x)$ up to the discontinuity x_s from right. We thus extend $p_{i-1}(x)$ from the left into the cell I_i, and extend $p_{i+1}(x)$ from the right into the cell I_i, and require that the cell average $\bar{v}_i$ be preserved:

$$\int_{x_{i-\frac{1}{2}}}^{x_s} p_{i-1}(x)\,dx + \int_{x_s}^{x_{i+\frac{1}{2}}} p_{i+1}(x)\,dx = \Delta x_i \bar{v}_i. \tag{4.1}$$

It can be proven that under very general conditions, (4.1) has only one root x_s inside the cell I_i, hence one could use Newton iterations to find this root.

Subcell resolution can be applied to both finite volume and finite difference ENO and WENO schemes [35], [70], However, it should be applied only to sharpen contact discontinuities. It is quite dangerous to apply the subcell resolution to a shock, since it might generate entropy violating expansion shocks in the numerical solution.

Another very serious restriction about subcell resolution is that it is very difficult to be applied to 2D. However, see Siddiqi, Kimia and Shu [73], where a geometrical ENO is used to extend the subcell resolution idea to 2D for image processing problems (we termed it geometric ENO, or GENO).

Artificial compression. Another very useful idea to sharpen a contact discontinuity is the artificial compression, first developed by Harten [32] and further improved by Yang [83]. The idea is to *increase* the magnitude of the slope of a reconstruction, of course subject to certain monotonicity restrictions, near such a discontinuity. Notice that this goes against the idea of limiting, which typically *decreases* the magnitude of the slope of a reconstruction.

Artificial compression can be applied both to finite volume and to finite difference ENO and WENO schemes [83], [70], [43]. Unlike subcell resolution, artificial compression can also be applied easily to multi space dimensions, at least in principle.

Other building blocks. It is not necessary to stay within polynomial building blocks, although polynomials are the most natural functions to work with. For some applications, other building blocks, such as rational functions, trigonometric polynomials, exponential functions, radial functions, etc., may be more appropriate. The idea of ENO or WENO can be applied also in such situations. The key idea is to find suitable "smooth indicators", similar to the Newton divided differences for the polynomial case, for applying the ENO or WENO idea. See [16] and [40] for some examples.

4.2   Time Discretization

Up to now we have only considered spatial discretizations, leaving the time variable continuous (method of lines). In this section we consider the issue of time discretization. The techniques discussed in this section can also be applied to other types of spatial discretizations using the method of lines approach, such as various TVD and TVB schemes [52,78,65] and discontinuous Galerkin methods [18–21,17].

TVD Runge-Kutta methods. A class of TVD (total variation diminishing) high order Runge-Kutta methods is developed in [69] and further in [29].

These Runge-Kutta methods are used to solve a system of initial value problems of ODEs written as:

$$u_t = L(u), \qquad (4.2)$$

resulting from a method of lines spatial approximation to a PDE such as:

$$u_t = -f(u)_x. \qquad (4.3)$$

We have written the equation in (4.3) as a 1D conservation law, but the discussion which follows apply to general initial value problems of PDEs in any spatial dimensions. Clearly, $L(u)$ in (4.2) is an approximation (e.g. ENO or WENO approximation in these lecture notes), to the derivative $-f(u)_x$ in the PDE (4.3).

If we *assume* that a first order Euler forward time stepping:

$$u^{n+1} = u^n + \Delta t L(u^n) \qquad (4.4)$$

is stable in a certain norm:

$$||u^{n+1}|| \leq ||u^n|| \qquad (4.5)$$

under a suitable restriction on Δt:

$$\Delta t \leq \Delta t_1, \qquad (4.6)$$

then we look for higher order in time Runge-Kutta methods such that the same stability result (4.5) holds, under a perhaps different restriction on Δt:

$$\Delta t \leq c \, \Delta t_1 \qquad (4.7)$$

where c is termed *the CFL coefficient* for the high order time discretization.

We remark that the stability condition (4.5) for the first order Euler forward in time (4.4) is easy to obtain in many cases, such as various TVD and TVB schemes in 1D (where the norm is the total variation norm) and in multi dimensions (where the norm is the L^∞ norm), see, e.g. [52,78,65].

Originally in [69,66] the norm in (4.5) was chosen to be the total variation norm, hence the terminology "TVD time discretization".

As it stands, the TVD high order time discretization defined above maintains stability in whatever norm, of the Euler forward first order time stepping, for the high order time discretization, under the time step restriction (4.7). For example, if it is used for multi dimensional scalar conservation laws, for which TVD is not possible but maximum norm stability can be maintained for high order spatial discretizations plus forward Euler time stepping (e.g. [20]), then the same maximum norm stability can be maintained if TVD high order time discretization is used. As another example, if an entropy inequality can be proved for the Euler forward, then the same entropy inequality is valid under a high order TVD time discretization.

In [69], a general Runge-Kutta method for (4.2) is written in the form:

$$u^{(i)} = \sum_{k=0}^{i-1} \left(\alpha_{ik} u^{(k)} + \Delta t \beta_{ik} L(u^{(k)}) \right), \qquad i = 1, ..., m \qquad (4.8)$$

$$u^{(0)} = u^n, \qquad u^{(m)} = u^{n+1}.$$

Clearly, if all the coefficients are nonnegative $\alpha_{ik} \geq 0$, $\beta_{ik} \geq 0$, then (4.8) is just a convex combination of the Euler forward operators, with Δt replaced by $\frac{\beta_{ik}}{\alpha_{ik}} \Delta t$, since by consistency $\sum_{k=0}^{i-1} \alpha_{ik} = 1$. We thus have

Lemma 4.1. [69] The Runge-Kutta method (4.8) is TVD under the CFL coefficient (4.7):

$$c = \min_{i,k} \frac{\alpha_{ik}}{\beta_{ik}}, \qquad (4.9)$$

provided that $\alpha_{ik} \geq 0$, $\beta_{ik} \geq 0$.

$\square$

In [69], schemes up to third order were found to satisfy the conditions in Lemma 4.1 with CFL coefficient equal to 1.

The optimal second order TVD Runge-Kutta method is given by [69,8]:

$$u^{(1)} = u^n + \Delta t L(u^n) \qquad (4.10)$$

$$u^{n+1} = \frac{1}{2} u^n + \frac{1}{2} u^{(1)} + \frac{1}{2} \Delta t L(u^{(1)}),$$

with a CFL coefficient $c = 1$ in (4.9).

The optimal third order TVD Runge-Kutta method is given by [69,8]:

$$u^{(1)} = u^n + \Delta t L(u^n)$$

$$u^{(2)} = \frac{3}{4} u^n + \frac{1}{4} u^{(1)} + \frac{1}{4} \Delta t L(u^{(1)}) \qquad (4.11)$$

$$u^{n+1} = \frac{1}{3} u^n + \frac{2}{3} u^{(2)} + \frac{2}{3} \Delta t L(u^{(2)}),$$

with a CFL coefficient $c = 1$ in (4.9).

Unfortunately, it is proven in [29] that no four stage, fourth order TVD Runge-Kutta method exists with nonnegative α_{ik} and β_{ik}. We thus have to consider the situation where $\alpha_{ik} \geq 0$ but β_{ik} might be negative. In such situations we need to introduce an adjoint operator $\tilde{L}$. The requirement for $\tilde{L}$ is that it approximates the same spatial derivative(s) as L, but is TVD (or stable in another relevant norm) for first order Euler, backward in time:

$$u^{n+1} = u^n - \Delta t \tilde{L}(u^n) \qquad (4.12)$$

This can be achieved, for hyperbolic conservation laws, by solving the backward in time version of (4.3):

$$u_t = f(u)_x. \tag{4.13}$$

Numerically, the only difference is the change of upwind direction. Clearly, $\tilde{L}$ can be computed with the same cost as that of computing L. We then have the following lemma:

Lemma 4.2. [69] The Runge-Kutta method (4.8) is TVD under the CFL coefficient (4.7):

$$c = \min_{i,k} \frac{\alpha_{ik}}{|\beta_{ik}|}, \tag{4.14}$$

provided that $\alpha_{ik} \geq 0$, and L is replaced by $\tilde{L}$ for negative β_{ik}.

□

Notice that, if for the same k, both $L(u^{(k)})$ and $\tilde{L}(u^{(k)})$ must be computed, the cost as well as storage requirement for this k is doubled. For this reason, we would like to avoid negative β_{ik} as much as possible.

An extensive search performed in [29] gives the following preferred four stage, fourth order TVD Runge-Kutta method:

$$
\begin{aligned}
u^{(1)} &= u^n + \frac{1}{2}\Delta t L(u^n) \\
u^{(2)} &= \frac{649}{1600}u^{(0)} - \frac{10890423}{25193600}\Delta t \tilde{L}(u^n) + \frac{951}{1600}u^{(1)} + \frac{5000}{7873}\Delta t L(u^{(1)}) \\
u^{(3)} &= \frac{53989}{2500000}u^n - \frac{102261}{5000000}\Delta t \tilde{L}(u^n) + \frac{4806213}{20000000}u^{(1)} \\
&\quad - \frac{5121}{20000}\Delta t \tilde{L}(u^{(1)}) + \frac{23619}{32000}u^{(2)} + \frac{7873}{10000}\Delta t L(u^{(2)}) \\
u^{n+1} &= \frac{1}{5}u^n + \frac{1}{10}\Delta t L(u^n) + \frac{6127}{30000}u^{(1)} + \frac{1}{6}\Delta t L(u^{(1)}) + \frac{7873}{30000}u^{(2)} \\
&\quad + \frac{1}{3}u^{(3)} + \frac{1}{6}\Delta t L(u^{(3)})
\end{aligned} \tag{4.15}
$$

with a CFL coefficient $c = 0.936$ in (4.14). Notice that two $\tilde{L}$'s must be computed. The effective CFL coefficient, comparing with an ideal case without $\tilde{L}$'s, is $0.936 \times \frac{4}{6} = 0.624$. Since it is difficult to solve the global optimization problem, we do not claim that (4.15) is the optimal 4 stage, 4th order TVD Runge-Kutta method.

A fifth order TVD Runge-Kutta method is also given in [69].

For large scale scientific computing in three space dimensions, storage is usually a paramount consideration. There are therefore discussions about low storage Runge-Kutta methods [81], [10], which only require 2 storage units per ODE equation. In [29], we considered the TVD properties among such low storage Runge-Kutta methods and found third order low storage TVD Runge-Kutta methods.

The general low-storage Runge-Kutta schemes can be written in the form [81], [10]:

$$du^{(i)} = A_i du^{(i-1)} + \Delta t L(u^{(i-1)})$$
$$u^{(i)} = u^{(i-1)} + B_i du^{(i)}, \qquad i = 1, ..., m \qquad (4.16)$$
$$u^{(0)} = u^n, \ \ u^{(m)} = u^{n+1}, \ \ A_0 = 0$$

Only u and du must be stored, resulting in two storage units for each variable.

Carpenter and Kennedy [10] have classified all the three stage, third order (m=3) low storage Runge-Kutta methods, obtaining the following one parameter family:

$$z_1 = \sqrt{36c_2^4 + 36c_2^3 - 135c_2^2 + 84c_2 - 12}$$
$$z_2 = 2c_2^2 + c_2 - 2$$
$$z_3 = 12c_2^4 - 18c_2^3 + 18c_2^2 - 11c_2 + 2$$
$$z_4 = 36c_2^4 - 36c_2^3 + 13c_2^2 - 8c_2 + 4$$
$$z_5 = 69c_2^3 - 62c_2^2 + 28c_2 - 8$$
$$z_6 = 34c_2^4 - 46c_2^3 + 34c_2^2 - 13c_2 + 2$$
$$B_1 = c_2 \qquad (4.17)$$
$$B_2 = \frac{12c_2(c_2 - 1)(3z_2 - z_1) - (3z_2 - z_1)^2}{144c_2(3c_2 - 2)(c_2 - 1)^2}$$
$$B_3 = \frac{-24(3c_2 - 2)(c_2 - 1)^2}{(3z_2 - z_1)^2 - 12c_2(c_2 - 1)(3z_2 - z_1)}$$
$$A_2 = \frac{-z_1(6c_2^2 - 4c_2 + 1) + 3z_3}{(2c_2 + 1)z_1 - 3(c_2 + 2)(2c_2 - 1)^2}$$
$$A_3 = \frac{-z_4 z_1 + 108(2c_2 - 1)c_2^5 - 3(2c_2 - 1)z_5}{24z_1 c_2(c_2 - 1)^4 + 72c_2 z_6 + 72c_2^6(2c_2 - 13)}$$

In [29] we converted this form into the form (4.8), by introducing three new parameters. Then we searched for values of these parameters that would maximize the CFL restriction, by a computer program. The result seems to indicate that

$$c_2 = 0.924574 \qquad (4.18)$$

gives an almost best choice, with CFL coefficient $c = 0.32$ in (4.9). This is of course less optimal than (4.11) in terms of CFL coefficients, however the low storage form is useful for large scale calculations.

We end this subsection by quoting the following numerical example [29], which shows that, even with a very nice second order TVD spatial discretization, if the time discretization is by a non-TVD but linearly stable Runge-Kutta method, the result may be oscillatory. Thus it would always be safer to use TVD Runge-Kutta methods for hyperbolic problems.

The numerical example uses the standard minmod based MUSCL second order spatial discretization [79]. We will compare the results of a TVD versus a non-TVD second order Runge-Kutta time discretizations. The PDE is the simple Burgers equation

$$u_t + \left(\frac{1}{2}u^2\right)_x = 0 \tag{4.19}$$

with a Riemann initial data:

$$u(x,0) = \begin{cases} 1, & \text{if } x \le 0 \\ -0.5, & \text{if } x > 0 \end{cases} \tag{4.20}$$

The nonlinear flux $\left(\frac{1}{2}u^2\right)_x$ in (4.19) is approximated by the conservative difference

$$\frac{1}{\Delta x}\left(\hat{f}_{i+\frac{1}{2}} - \hat{f}_{i-\frac{1}{2}}\right),$$

where the numerical flux $\hat{f}_{i+\frac{1}{2}}$ is defined by

$$\hat{f}_{i+\frac{1}{2}} = h\left(u^-_{i+\frac{1}{2}}, u^+_{i+\frac{1}{2}}\right)$$

with

$$u^-_{i+\frac{1}{2}} = u_i + \frac{1}{2}minmod(u_{i+1} - u_i, u_i - u_{i-1}),$$

$$u^+_{i+\frac{1}{2}} = u_{i+1} - \frac{1}{2}minmod(u_{i+2} - u_{i+1}, u_{i+1} - u_i)$$

The monotone flux h is the Godunov flux defined by (2.70), and the *minmod* function is given by

$$minmod(a,b) = \frac{sign(a) + sign(b)}{2}\min(|a|,|b|).$$

It is easy to prove, by using Harten's Lemma [33], that the Euler forward time discretization with this second order MUSCL spatial operator is TVD under the CFL condition (4.6):

$$\Delta t \le \frac{\Delta x}{2\max_j |u^n_j|} \tag{4.21}$$

Thus $\Delta t = \frac{\Delta x}{2\max_j |u^n_j|}$ will be used in all our calculations. Actually, apart from a slight difference (the *minmod* function is replaced by a minimum-in-absolute-value function), this MUSCL scheme is the same as the second order ENO scheme discussed in Sect.2.3.

The TVD second order Runge-Kutta method we consider is the optimal one (4.10). The non-TVD method we use is:

$$u^{(1)} = u^n - 20\Delta t L(u^n) \qquad (4.22)$$
$$u^{n+1} = u^n + \frac{41}{40}\Delta t L(u^n) - \frac{1}{40}\Delta t L(u^{(1)}).$$

It is easy to verify that both methods are second order accurate in time. The second one (4.22) is however clearly non-TVD, since it has negative β's in both stages (i.e. it partially simulates backward in time with wrong upwinding).

If the operator L is linear (for example the first order upwind scheme applied to a linear PDE), then both Runge-Kutta methods (actually all the two stage, second order Runge-Kutta methods) yield identical results (the two stage, second order Runge-Kutta method for a linear ODE is unique). However, since our L is nonlinear, we may and do observe different results when the two Runge-Kutta methods are used.

In Fig.4.1 we show the result of the TVD Runge-Kutta method (4.10) and the non-TVD method (4.22), after the shock moves about 50 grids (400 time steps for the TVD method, 528 time steps for the non-TVD method). We can clearly see that the non-TVD result is oscillatory (there is an overshoot).

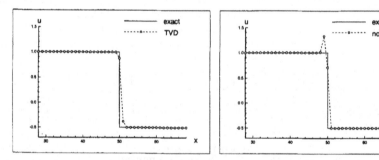

Fig. 4.1: Second order TVD MUSCL spatial discretization. Solution after 500 time steps. Left: TVD time discretization (4.10); Right: non-TVD time discretization (4.22).

Such oscillations are also observed when the non-TVD Runge-Kutta method coupled with a second order TVD MUSCL spatial discretization is applied to a linear PDE ($u_t + u_x = 0$) (the scheme is still nonlinear due to the *minmod* functions). Moreover, for some Runge-Kutta methods, if one looks at the intermediate stages, i.e. $u^{(i)}$ for $1 \leq i < m$ in (4.8), one observes even bigger oscillations. Such oscillations may render difficulties when physical problems are solved, such as the appearance of negative density and pressure for Euler equations of gas dynamics. On the other hand, TVD Runge-Kutta method guarantees that each middle stage solution is also TVD.

This simple numerical test convinces us that it is much safer to use a TVD Runge-Kutta method for solving hyperbolic problems.

TVD multi-step methods. If one prefers multi-step methods rather than Runge-Kutta methods, one can use the TVD high order multi-step methods developed in [66]. The philosophy is very similar to the TVD Runge-Kutta methods discussed in the previous subsection. One starts with a method of lines approximation (4.2) to the PDE (4.3), and an assumption that the first order Euler forward in time discretization (4.4) is stable under a certain norm (4.5), with the time step restriction (4.6). One then looks for higher order in time multi-step methods such that the same stability result (4.5) holds, under a perhaps different restriction on Δt in (4.7), where c is again termed *the CFL coefficient* for the high order time discretization.

The general form of the multi-step methods studied in [66] is:

$$u^{n+1} = \sum_{k=0}^{m} \left(\alpha_k u^{n-k} + \Delta t \beta_k L(u^{n-k}) \right). \tag{4.23}$$

Similar to the Runge-Kutta methods in the previous subsection, if all the coefficients are nonnegative $\alpha_k \geq 0$, $\beta_k \geq 0$, then (4.23) is just a convex combination of the Euler forward operators, with Δt replaced by $\frac{\beta_k}{\alpha_k} \Delta t$, since by consistency $\sum_{k=0}^{m} \alpha_k = 1$. We thus have

Lemma 4.3. [66] The multi-step method (4.23) is TVD under the CFL coefficient (4.7):

$$c = \min_k \frac{\alpha_k}{\beta_k}, \tag{4.24}$$

provided that $\alpha_k \geq 0$, $\beta_k \geq 0$.

$\square$

In [66], schemes up to third order were found to satisfy the conditions in Lemma 4.3. Here we list a few examples.

The following three step ($m = 2$) scheme is second order and TVD

$$u^{n+1} = \frac{3}{4}u^n + \frac{3}{2}\Delta t L(u^n) + \frac{1}{4}u^{n-2} \tag{4.25}$$

with a CFL coefficient $c = 0.5$ in (4.24). This translates to the same efficiency as the optimal second order TVD Runge-Kutta scheme (4.10), as here only one residue evaluation is needed per time step. Of course, the storage requirement is bigger here. There is also the problem of the starting values u^1 and u^2.

The following five step ($m = 4$) scheme is third order and TVD

$$u^{n+1} = \frac{25}{32}u^n + \frac{25}{16}\Delta t L(u^n) + \frac{7}{32}u^{n-4} + \frac{5}{16}\Delta t L(u^{n-4}) \qquad (4.26)$$

with a CFL coefficient $c = 0.5$ in (4.24). This translates to a better efficiency than the optimal third order TVD Runge-Kutta scheme (4.11), as here only one residue evaluation is needed per time step. Of course, the storage requirement is much bigger here. There is also the problem of the starting values u^1, u^2, u^3 and u^4.

There are many other TVD multi-step methods satisfying the conditions in Lemma 4.3 listed in [66]. It seems that if one uses more storage (larger m) one could get better CFL coefficients.

In [66] we have been unable to find multi-step schemes of order four or higher satisfying the condition of Lemma 4.3. As in the Runge-Kutta case, we can relax the condition $\beta_k \geq 0$ by introducing the adjoint operator $\tilde{L}$. We thus have

Lemma 4.4. [66] The multi-step method (4.23) is TVD under the CFL coefficient (4.7):

$$c = \min_k \frac{\alpha_k}{|\beta_k|}, \qquad (4.27)$$

provided that $\alpha_k \geq 0$, and L is replaced by $\tilde{L}$ for negative β_k.

□

Again, notice that, if we have both positive and negative β_k's, then both $L(u^n)$ and $\tilde{L}(u^n)$ must be computed, the cost as well as storage requirement will thus be doubled.

We list here a six step ($m = 5$), fourth order multi-step method which is TVD with a CFL coefficient $c = 0.245$ in (4.24) [66]:

$$u^{n+1} = \frac{747}{1280}u^n + \frac{237}{128}\Delta t L(u^n) \qquad (4.28)$$
$$+ \frac{81}{256}u^{n-4} + \frac{165}{128}\Delta t L(u^{n-4}) + \frac{1}{10}u^{n-5} - \frac{3}{8}\Delta t \tilde{L}(u^{n-5})$$

The Lax-Wendroff procedure. Another way to discretize the time variable is by the Lax-Wendroff procedure [51]. This is also referred to as the Taylor series method for discretizing the ODE (4.2). We will again use the simple 1D scalar conservation law (4.3) as an example to illustrate the procedure, however it applies to more general multidimensional systems.

Starting from a Taylor series expansion in time:

$$u(x, t + \Delta t) = u(x, t) + u_t(x, t)\Delta t + u_{tt}(x, t)\frac{\Delta t^2}{2} + ... \qquad (4.29)$$

The expansion is carried out to the desired order of accuracy in time. For example, a second order in time would need the three terms written out in (4.29). We then use the PDE (4.3) to replace the time derivatives by the spatial derivatives:

$$
\begin{aligned}
u_t(x,t) &= -f(u(x,t))_x = -f'(u(x,t))\,u_x(x,t); \\
u_{tt}(x,t) &= -(f(u(x,t))_t)_x = -(f'(u(x,t))\,u_t(x,t))_x \\
&= ((f'(u(x,t))^2 u_x(x,t))_x \\
&= 2f'(u(x,t))\,f''(u(x,t)\,(u_x(x,t))^2 + (f'(u(x,t)))^2\,u_{xx}(x,t).
\end{aligned}
\tag{4.30}
$$

This little exercise in (4.30) should convince us that it is always possible to write all the time derivatives as functions of the $u(x,t)$ and its spatial derivatives. But the expression could be terribly complicated, especially for multidimensional systems.

Once this is done, we substitute (4.30) into (4.29), and then discretize the spatial derivatives of $u(x,t)$ by whatever methods we use. For example, in the cell averaged (finite volume) ENO schemes discussed in Sect.2.3, we proceed as follows. We first integrate the PDE (2.65) in space-time over the region $[x_{i-\frac{1}{2}}, x_{i+\frac{1}{2}}] \times [t^n, t^{n+1}]$ to obtain

$$
\bar{u}_i^{n+1} = \bar{u}_i^n - \frac{1}{\Delta x_i}\left(\int_{t^n}^{t^{n+1}} f(u(x_{i+\frac{1}{2}},t))dt - \int_{t^n}^{t^{n+1}} f(u(x_{i-\frac{1}{2}},t))dt \right)
\tag{4.31}
$$

Then, we use a suitable Gaussian quadrature to discretize the time integration for the flux in (4.31):

$$
\frac{1}{\Delta t}\int_{t^n}^{t^{n+1}} f(u(x_{i+\frac{1}{2}},t))dt \approx \sum_\alpha w_\alpha f(u(x_{i+\frac{1}{2}}, t^n + \beta_\alpha \Delta t),
\tag{4.32}
$$

where β_α and w_α are Gaussian quadrature nodes and weights. Next we replace each

$$
f(u(x_{i+\frac{1}{2}}, t^n + \beta_\alpha \Delta t)
$$

by a monotone flux:

$$
f(u(x_{i+\frac{1}{2}}, t^n + \beta_\alpha \Delta t) \approx h\left(u(x_{i+\frac{1}{2}}^-, t^n + \beta_\alpha \Delta t), u(x_{i+\frac{1}{2}}^+, t^n + \beta_\alpha \Delta t) \right),
\tag{4.33}
$$

and use the Lax-Wendroff procedure (4.29)-(4.30) to convert

$$
u(x_{i+\frac{1}{2}}^\pm, t^n + \beta_\alpha \Delta t)
$$

to $u(x_{i+\frac{1}{2}}^\pm, t^n)$ and its spatial derivatives also at t^n, which can then be obtained by the reconstructions $p(x)$ inside I_i and I_{i+1}. Notice that the accuracy is just enough in this procedure, as each derivative of the reconstruction $p(x)$ will be one order lower in accuracy, but this is compensated by the Δt in front of it in (4.29).

This Lax-Wendroff procedure, comparing with the method of lines approach coupled with TVD Runge-Kutta or multi-step time discretizations, has the following advantages and disadvantages.

Advantages:

1. This is a truly one step method, hence it is quite compact (a second order method in space and time uses only three cells on time level n to advance to time level $n + 1$ for one cell), and there are no complications such as boundary conditions needed in middle stages;
2. It utilizes the PDE more extensively than the method of lines approach. This is also one reason that it can be so compact.

Disadvantages:

1. The algebra is very, very complicated for multi dimensional systems. This also increases operation counts for complicated nonlinear systems;
2. It is more difficult to prove stability properties (e.g. TVD) for higher order methods in this framework;
3. It is difficult and costly to apply this procedure to the conservative finite difference framework established in Sections 2.3 and 3.3.

4.3 Formulation of the ENO and WENO Schemes for the Hamilton-Jacobi Equations

In this section we describe high order ENO and WENO approximations to the Hamilton-Jacobi equation:

$$\begin{cases} \varphi_t + H(\varphi_x, \varphi_y) = 0 \\ \varphi(x, y, 0) = \varphi^0(x, y) \end{cases} \tag{4.34}$$

where H is a locally Lipschitz continuous Hamiltonian and the initial condition $\varphi^0(x, y)$ is locally Lipschitz continuous. We have written the equation (4.34) in two space dimensions, but the discussion is valid for other space dimensions as well.

As is well known, solutions to (4.34) are Lipschitz continuous but may have discontinuous derivatives, regardless of the smoothness of $\varphi^0(x, y)$. The non-uniqueness of such generalized solutions also necessitates the definition of viscosity solutions, to single out a unique, practically relevant solution. The viscosity solution to (4.34) is a locally Lipschitz continuous function $\varphi(x, y, t)$, which satisfies the initial condition and the following property: for any smooth function $\psi(x, y, t)$, if (x_0, y_0, t_0) is a local maximum point of $\varphi - \psi$, then

$$\psi_t(x_0, y_0, t_0) + H(\psi_x(x_0, y_0, t_0) + \psi_y(x_0, y_0, t_0)) \leq 0, \tag{4.35}$$

and, if (x_0, y_0, t_0) is a local minimum point of $\varphi - \psi$, then

$$\psi_t(x_0, y_0, t_0) + H(\psi_x(x_0, y_0, t_0) + \psi_y(x_0, y_0, t_0)) \geq 0. \qquad (4.36)$$

Of course, the above definition means that whenever $\varphi(x, y, t)$ is differentiable, (4.34) is satisfied in the classical sense. Viscosity solution defined this way exists and is unique. For details and equivalent definitions of viscosity solutions, see Crandall and Lions [22].

Hamilton-Jacobi equations are actually easier to solve than conservation laws, because the solutions are typically continuous (only the derivatives are discontinuous).

As before, given mesh sizes Δx, Δy and Δt, we denote the mesh points as $(x_i, y_j, t_n) = (i\Delta x, j\Delta y, n\Delta t)$. The numerical approximation to the viscosity solution $\varphi(x_i, y_j, t_n)$ of (4.34) at the mesh point (x_i, y_j, t_n) is denoted by φ_{ij}^n. We again use a semi-discrete (discrete in the spatial variables only) formulation as a middle step in designing algorithms. In such cases, the numerical approximation to the viscosity solution $\varphi(x_i, y_j, t)$ of (4.34) at the mesh point (x_i, y_j, t) is denoted by $\varphi_{ij}(t)$, the temporal variable t is not discretized. We will also use the notations $D_\pm^x \varphi_{ij} = \frac{\pm(\varphi_{i\pm 1, j} - \varphi_{ij})}{\Delta x}$ and $D_\pm^y \varphi_{ij} = \frac{\pm(\varphi_{i, j\pm 1} - \varphi_{ij})}{\Delta y}$ to denote the first order forward/backward difference approximations to the left and right derivatives of $\varphi(x, y)$ at the location (x_i, y_j).

Since the viscosity solution to (4.34) is usually only Lipschitz continuous but not everywhere differentiable, the *formal* order of accuracy of a numerical scheme is again defined as that determined by the local truncation error in the smooth regions of the solution. Thus, a monotone scheme of the form

$$\varphi_{ij}^{n+1} = G(\varphi_{i-p, j-r}^n, \cdots, \varphi_{i+q, j+s}^n) \qquad (4.37)$$

where G is a non-decreasing function of each argument, is called a first order scheme, although the provable order of accuracy in the L_∞ norm is just $\frac{1}{2}$ [23]. In the semi-discrete formulation, a five point monotone scheme (it does not pay to use more points for a monotone scheme because the order of accuracy of a monotone scheme is at most one [36]) is of the form

$$\frac{d}{dt}\varphi_{ij}(t) = -\hat{H}(D_+^x \varphi_{ij}(t), D_-^x \varphi_{ij}(t), D_+^y \varphi_{ij}(t), D_-^y \varphi_{ij}(t)). \qquad (4.38)$$

The numerical Hamiltonian $\hat{H}$ is assumed to be locally Lipschitz continuous, consistent with H: $\hat{H}(u, u, v, v) = H(u, v)$, and is non-increasing in its first and third arguments and non-decreasing in the other two. Symbolically $\hat{H}(\downarrow, \uparrow, \downarrow, \uparrow)$. It is easy to see that, if the time derivative in (4.38) is discretized by Euler forward differencing, the resulting fully discrete scheme, in the form of (4.37), will be monotone when Δt is suitably small. We have chosen the semi-discrete formulation (4.38) in order to apply suitable nonlinearly stable high order Runge-Kutta type time discretization, see Sect.4.2.

Semi-discrete or fully discrete monotone schemes (4.38) and (4.37) are both convergent towards the viscosity solution of (4.34) [23]. However, monotone schemes are at most first order accurate. As before, we will use the monotone schemes as building blocks for higher order ENO and WENO schemes.

ENO schemes were adapted to the Hamilton-Jacobi equations (4.34) by Osher and Sethian [59] and Osher and Shu [60]. As we know now, the key feature of the ENO algorithm is an adaptive stencil high order interpolation which tries to avoid shocks or high gradient regions whenever possible. Since the Hamilton-Jacobi equation (4.34) is closely related to the conservation law (3.22), in fact in one space dimension they are exactly the same if one takes $u = \varphi_x$, it is not surprising that successful numerical schemes for the conservation laws (3.22), such as ENO and WENO, can be applied to the Hamilton-Jacobi equation (4.34). ENO and WENO schemes, when applied to Hamilton-Jacobi equations (4.34), can produce high order accuracy in the smooth regions of the solution, and sharp, non-oscillatory corners (discontinuities in derivatives).

There are many monotone Hamiltonians [23], [59], [60]. In this section we mainly discuss the following two:

1. For the special case $H(u, v) = f(u^2, v^2)$ where f is a monotone function of both arguments, such as the example $H(u, v) = \sqrt{u^2 + v^2}$, we can use the Osher-Sethian monotone Hamiltonian [59]:

$$\hat{H}^{OS}(u^+, u^-, v^+, v^-) = f(u^2, v^2) \qquad (4.39)$$

where, if f is a non-increasing function of u^2, u^2 is implemented by

$$u^2 = (\min(u^-, 0))^2 + (\max(u^+, 0))^2 \qquad (4.40)$$

and, if f is a non-decreasing function of u^2, u^2 is implemented by

$$u^2 = (\min(u^+, 0))^2 + (\max(u^-, 0))^2 \qquad (4.41)$$

Similarly for v^2. This Hamiltonian is purely upwind (i.e. when $H(u, v)$ is monotone in u in the relevant domain $[u^-, u^+] \times [v^-, v^+]$, only u^- or u^+ is used in the numerical Hamiltonian according to the wind direction), and simple to program. Whenever applicable it should be used. This flux is similar to the Engquist-Osher monotone flux (2.71) for the conservation laws.

2. For the general H we can always use the Godunov type Hamiltonian [5], [60]:

$$\hat{H}^G(u^+, u^-, v^+, v^-) = ext_{u \in I(u^-, u^+)} \, ext_{v \in I(v^-, v^+)} \, H(u, v) \qquad (4.42)$$

where the extrema are defined by

$$ext_{u \in I(a,b)} = \begin{cases} \min_{a \leq u \leq b} & \text{if } a \leq b \\ \max_{b \leq u \leq a} & \text{if } a > b \end{cases} \qquad (4.43)$$

Godunov Hamiltonian is obtained by attempting to solve the Riemann problem of the equation (4.34) exactly with piecewise linear initial condition determined by $u^{\pm}$ and $v^{\pm}$. It is in general not unique, because $\min_u \max_v H(u,v) \neq \max_v \min_u H(u,v)$ and interchanging the order of the two ext's in (4.42) can produce a different monotone Hamiltonian.

Godunov Hamiltonian is purely upwind and is the least dissipative among all monotone Hamiltonians [57]. However, it might be extremely difficult to program, since in general analytical expressions for things like $\min_u \max_v H(u,v)$ can be quite complicated. The readers will be convinced by doing the exercise of obtaining the analytical expression and programming H^G for the ellipse in ellipse case in image processing where $H(u,v) = \sqrt{au^2 + 2buv + cv^2}$. For this case the Osher-Sethian Hamiltonian H^{OS} does not apply.

We are now ready to discuss about higher order ENO or WENO schemes for (4.34). The framework is quite simple: we simply replace the first order scheme (4.38) by:

$$\frac{d}{dt}\varphi_{ij}(t) = -\hat{H}(u_{ij}^+(t), u_{ij}^-(t), v_{ij}^+(t), v_{ij}^-(t)) \qquad (4.44)$$

where $u_{ij}^{\pm}(t)$ are high order approximations to the left and right x-derivatives of $\varphi(x,y,t)$ at (x_i, y_j, t):

$$u_{ij}^{\pm}(t) = \frac{\partial \varphi}{\partial x}(x_i^{\pm}, y_j, t) + O(\Delta x^r) \qquad (4.45)$$

Similarly for $v_{ij}^{\pm}(t)$. Notice that there is no cell-averaged version now.

The key feature of ENO to avoid numerical oscillations is through the following interpolation procedure to obtain $u_{ij}^{\pm}(t)$ and $v_{ij}^{\pm}(t)$. These are just the same ENO procedure we discussed before in Sect.2.2. We repeat it here with its own notations:

ENO Interpolation Algorithm: Given point values $f(x_j)$, $j = 0, \pm1, \pm2, \cdots$ of a (usually piecewise smooth) function $f(x)$ at discrete nodes x_j, we associate an r-th degree polynomial $P_{j+1/2}^{f,r}(x)$ with each interval $[x_j, x_{j+1}]$, with the left-most point in the stencil as $x_{k_{min}^{(r)}}$, constructed inductively as follows:

(1) $P_{j+1/2}^{f,1}(x) = f[x_j] + f[x_j, x_{j+1}](x - x_j)$, $k_{min}^{(1)} = j$;

(2) If $k_{min}^{(l-1)}$ and $P_{j+1/2}^{f,l-1}(x)$ are both defined, then let

$$a^{(l)} = f[x_{k_{min}^{(l-1)}}, \cdots, x_{k_{min}^{(l-1)}+l}]$$

$$b^{(l)} = f[x_{k_{min}^{(l-1)}-1}, \cdots, x_{k_{min}^{(l-1)}+l-1}]$$

and

(i) If $|a^{(l)}| \geq |b^l|$, then $c^{(l)} = b^{(l)}$ and $k_{min}^{(l)} = k_{min}^{(l-1)} - 1$; otherwise $c^{(l)} = a^{(l)}$ and $k_{min}^{(l)} = k_{min}^{(l-1)}$;

(ii) $P^{f,l}_{j+1/2}(x) = P^{f,l-1}_{j+1/2}(x) + c^{(l)} \prod_{i=k^{(l-1)}_{min}}^{k^{(l-1)}_{min}+l-1} (x - x_i)$.

$\square$

In the above procedure $f[\cdot, \cdots, \cdot]$ are the standard Newton divided differences, inductively defined as

$$f[x_1, x_2, \cdots, x_{k+1}] = \frac{f[x_2, \cdots, x_{k+1}] - f[x_1, \cdots, x_k]}{x_{k+1} - x_1}$$

with $f[x_1] = f(x_1)$.

ENO Interpolation Algorithm starts with a first degree polynomial $P^{f,1}_{j+1/2}(x)$ interpolating the function $f(x)$ at the two grid points x_j and x_{j+1}. If we stop here, we would obtain the first order monotone scheme. When higher order is desired, we will in each step add just one point to the existing stencil, chosen from the two immediate neighbors by the size of the two relevant divided differences, which measures the local smoothness of the function $f(x)$.

The approximations to the left and right x-derivatives of φ are then taken as

$$u^{\pm}_{ij} = \frac{\partial}{\partial x} P^{\varphi,r}_{i\pm1/2,j}(x_i) \qquad (4.46)$$

where $P^{\varphi,r}_{i\pm1/2,j}(x)$ is obtained by the ENO Interpolation Algorithm in the x-direction, with $y = y_j$ and t both fixed. $v^{\pm}_{ij}$ are obtained in a similar fashion. The resulting ODE (4.44) is then discretized by an r-th order TVD Runge-Kutta time discretization in Sect.4.2 to guarantee nonlinear stability. More specifically, the high order Runge-Kutta method we use in Sect.4.2 will maintain TVD (total-variation-diminishing) or other stability properties, if these properties are valid for the simple first order Euler forward time discretization of the ODE (4.44). Notice that this is different from the usual linear stability requirement for the ODE solver. We thus obtain both nonlinear stability and high order accuracy in time. The second order ($r = 2$) and third order ($r = 3$) methods we use which has this stability property are given by (4.10) and (4.11), respectively.

Time step restriction is taken as

$$\Delta t \left(\frac{1}{\Delta x} \max_{u,v} \left| \frac{\partial}{\partial u} H(u,v) \right| + \frac{1}{\Delta y} \max_{u,v} \left| \frac{\partial}{\partial v} H(u,v) \right| \right) \leq 0.6 \qquad (4.47)$$

where the maximum is taken over the relevant ranges of u, v. Here 0.6 is just a convenient number used in practice. This number should be chosen between 0.5 and 0.7 according to our numerical experience.

WENO schemes can be used in a similar fashion for Hamilton-Jacobi equations [45]. We will not present the details here.

5 Applications

5.1 Applications to Compressible Gas Dynamics

One of the main application areas of ENO and WENO schemes is compressible gas dynamics.

In 3D, Euler equations of gas dynamics are written as

$$U_t + f(U)_x + g(U)_y + h(U)_z = 0$$

where

$$U = (\rho, \rho u, \rho v, \rho w, E),$$

$$f(U) = (\rho u, \rho u^2 + P, \rho u v, \rho u w, u(E + P)),$$

$$g(U) = (\rho v, \rho u v, \rho v^2 + P, \rho v w, v(E + P)),$$

$$h(U) = (\rho w, \rho u w, \rho v w, \rho w^2 + P, w(E + P)).$$

Here ρ is density, (u, v, w) is the velocity, E is the total energy, P is the pressure, related to the total energy E by

$$E = \frac{P}{\gamma - 1} + \frac{1}{2}\rho(u^2 + v^2 + w^2)$$

with $\gamma = 1.4$ for air.

For the form of Navier-Stokes equations, for the eigenvalues and eigenvectors needed for the characteristic-wise ENO and WENO schemes, and for those equations appearing in curvilinear coordinates, see, e.g. [71].

We mention the following applications of ENO and WENO schemes for compressible flow calculations:

1. Shock tube problem. This is a standard problem for testing codes for shock calculations. However, it is not the best test case for high order methods, as the solution structure is relatively simple (basically piecewise linear). The set-up is a Riemann type initial data:

$$U(x, 0) = \begin{cases} U_L & \text{if } x \leq 0 \\ U_R & \text{if } x > 0 \end{cases}$$

The two standard test cases are the Sod's problem:

$$(\rho_L, q_L, P_L) = (1, 0, 1); \quad (\rho_R, q_R, P_R) = (0.125, 0, 0.1)$$

and the Lax's problem:

$$(\rho_L, q_L, P_L) = (0.445, 0.698, 3.528); \quad (\rho_R, q_R, P_R) = (0.5, 0, 0.571).$$

We show the results of WENO (third order and fifth order) schemes for the Lax problem, in Fig.5.1. Notice that "PS" in the pictures means a

400

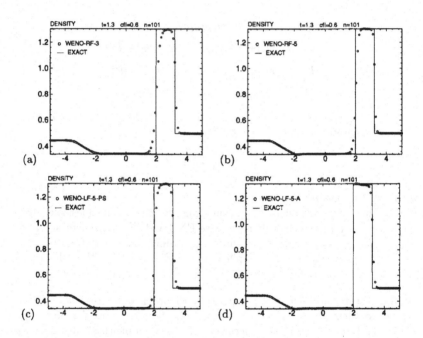

Fig. 5.1: Shock tube, Lax problem, density. (a): third order WENO; (b): fifth order WENO; (c): fifth order WENO with cheaper characteristic decomposition; (d): fifth order WENO with artificial compression.

way of treating the system cheaper than the local characteristic decompositions (for details, see [43]). "A" stands for Yang's artificial compression [83] applied to these cases [43].

We can see from Fig.5.1 that WENO perform reasonably well for these shock tube problem. The contact discontinuity is smeared more than the shock, as expected. Artificial compression helps sharpening contacts. For this problem, which is not the most demanding, the less expensive "PS" version of WENO work quite well.

ENO schemes on this test case perform similarly. We will not give the pictures here. See [70].

2. Shock entropy wave interactions. This problem is very suitable for high order ENO and WENO schemes, because both shocks and complicated smooth flow feature co-exist. In this example, a moving shock interacting with an entropy wave of small amplitude. On a domain $[0,5]$, the initial condition is:

$$\rho = 3.85714; \quad u = 2.629369; \ P = 10.33333; \text{ when } x < 0.5$$
$$\rho = e^{-\epsilon \sin(kx)}; \ u = 0; \qquad P = 1; \qquad \text{when } x \geq 0.5$$

where ϵ and k are the amplitude and wave number of the entropy wave, respectively. The mean flow is a pure right moving Mach 3 shock. If ϵ is small compared to the shock strength, the shock will march to the right at approximately the non-perturbed shock speed and generate a sound wave which travels along with the flow behind the shock. At the same time, the perturbing entropy wave, after "going through" the shock, is compressed and amplified and travels approximately at the speed of $u + c$ where u and c are the velocity and speed of the sound of the mean flow left to the shock. The amplification factor for the entropy wave can be obtained by linear analysis.

Since the entropy wave here is set to be very weak relative to the shock, any numerical oscillation might pollute the generated waves (e.g. the sound waves) and the amplified entropy waves. In our tests, we take $\epsilon = 0.01$ and $k = 13$. The amplitude of the amplified entropy waves predicted by the linear analysis is 0.08690716 (shown in the following figures as horizontal solid lines).

In Fig.5.2, we show the result (entropy) when 12 waves have passed through the shock. It is clear that a lower order method (more dissipative) will damp the magnitude of the transmitted wave more seriously, especially when the waves are traveling more and more away from the shock. We can see that, while fifth order WENO with 800 points already resolves the passing waves well, and with 1200 points resolves the waves excellently, a second order TVD scheme (which is a good one among second order schemes) with 2000 points still shows excessive dissipation downstream. If we agree that fifth order WENO with 800 points behaves similarly as second order TVD with 2000 points, then there is a saving of a factor of 2.5 in grid points. This factor is *per dimension*, hence for a

3D time dependent problem the saving of the number of space-time grids will be a factor of $2.5^4 \approx 40$, a significant saving even after factoring in the extra cost per grid point for the higher order WENO method.

ENO schemes behave similarly for this problem.

There is a two dimensional version of this problem, when the entropy wave can make an angle with the shock. The simulation results again show an advantage in using a higher order method, in Fig.5.3. Several curves are clustered in Fig.5.3 around the exact solution, belonging to various fourth and fifth order ENO or WENO schemes. The circles correspond to a second order TVD scheme, which dissipates the amplitude of the transmitted entropy wave much more rapidly.

3. Steady state calculations. This is important both in gas dynamics and in other fields of applications, such as in semiconductor device simulation, Sect.5.3. For ENO or TVD schemes, the residue does not settle down to machine zero during the time evolution. It will decay first and then hang at the level of the local truncation errors. Presumably this is due to the fact that the numerical flux is not smooth enough (it is only Lipschitz continuous but not C^1). Although this is not satisfactory, it does not seem to affect the final solution (up to the truncation error level, which is how accurate the solution will be anyway).

WENO schemes are much better in getting the residues to settle down to machine zeroes, due to the smoothness of their fluxes.

In Fig.5.4 we show the result of a one dimensional nozzle calculation. The residue in this case settles down nicely to machine zeros. Both fourth and fifth order WENO results are shown.

4. Forward facing step problem. This is a standard test case for high resolution schemes [82]. However, second order methods usually already work well. High order methods might have some advantage in resolving the slip lines.

The set up of the problem is the following: the wind tunnel is 1 length unit wide and 3 length units long. The step is 0.2 length units high and is located 0.6 length units from the left-hand end of the tunnel. The problem is initialized by a right-going Mach 3 flow. Reflective boundary conditions are applied along the walls of the tunnel and in-flow and out-flow boundary conditions are applied at the entrance (left-hand end) and the exit (right-hand end). For the treatment of the singularity at the corner of the step, we adopt the same technique used in [82], which is based on the assumption of a nearly steady flow in the region near the corner.

In Fig.5.5 we present the results of fifth order WENO and fourth order ENO with 242×79 grid points.

5. Double Mach reflection. This is again a standard test case for high resolution schemes [82]. However, second order methods usually again already work well. High order methods might have some advantage in resolving the flow below the Mach stem.

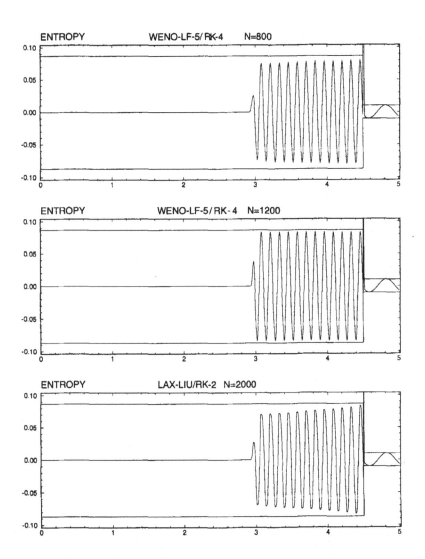

Fig. 5.2: 1D shock entropy wave interaction. Entropy. Top: fifth order WENO with 800 points; middle: fifth order WENO with 1200 points; bottom: second order TVD with 2000 points.

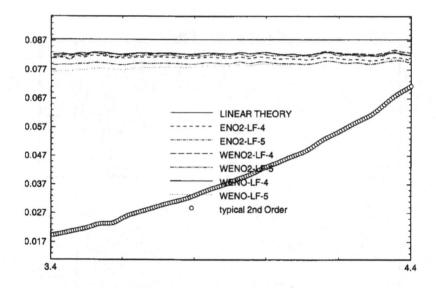

Fig. 5.3: 2D shock entropy wave interaction. Amplitude of amplified entropy waves. 800 points (about 20 points per entropy wave length).

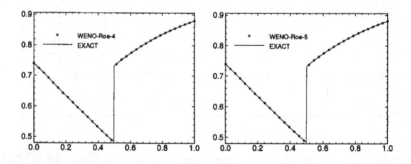

Fig. 5.4: Density. Steady quasi-1D nozzle flow. 34 points. Left: fourth order WENO; right: fifth order WENO.

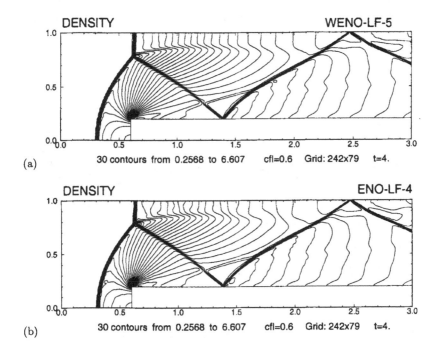

DENSITY   WENO-LF-5

30 contours from 0.2568 to 6.607   cfl=0.6   Grid: 242x79   t=4.

(a)

DENSITY   ENO-LF-4

30 contours from 0.2568 to 6.607   cfl=0.6   Grid: 242x79   t=4.

(b)

Fig. 5.5: Flow past a forward facing step. Density: 242×79 grid points. Top: fifth order WENO; bottom: fourth order ENO.

The computational domain for this problem is chosen to be $[0, 4] \times [0, 1]$, although only part of it, $[0, 3] \times [0, 1]$, is shown [82]. The reflecting wall lies at the bottom of the computational domain starting from $x = \frac{1}{6}$. Initially a right-moving Mach 10 shock is positioned at $x = \frac{1}{6}, y = 0$ and makes a 60° angle with the x-axis. For the bottom boundary, the exact post-shock condition is imposed for the part from $x = 0$ to $x = \frac{1}{6}$ and a reflective boundary condition is used for the rest. At the top boundary of our computational domain, the flow values are set to describe the exact motion of the Mach 10 shock. See [82] for a detailed description of this problem.

In Fig.5.6 we present the results of fifth order WENO and fourth order ENO with 480×119 grid points.

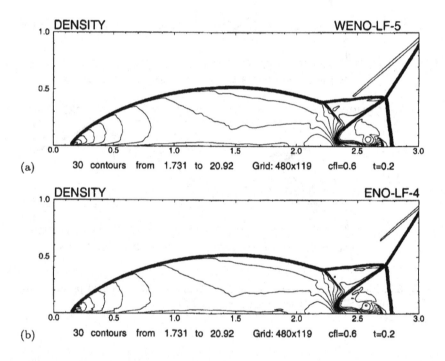

Fig. 5.6: Double Mach reflection. Density: 480×119 grid points. Top: fifth order WENO; bottom: fourth order ENO.

6. 2D shock vortex interactions. High order methods have some advantages in this case, as it resolves the vortex and the interaction better.

The model problem we use describes the interaction between a stationary shock and a vortex. The computational domain is taken to be $[0, 2] \times [0, 1]$. A stationary Mach 1.1 shock is positioned at $x = 0.5$ and normal to the x-axis. Its left state is $(\rho, u, v, P) = (1, \sqrt{\gamma}, 0, 1)$. A small vortex is superposed to the flow left to the shock and centers at $(x_c, y_c) = (0.25, 0.5)$. We describe the vortex as a perturbation to the velocity (u, v), temperature $(T = \frac{P}{\rho})$ and entropy $(S = \ln \frac{P}{\rho^\gamma})$ of the mean flow and denote it by the tilde values:

$$\tilde{u} = \epsilon \tau e^{\alpha(1-\tau^2)} \sin \theta \tag{5.1}$$

$$\tilde{v} = -\epsilon \tau e^{\alpha(1-\tau^2)} \cos \theta \tag{5.2}$$

$$\tilde{T} = -\frac{(\gamma - 1)\epsilon^2 e^{2\alpha(1-\tau^2)}}{4\alpha\gamma} \tag{5.3}$$

$$\tilde{S} = 0 \tag{5.4}$$

where $\tau = \frac{r}{r_c}$ and $r = \sqrt{(x - x_c)^2 + (y - y_c)^2}$. Here ϵ indicates the strength of the vortex, α controls the decay rate of the vortex and r_c is the critical radius for which the vortex has the maximum strength. In our tests, we choose $\epsilon = 0.3, r_c = 0.05$ and $\alpha = 0.204$. The above defined vortex is a steady state solution to the 2D Euler equation.

We use a grid of 251×100 which is uniform in y but refined in x around the shock. The upper and lower boundaries are intentionally set to be reflective. The results (pressure contours) are shown in Fig.5.7 for a fifth order WENO with the cheap "PS" way of treating characteristic decomposition for the system.

In [27], interaction of a shock with a longitudinal vortex is also investigated by the ENO method.

7. How does the finite difference version of ENO and WENO handle non-rectangular domain? As we mentioned before, as long as the domain can be *smoothly* transformed to a rectangle, the schemes can be handily applied.

We consider, as an example, the problem of a supersonic flow past a cylinder. In the physical space, a cylinder of unit radius is positioned at the origin on a $x - y$ plane. The computational domain is chosen to be $[0, 1] \times [0, 1]$ on $\xi - \eta$ plane. The mapping between the computational domain and the physical domain is:

$$x = (R_x - (R_x - 1)\xi) \cos(\theta(2\eta - 1)) \tag{5.5}$$
$$y = (R_y - (R_y - 1)\xi) \sin(\theta(2\eta - 1)) \tag{5.6}$$

where we take $R_x = 3, R_y = 6$ and $\theta = \frac{5\pi}{12}$. Fifth order WENO and a uniform mesh of 60×80 in the computational domain are used.

The problem is initialized by a Mach 3 shock moving toward the cylinder from the left. Reflective boundary condition is imposed at the surface of

408

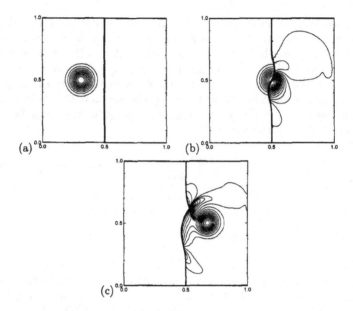

Fig. 5.7: 2D shock vortex interaction. Pressure. Fifth order WENO-LF-5-PS. 30 contours. (a) t=0.05. (b) t=0.20. (c) t=0.35.

the cylinder, i.e. $\xi = 1$, inflow boundary condition is applied at $\xi = 0$ and outflow boundary condition is applied at $\eta = 0, 1$,

We present an illustration of the mesh in the physical space (drawing every other grid line), and the pressure contour, in Fig. 5.8. Similar results are obtained by the ENO schemes but are not shown here.

8. Finally, we use the following problem to illustrate more clearly the power of high order methods. Consider the following idealized problem for the Euler equations in 2D: The mean flow is $\rho = 1$, $P = 1$, and $(u, v) = (1, 1)$ (diagonal flow). We add, to this mean flow, an isentropic vortex (perturbations in (u, v) and the temperature $T = \frac{P}{\rho}$, no perturbation in the entropy $S = \frac{P}{\rho^\gamma}$):

$$(\delta u, \delta v) = \frac{\epsilon}{2\pi} e^{0.5(1-r^2)} (-\bar{y}, \bar{x})$$

$$\delta T = -\frac{(\gamma - 1)\epsilon^2}{8\gamma\pi^2} e^{1-r^2}, \qquad \delta S = 0,$$

where $(\bar{x}, \bar{y}) = (x - 5, y - 5)$, $r^2 = \bar{x}^2 + \bar{y}^2$, and the vortex strength $\epsilon = 5$. Since the mean flow is in the diagonal direction, the vortex movement is not aligned with the mesh direction.

The computational domain is taken as $[0,10] \times [0,10]$, *extended periodically in both directions*. This allows us to perform long time simulation without

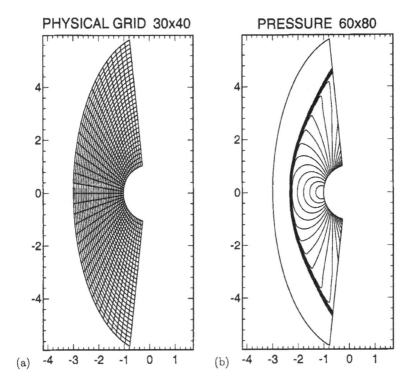

PHYSICAL GRID 30x40 PRESSURE 60x80

Fig. 5.8: Flow past a cylinder. (a) Physical grid. (b) Pressure. WENO-LF-5. 20 contours

having to deal with a large domain. As we will see, the advantage of the high order methods are more obvious for long time simulations.

It is clear that the exact solution of the Euler equation with the above initial and boundary conditions is just the passive convection of the vortex with the mean velocity.

A grid of 80^2 points is used. The simulation is performed until $t = 100$ (10 periods in time). As can be seen from Fig.5.9, fifth order WENO has a much better resolution than a second order TVD scheme, especially for the larger time $t = 100$.

5.2 Applications to Incompressible Flows

In this section we consider numerically solving the incompressible Navier-Stokes or Euler equations

$$u_t + uu_x + vu_y = \mu(u_{xx} + u_{yy}) - p_x$$

410

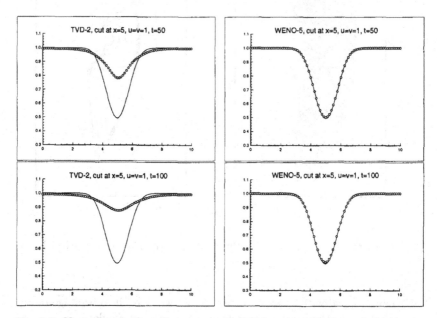

Fig. 5.9: Vortex evolution. Cut at $x = 5$. Solid: exact solution; circles: computed solution. Top: $t = 50$ (after 5 time periods); bottom: $t = 100$ (after 10 time periods). Left: second order TVD scheme; right: fifth order WENO scheme.

$$v_t + uv_x + vv_y = \mu(v_{xx} + v_{yy}) - p_y \qquad (5.7)$$
$$u_x + v_y = 0$$

or their equivalent conservative form

$$u_t + (u^2)_x + (uv)_y = \mu(u_{xx} + u_{yy}) - p_x$$
$$v_t + (uv)_x + (v^2)_y = \mu(v_{xx} + v_{yy}) - p_y \qquad (5.8)$$
$$u_x + v_y = 0$$

where (u, v) is the velocity vector, p is the pressure, $\mu > 0$ for the Navier-Stokes equations and $\mu = 0$ for the Euler equations, using ENO and WENO schemes. We do not discuss the issue of boundary conditions here, thus the equation is defined on the box $[0, 2\pi] \times [0, 2\pi]$ with periodic boundary conditions in both directions. We choose two space dimensions for easy presentation, although our method is also applicable for three space dimensions.

In some sense equations (5.7) are easier to solve numerically than their compressible counter-parts in Sect.5.1, because the latter have solutions containing possible discontinuities (for example shocks and contact discontinuities). However, the solution to (5.7), even if for most cases smooth mathematically, may evolve rather rapidly with time t and may easily become too complicated to be fully resolved on a feasible grid. Traditional linearly stable schemes, such as spectral methods and high-order central difference methods, are suitable for the cases where the solution can be fully resolved, but typically produce signs of instability such as oscillations when small scale features of the flow, such as shears and roll-ups, cannot be adequately resolved on the computational grid. Although in principle one can always overcome this difficulty by refining the grid, today's computer capacity seriously restricts the largest possible grid size.

As we know, the high resolution "shock capturing" schemes such as ENO and WENO are based on the philosophy of giving up fully resolving rapid transition regions or shocks, just to "capture" them in a stable and somehow globally correct fashion (e.g., with correct shock speed), but at the same time to require a high resolution for the smooth part of the solution. The success of such an approach for the conservation laws is documented by many examples in these lecture notes and the references. One example is the one and two dimensional shock interaction with vorticity or entropy waves [70], [71]. The shock is captured sharply and certain key quantities related to the interaction between the shock and the smooth part of the flow, such as the amplification and generation factors when a wave passes through a shock, are well resolved. Another example is the homogeneous turbulence for compressible Navier-Stokes equations studied in [71]. In one of the test cases, the spectral method can resolve all the scales using a 256^2 grid, while third order ENO with just 64^2 points can adequately resolve certain interesting quantities although it cannot resolve local quantities achieved inside the rapid transition region such as the minimum divergence. The conclusion seems to be that, when fully resolving the flow is either impossible or too costly, a "capturing" scheme such as ENO can be used on a coarse grid to obtain at least some partial information about the flow.

We thus expect that, also for the incompressible flow, we can use high-order ENO or WENO schemes on a coarse grid, without fully resolving the flow, but still get back some useful information.

A pioneer work in applying shock capturing compressible flow techniques to incompressible flow is by Bell, Colella and Glaz [6], in which they considered a second order Godunov type discretization, investigated the projection into divergence-free velocity fields for general boundary conditions, and discussed accuracy of time discretizations. Higher order ENO and WENO schemes for incompressible flows are extensions of such methods.

We solve (5.8) in its equivalent projection form

$$\begin{pmatrix} u \\ v \end{pmatrix}_t = \mathbf{P}\left[-\begin{pmatrix} u^2 \\ uv \end{pmatrix}_x - \begin{pmatrix} uv \\ v^2 \end{pmatrix}_y + \mu\left(\begin{pmatrix} u \\ v \end{pmatrix}_{xx} + \begin{pmatrix} u \\ v \end{pmatrix}_{yy} \right) \right] \qquad (5.9)$$

where $\mathbf{P}$ is the Hodge projection into divergence-free fields, i.e., if $\begin{pmatrix} \tilde{u} \\ \tilde{v} \end{pmatrix} = \mathbf{P}\begin{pmatrix} u \\ v \end{pmatrix}$, then $\tilde{u}_x + \tilde{v}_y = 0$ and $\tilde{v}_y - \tilde{u}_x = v_y - u_x$. See, e.g., [6]. For the current periodic case the additional condition to obtain a unique projection $\mathbf{P}$ is that the mean values of u and v are preserved, i.e., $\int_0^{2\pi} \int_0^{2\pi} \tilde{u}(x,y)dxdy = \int_0^{2\pi} \int_0^{2\pi} u(x,y)dxdy$ and $\int_0^{2\pi} \int_0^{2\pi} \tilde{v}(x,y)dxdy = \int_0^{2\pi} \int_0^{2\pi} v(x,y)dxdy$.

We use N_x and N_y (even numbers) equally spaced grid points in x and y, respectively. The grid sizes are denoted by $\Delta x = \frac{N_x}{2\pi}$ and $\Delta y = \frac{N_y}{2\pi}$, and the grid points are denoted by $x_i = i\Delta x$ and $y_j = j\Delta y$. The approximated numerical values of u and v at the grid point (x_i, y_j) are denoted by u_{ij} and v_{ij}.

We first describe the numerical implementation of the projection $\mathbf{P}$. In the periodic case this is easily achieved in the Fourier space. We first expand u and v using Fourier collocation:

$$u_N(x,y) = \sum_{l=-\frac{N_y}{2}}^{\frac{N_y}{2}} \sum_{k=-\frac{N_x}{2}}^{\frac{N_x}{2}} \hat{u}_{kl}\, e^{I(kx+ly)},$$

$$v_N(x,y) = \sum_{l=-\frac{N_y}{2}}^{\frac{N_y}{2}} \sum_{k=-\frac{N_x}{2}}^{\frac{N_x}{2}} \hat{v}_{kl}\, e^{I(kx+ly)} \qquad (5.10)$$

where $I = \sqrt{-1}$, $\hat{u}_{kl}$ and $\hat{v}_{kl}$ are the Fourier collocation coefficients which can be computed from the point values u_{ij} and v_{ij}, using either FFT or matrix-vector multiplications. The detail can be found in, e.g., [9]. Derivatives, either by spectral method or by central differences, involve only multiplications by factors d_k^x or d_l^y in (5.10) because $e^{I(kx+ly)}$ are eigenfunctions of such derivative operators. For example,

$$d_k^x = Ik, \qquad d_l^y = Il \tag{5.11}$$

for spectral derivatives;

$$d_k^x = \frac{2I\sin(\frac{k\Delta x}{2})}{\Delta x}, \qquad d_l^y = \frac{2I\sin(\frac{l\Delta y}{2})}{\Delta y} \tag{5.12}$$

for the second order central differences which, when used twice, will produce the second order central difference approximation for w_{xx}

$$\frac{w_{i+1} - 2w_i + w_{i-1}}{\Delta x^2}$$

and

$$
\begin{aligned}
d_k^x &= \frac{2I\sqrt{(1-\cos(k\Delta x))(7-\cos(k\Delta x))}}{\Delta x}, \\
d_l^y &= \frac{2I\sqrt{(1-\cos(l\Delta y))(7-\cos(l\Delta y))}}{\Delta y}
\end{aligned}
\tag{5.13}
$$

for the fourth order central differences which, when used twice, will produce the fourth order central difference approximation for w_{xx}

$$\frac{16(w_{i+1} + w_{i-1}) - (w_{i+2} + w_{i-2}) - 30w_i}{12\Delta x^2}.$$

High order filters, such as the exponential filter [55], [46]:

$$\sigma_k^x = e^{-\alpha(\frac{k}{N_x})^{2p}}, \qquad \sigma_l^y = e^{-\alpha(\frac{l}{N_y})^{2p}} \tag{5.14}$$

where $2p$ is the order of the filter and α is chosen so that $e^{-\alpha}$ is machine zero, can be used to enhance the stability while keeping at least $2p$-th order of accuracy. This is especially helpful when the projection $\mathbf{P}$ is used for the under-resolved coarse grid with ENO methods. We use the fourth order projection (5.13) and the filter (5.14) with $2p = 8$ in our calculations. This will guarantee third order accuracy (fourth order in L_1) of the ENO scheme. We will denote this combination (the fourth order projection plus the eighth order filtering) by $\mathbf{P}_4$. To be precise, if $\begin{pmatrix} \tilde{u} \\ \tilde{v} \end{pmatrix} = \mathbf{P}_4 \begin{pmatrix} u \\ v \end{pmatrix}$ and $\hat{u}_{kl}$ and $\hat{v}_{kl}$ are Fourier collocation coefficients of u and v, then the Fourier collocation coefficients of $\tilde{u}$ and $\tilde{v}$ are given by

$$\hat{\tilde{u}} = \sigma_k^x \sigma_l^y \frac{d_l^y(d_l^y\hat{u} - d_k^x\hat{v})}{(d_k^x)^2 + (d_l^y)^2}, \qquad \hat{\tilde{v}} = \sigma_k^x \sigma_l^y \frac{-d_k^x(d_l^y\hat{u} - d_k^x\hat{v})}{(d_k^x)^2 + (d_l^y)^2} \tag{5.15}$$

where σ_k^x and σ_l^y are defined by (5.14) with $2p = 8$, and d_k^x and d_l^y are defined by (5.13).

Next we shall describe the ENO scheme for (5.8). Since (5.8) is equivalent to the non-conservative form (5.7), it is natural to implement upwinding by the signs of u and v, and to implement ENO equation by equation (the component version described in Sect.2.3). The r-th order ENO approximation of, e.g., $(u^2)_x$ is thus carried out using the ENO Procedure 4.2. We mention a couple of facts needing attention:

1. Take $f(x) = u^2(x, y)$ with y fixed. We start with the point values $f_i = f(x_i)$;
2. The stencil of the reconstruction is determined adaptively by upwinding and smoothness of $f(x)$. It starts with either x_j or x_{j+1} according to whether $u \geq 0$ or $u < 0$.

There are two ways to handle the second derivative terms for the Navier-Stokes equations. One can absorb them into the convection part and treat them using ENO. For example, $f(x) = u^2(x, y)$ can be replaced by $f(x) = u^2(x, y) - \mu u(x, y)_x$, where $u(x, y)_x$ itself can be obtained using either ENO or central difference of a suitable order. The remaining procedure for computing $f(x)_x$ would be the same as described above. Another simpler possibility is just to use standard central differences (of suitable order) to compute the double derivative terms. Our experience with compressible flow is that there is little difference between the two approaches, especially when the viscosity μ is small.

In the above we have described the discretization for the spatial derivatives

$$
L_{ij} \approx \left[-\left(\begin{array}{c} u^2 \\ uv \end{array} \right)_x - \left(\begin{array}{c} uv \\ v^2 \end{array} \right)_y + \mu \left(\left(\begin{array}{c} u \\ v \end{array} \right)_{xx} + \left(\begin{array}{c} u \\ v \end{array} \right)_{yy} \right) \right] \begin{array}{c} x = x_i \\ y = y_j \end{array} . \tag{5.16}
$$

We then use the third order TVD (total variation diminishing) Runge-Kutta method (4.11) to discretize the resulting ODE:

$$
\left(\begin{array}{c} u \\ v \end{array} \right)_t = \mathbf{P}_4 L_{ij} \tag{5.17}
$$

obtaining:

$$
\begin{aligned}
\left(\begin{array}{c} u \\ v \end{array} \right)^{(1)} &= \mathbf{P}_4 \left[\left(\begin{array}{c} u \\ v \end{array} \right)^n + \Delta t L_{ij}^n \right] \\
\left(\begin{array}{c} u \\ v \end{array} \right)^{(2)} &= \mathbf{P}_4 \left[\frac{3}{4} \left(\begin{array}{c} u \\ v \end{array} \right)^n + \frac{1}{4} \left(\begin{array}{c} u \\ v \end{array} \right)^{(1)} + \frac{1}{4} \Delta t L_{ij}^{(1)} \right] \\
\left(\begin{array}{c} u \\ v \end{array} \right)^{n+1} &= \mathbf{P}_4 \left[\frac{1}{3} \left(\begin{array}{c} u \\ v \end{array} \right)^n + \frac{2}{3} \left(\begin{array}{c} u \\ v \end{array} \right)^{(2)} + \frac{2}{3} \Delta t L_{ij}^{(2)} \right]
\end{aligned} \tag{5.18}
$$

Notice that we have used the property $P_4 \circ P_4 = P_4$ in obtaining the discretization (5.18) from (5.17).

This explicit time discretization is expected to be nonlinearly stable under the CFL condition

$$\Delta t \left[\max_{i,j} \left(\frac{|u_{ij}|}{\Delta x} + \frac{|v_{ij}|}{\Delta y} \right) + 2\mu \left(\frac{1}{\Delta x^2} + \frac{1}{\Delta y^2} \right) \right] \leq 1 \qquad (5.19)$$

For small μ (which is the case we are interested in) this is not a serious restriction on Δt.

We present some numerical examples in the following.

Example 5.1: This example is used to check the third order accuracy of our ENO scheme for smooth solutions. We first take the initial condition as

$$u(x, y, 0) = -\cos(x)\sin(y), \qquad v(x, y, 0) = \sin(x)\cos(y) \qquad (5.20)$$

which was used in [6]. The exact solution for this case is known:

$$u(x, y, t) = -\cos(x)\sin(y)e^{-2\mu t}, \qquad v(x, y, t) = \sin(x)\cos(y)e^{-2\mu t} \qquad (5.21)$$

We take $\Delta x = \Delta y = \frac{1}{N}$ with $N = 32, 64, 128$ and 256. The solution is computed up to $t = 2$ and the L_2 error and numerical order of accuracy are listed in Table 5.1. For the $\mu = 0.05$ case, we list results both with fourth order central approximation to the double derivative terms (central) and with ENO to handle the double derivative terms by absorbing them into the convection part (ENO). We can clearly observe fully third order accuracy (actually better in many cases because the spatial ENO is fourth order in the L_1 sense) in this table.

Example 5.2: This is our test example to study resolution of ENO schemes when the grid is coarse. It is a double shear layer taken from [6]:

$$u(x, y, 0) = \begin{cases} \tanh((y - \pi/2)/\rho) & y \leq \pi \\ \tanh((3\pi/2 - y)/\rho) & y > \pi \end{cases} \qquad v(x, y, 0) = \delta \sin(x) \qquad (5.22)$$

where we take $\rho = \pi/15$ and $\delta = 0.05$. The Euler equations ($\mu = 0$) are used for this example. The solution quickly develops into roll-ups with smaller and smaller scales, so on any fixed grid the full resolution is lost eventually. For example, the expensive run we performed using 512^2 points for the spectral collocation code (with a 18-th order filter (5.14)) is able to resolve the solution fully up to $t = 8$, Fig.5.10, top left, as verified by the spectrum of the solution (not shown here), but begins to lose resolution as indicated by the wriggles in the vorticity contour at $t = 10$ (not shown here). On the other hand, the ENO runs with 64^2 and 128^2 points produces smooth, stable results Fig.5.10, top right and bottom left. In Fig.5.10, bottom right, we show a cut at $x = \pi$ for v at $t = 8$. This gives a better feeling about the resolution in physical space. Apparently with these coarse grids the full structure of the roll-up is not resolved. However, when we compute the total circulation

$$c_\Omega = \int_\Omega \omega(x, y) dx dy = \int_{\partial\Omega} u dx + v dy \qquad (5.23)$$

Table 5.1. Accuracy of ENO Schemes for (12.2).

N	$\mu = 0$		$\mu = 0.05$, central		$\mu = 0.05$, ENO	
	L_2 error	order	L_2 error	order	L_2 error	order
32	9.10(-4)		5.28(-4)		4.87(-4)	
64	5.73(-5)	3.99	3.20(-5)	4.04	3.09(-5)	3.98
128	3.62(-6)	3.98	1.93(-6)	4.05	1.89(-6)	4.03
256	2.28(-7)	3.99	1.18(-7)	4.03	1.16(-7)	4.03

N	$\mu = 0$		
	L_2 diff	order	error
32	1.14(-1)		
64	1.40(-2)	3.02	1.96(-3)
128	1.46(-3)	3.26	1.69(-4)
256	1.11(-4)	3.77	8.78(-6)
	$\mu = 0.05$, central		
	L_2 diff	order	error
32	3.20(-2)		
64	2.78(-3)	3.52	2.66(-4)
128	1.81(-4)	3.94	1.26(-5)
256	1.09(-5)	4.06	6.91(-7)
	$\mu = 0.05$, ENO		
	L_2 diff	order	error
32	3.60(-2)		
64	2.93(-3)	3.62	2.60(-4)
128	1.80(-4)	4.02	1.18(-5)
256	1.10(-5)	4.04	7.15(-7)

around the roll-up by taking $\Omega = [\frac{\pi}{2}, \frac{3\pi}{2}] \times [0, 2\pi]$ and using the rectangular rule (which is infinite order accurate for the periodic case) on the line integrals at the right-hand-side of (5.13), we can see that this number is resolved much better than the roll-up itself, Table 5.2.

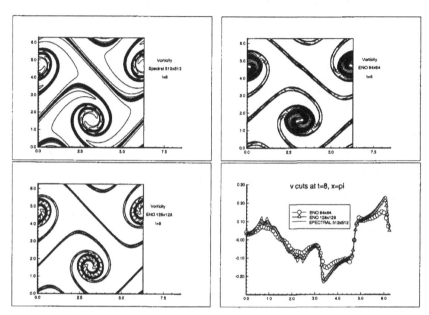

Fig. 5.10: Double shear layer. Contours of vorticity. $t = 8$. Top left: spectral with 512^2 points; top right: ENO with 64^2 points; bottom left: ENO with 128^2 points; bottom right: the cut at $x = \pi$ of v, spectral method with 512^2 points, ENO method with 64^2 and with 128^2 points.

As an application of ENO scheme for incompressible flow, we consider the motion of an incompressible fluid, in two and three dimensions, in which the vorticity is concentrated on a lower dimensional set [31]. Prominent examples are vortex sheets and vortex filaments in three dimensions, and vortex sheets, vortex dipole sheets and point vortices in two dimensions.

In three dimensions, the equations are written in the form

$$\xi_t + v\nabla\xi - \nabla v \ \xi = 0$$
$$\nabla \times v = \xi \qquad (5.24)$$
$$\nabla \cdot v = 0$$

where $\xi(x, y, z, t)$ is the vorticity vector, and $v(x, y, z, t)$ is the velocity vector.

Table 5.2. Resolution of the Total Circulation.

t	2	4	6	8	10
ENO 64^2	0.87300	3.07100	7.16889	9.88063	10.90122
ENO 128^2	0.87452	2.97810	7.30999	10.34414	11.79418
spectral 512^2	0.87433	2.98029	7.28308	10.46212	11.85875

In a vortex sheet, ξ is a singular measure concentrated on a two dimensional surface, while in a vortex filament, ξ is a function concentrated on a tubular neighborhood of a curve.

We use an Eulerian, fixed grid, approach, that works in general in two and three dimensions. In the particular case of the two dimensional vortex sheet problem in which the vorticity does not change sign, the approach yields a very simple and elegant formulation.

The basic observation involves a variant of the level set method for capturing fronts, developed in [59].

The formulation we use here regularizes general ill-posed problems via the level set approach, using the idea that a simple closed curve which is the level set of a function cannot change its index, i.e. there is an automatic topological regularization. This is very helpful for numerical calculations. The regularization is automatically accomplished through the use of dissipative schemes, which has the effect of adding a small curvature term (which vanishes as the grid size goes to zero) to the evolution of the interface. The formulation allows for topological changes, such as merging of surfaces.

The main idea is to decompose ξ into a product of the form

$$\xi = P(\varphi)\eta \tag{5.25}$$

where P is a scalar function, typically an approximate δ function. The variable φ is a scalar function whose zero level set represents the points where vorticity concentrates, and η represents the vorticity strength vector. This decomposition is performed at time zero and is of course not unique.

The observation is that once a decomposition is found, the following system of equations yields a solution to the Euler equations, replacing the original set of equations (5.24).

$$\varphi_t + v\nabla\varphi = 0$$
$$\eta_t + v\nabla\eta - \nabla v\ \eta = 0$$
$$\nabla \times v = P(\varphi)\eta \tag{5.26}$$
$$\nabla \cdot v = 0$$

These equations have initial conditions

$$\varphi(0, \cdot) = \varphi_0$$

$$\eta(0, \cdot) = \eta_0$$

where φ_0, η_0 and P are chosen so that (5.25) holds at time $t = 0$. Notice that (5.25) and (5.26) imply that $\nabla\varphi$ is orthogonal to η, and $div(\eta) = 0$. This is enforced in the initial condition and is maintained automatically by (5.25) and (5.26).

When P is a distribution, such as a δ function, approaching P with a sequence of smooth mollifiers P_ϵ yields a sequence of approximating solutions. This is the approach used in numerical calculations, since the δ function can only be represented approximately on a finite grid. The parameter ϵ is usually chosen to be proportional to the mesh size.

The advantage of this formulation, is that it replaces a possibly singular and unbounded vorticity function ξ, by bounded, smooth (at least uniformly Lipschitz) functions φ and η. Therefore, while it is not feasible to compute solutions of (5.24) directly, it is very easy to compute solutions of (5.26).

In two dimensions, the vorticity is given by

$$\xi = \begin{pmatrix} 0 \\ 0 \\ \omega(t, x, y) \end{pmatrix}$$

and hence the Euler equations are given by

$$\omega_t + v\nabla\omega = 0$$
$$curl(v) = \omega \tag{5.27}$$
$$div(v) = 0 \tag{5.28}$$

Our formulation (5.26), becomes

$$\varphi_t + v\nabla\varphi = 0$$
$$\eta_t + v\nabla\eta = 0 \tag{5.29}$$
$$curl(v) = P(\varphi)\eta$$
$$div(v) = 0$$

where η is now a scalar.

If the vortex sheet strength η does not change sign along the curve, it can be normalized to $\eta \equiv 1$ and the equations take on a particularly simple and elegant form:

$$\varphi_t + v(\varphi)\nabla\varphi = 0 \tag{5.30}$$

where the velocity $v(\varphi)$ is given by

$$v = - \begin{pmatrix} -\partial_y \\ \partial_x \end{pmatrix} \Delta^{-1} P(\varphi) \tag{5.31}$$

In this case, the vortex sheet strength along the curve is given by $\frac{1}{|\nabla\varphi|}$ (see (5.33)).

Example 5.3: *Vortex Sheets in 2D.* We consider the periodic vortex sheet in two dimensions, i.e. $P(\varphi) = \delta(\varphi)$ in (5.31). The three dimensional case is defined in detail later. The evolution of the vortex sheet in the Lagrangian framework has been considered by various authors. Krasny [47], [48] has computed vortex sheet roll-up using vortex blobs and point vortices with filtering. Baker and Shelley [4] have approximated the vortex sheet by a layer of constant vorticity which they computed by Lagrangian methods. In the context of our approach, their approximation corresponds to approximating the δ function by a step function.

In our framework, we use a fixed Eulerian grid, and approximate (5.30) by the third order upwind ENO finite difference scheme with a third order TVD Runge-Kutta time stepping. At every time step, the velocity v is first obtained by solving the Poisson equation for the stream function Ψ:

$$\Delta\Psi = -P(\varphi)$$

with boundary conditions

$$\Psi(x, \pm1) = 0$$

and periodic in x. This is done by using a second order elliptic solver FISH-PAK. Once Ψ is obtained, the velocity is recovered by $v = (-\Psi_y, \Psi_x)$ by using either ENO or central difference approximations (we do not observe major difference among the two: the results shown are those obtained by central difference). Once v is obtained, upwind biased ENO is easily applied to (5.30).

The initial conditions are similar to the ones in [48], i.e given by a sinusoidal perturbation of a flat sheet:

$$\varphi_0(x, y) = y + 0.05\sin(\pi x)$$

The boundary condition for φ are periodic, of the form:

$$\varphi(t, -1, y) = \varphi(t, 1, y)$$

$$\varphi(t, x, -1) = \varphi(t, x, 1) - 2$$

The δ function is approximated as in [61],[77] by

$$\delta_\epsilon(\varphi) = \begin{cases} \frac{1}{2\epsilon}\left(1 + \cos\left(\frac{\pi\varphi}{\epsilon}\right)\right) & \text{if } |\varphi| < \epsilon \\ 0 & \text{otherwise} \end{cases} \tag{5.32}$$

For fixed ϵ, there is convergence as $\Delta x \to 0$ to a smooth solution. One can then take $\epsilon \to 0$. This two step limit is very costly to implement numerically.

Our numerical results show that one can take ϵ to be proportional to Δx, but convergence is difficult to establish theoretically.

In Fig.5.11, top left, we present the result at $t = 4$, of using ENO with 128^2 grid points with the parameter ϵ in the approximate δ function chosen as $\epsilon = 12\Delta x$. We use the graphic package TECPLOT to draw the level curve of $\varphi = 0$. Next, we keep $\epsilon = 12\Delta x$ but double the grid points in each direction to 256^2, the result of $t = 4$ is shown in Fig.5.11, top right. Comparing with Fig.5.11, top left, we can see that there are more turns in the core at the same physical time when the grid size is reduced and the δ function width ϵ is kept proportional to Δx. One might wonder whether the core structure of Fig.5.11, top right, is distorted by numerical error. To verify that this is not the case, we keep $\epsilon = 12 \times \frac{2}{256} = \frac{3}{32}$ *fixed*, and reduce Δx, Fig.5.11, bottom left and right. The three pictures overlay very well, the bottom two pictures in Fig.5.11 are indistinguishable, indicating that the core structure is a resolved solution to the problem and convergence is obtained with fixed ϵ. By reducing ϵ for the more refined grids, more turns in the core can be obtained in shorter time (pictures not shown).

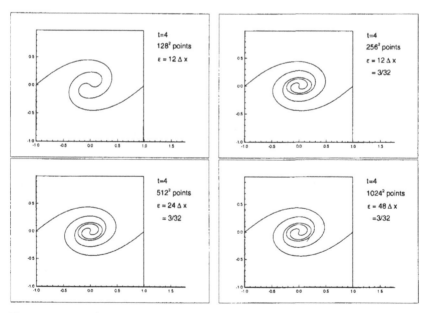

Fig. 5.11: Two dimensional vortex sheet simulation. $t = 4$. Top left: ENO with 128^2 points, δ function width $\epsilon = 12\Delta x = \frac{3}{16}$; Top right: ENO with 256^2 points, δ function width $\epsilon = 12\Delta x = \frac{3}{32}$; Bottom left: ENO with 512^2 points, δ function width $\epsilon = 24\Delta x = \frac{3}{32}$; Bottom right: ENO with 1024^2 points, δ function width $\epsilon = 48\Delta x = \frac{3}{32}$.

The smoothing of the δ function, and the third order truncation error in the advection step and the second order error in the inverse Laplacian are the only smoothing steps in our method.

We now give the same example in three dimensions. We first sketch the algorithm for initializing and computing a periodic 3D vortex sheet, using (5.26).

We let $P(\varphi) = \delta(\varphi)$ (in practice δ is replaced by an approximation). The zero level set of φ is the vortex sheet $\Gamma(s)$, parameterized by surface area s. The variable η_0 is chosen to fit the initial vortex sheet strength. For instance, given any smooth test function g

$$\langle \xi, g \rangle = \langle \eta_0 \delta(\varphi_0), g \rangle$$
$$= \int \eta_0(\Gamma_0(s)) g(\Gamma_0(s)) \frac{1}{|\nabla \varphi_0|} ds$$

Thus, the initial vortex sheet strength is given by

$$\frac{\eta_0}{|\nabla \varphi_0|} \tag{5.33}$$

To obtain the velocity vector, one introduces the vector potential A, where

$$v = \nabla \times A, \quad div(A) = 0$$

and solves the Poisson equation

$$\triangle A = -P(\varphi)\eta \tag{5.34}$$

To ensure that $div(A) = 0$, we require that $div(\eta) = 0$ and that $\nabla \varphi \cdot \eta = 0$ initially. It is easy to see that these equalities are maintained as t increases.

The boundary conditions for the velocity are $v_2(x, \pm 1, z) = 0$ and periodic in x and z. To obtain the boundary conditions for $A = (A_1, A_2, A_3)$, we use the divergence free condition on A in addition to the velocity boundary condition. Thus,

$$A_1(x, \pm 1, z) = A_3(x, \pm 1, z) = 0 \tag{5.35}$$
$$\partial_y A_2(x, \pm 1, z) = 0$$

and periodic in x, z. The Neumann condition requires the following compatibility condition

$$\int \xi_2(x, y, z, 0) dx dy dz = 0$$

Three dimensional runs are much more expensive than two dimensional runs, not only because the number of grid points increases, but also because there are now four evolution equations (for φ and η), and three potential equations. We still use the third order ENO scheme coupled with the second order elliptic solver FISHPAK, with 64^3 grid points, and ϵ is chosen as $6\Delta x$, which is the same in magnitude as that used in Fig.5.11 of Example 5.3. The boundary conditions for φ are similar to the ones in two dimensions: periodic in all directions (module the linear term in y). The vortex sheet strength vector η is periodic in all directions.

We first verify whether we can recover the two dimensional results with the three dimensional setting. We use the initial condition

$$\varphi_0(x, y, z) = y + 0.05\sin(\pi x)$$

which is the same as that for Example 5.3, and choose a constant initial condition for η as $\eta_0(x, y, z) = (0, 0, 1)$. We observe exact agreement with our two dimensional results in Example 5.3, Fig.5.11. Next, we consider the truly three dimensional problem with the initial condition chosen as

$$\varphi_0(x, y, z) = y + 0.05\sin(\pi x) + 0.1\sin(\pi z)$$

and η is chosen as $\eta_0(x, y, z) = (0, -0.1\pi\cos(\pi z), 1)$ which satisfies the divergence free condition as well as the condition to be orthogonal to $\nabla\varphi$. In Fig.5.12, left, we show the level set of $\varphi = 0$ for $t = 5$. We can clearly see the roll up process and the three dimensional features. The cut at the constants $z = 0$ plane is shown in Fig.5.12, right.

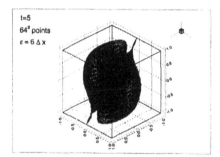 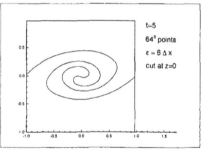

Fig. 5.12: Three dimensional vortex sheet simulation. $t = 5$. ENO with 64^3 points. δ function width $\epsilon = 6\Delta x$. Left: three dimensional level surface; Right: $z = 0$ plane cut.

5.3 Applications in Semiconductor Device Simulation

An interesting application area for ENO and WENO schemes is the equations in semiconductor device simulations. During the last decade, semiconductor device modeling has attempted to incorporate general carrier heating, velocity overshoot, and various small device features into carrier simulation. The popular wisdom emerging from such concentrated study holds that global dependence of critical quantities, such as mobilities, on energy and/or temperature, is essential if such phenomena are to be modeled adequately.

This gives rise to the various energy transport models, including the hydrodynamic model and the ET model, see, e.g. [41]. Unlike the earlier drift-diffusion models, which are basically parabolic, these new models contain significant transport effects [42], thus calling for discretization techniques suitable for hyperbolic problems.

In this section we present two of such models.

The first one is the hydrodynamic model. It is obtained by taking the first three moments of the Boltzmann equation. In the conservative format the hydrodynamic model is written as follows. Define the vector of dependent variables as

$$u = (n, \sigma, \tau, W), \tag{5.36}$$

where n is the electron concentration, $p = (\sigma, \tau)$ is the momenta, and W is the total energy. The equations, in two dimensions, take the form

$$u_t + f_1(u)_x + f_2(u)_y = c(u) + G(u, \varphi) + (0, 0, 0, \nabla \cdot (\kappa \nabla T)) \tag{5.37}$$

where

$$f_1(u) = (\frac{\sigma}{m}, \frac{2}{3}(\frac{\sigma^2}{mn} + W - \frac{\tau^2}{2mn}), \frac{\sigma\tau}{mn}, \frac{5\sigma W}{3mn} - \sigma\frac{\sigma^2 + \tau^2}{3m^2n^2}), \tag{5.38}$$

$$f_2(u) = (\frac{\tau}{m}, \frac{\sigma\tau}{mn}, \frac{2}{3}(\frac{\tau^2}{mn} + W - \frac{\sigma^2}{2mn}), \frac{5\tau W}{3mn} - \tau\frac{\sigma^2 + \tau^2}{3m^2n^2}), \tag{5.39}$$

$$c(u) = (0, -\frac{\sigma}{\tau_p}, -\frac{\tau}{\tau_p}, -\frac{W - W_0}{\tau_w}), \tag{5.40}$$

$$G(u) = (0, -enF_1, -enF_2, -enF \cdot v). \tag{5.41}$$

Here, F is the electric field, obtained by solving a Poisson's equation:

$$F = -\nabla\varphi, \tag{5.42}$$

$$\nabla \cdot (\epsilon\nabla\varphi) = -en - n_d \tag{5.43}$$

where n_d is the doping (a given function which is typically discontinuous).

The second model is the energy transport model, written as

$$u_t + f(u)_x = g(u)_{xx} + h(u). \tag{5.44}$$

In equation (5.44),

$$u = (en, \frac{nE}{m}), \tag{5.45}$$
$$f(u) = \varphi'n \, (e\mu(E), \, \mu^E(E) + D(E)), \tag{5.46}$$
$$g(u) = (nD(E), \, nD^E(E)), \tag{5.47}$$
$$h(u) = (0, \, en\mu(E)(\varphi')^2 + \frac{e}{\epsilon}(n - n_d)nD(E) - n\langle\frac{\partial E}{\partial t}\,|_{coll}\rangle) \tag{5.48}$$

It can be shown that the left hand side defines a hyperbolic system, since the eigenvalues of $f'(u)$ are real, for all positive n and T.

We first present one dimensional numerical results. The one dimensional n^+-n-n^+ channel we simulate is a standard silicon diode with a length of $0.6\mu m$, with a doping defined by $n_d = 5 \times 10^{17}cm^{-3}$ in $[0, 0.1]$ and in $[0.5, 0.6]$, and $n_d = 2 \times 10^{15}cm^{-3}$ in $[0.15, 0.45]$, joined by smooth junctions (Fig.5.13, left). The lattice temperature is taken as $T_0 = 300$ K. We apply a voltage bias of $vbias = 1.5$V. We use the full HD model; the relevant parameters can be found in [41]. In Fig.5.13, right, we present the simulated velocity using the HD model. The dashed line shows the result computed with a reduced HD model by ignoring the transport effects. This type of reduced HD models are used quite often in engineering, as they tend to reduce the numerical difficulty when standard (not high resolution) schemes are used. However, we can see here that there is significant difference in the simulated results.

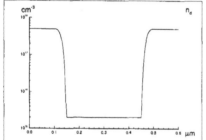 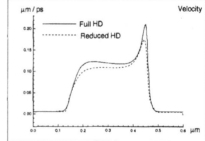

Fig. 5.13: The one dimensional n^+-n-n^+ channel. Left: the doping n_d; Right: the velocity v, comparison of the HD model and the reduced HD model.

We now present numerical simulation results for one carrier, two dimensional MESFET devices. The third order ENO shock-capturing algorithm with Lax-Friedrichs building blocks, as described elsewhere in these lecture notes, is applied to the hyperbolic part (the left hand side) of Equations (5.37) and (5.44). The TVD third order Runge-Kutta time discretization (4.11) is used for the time evolution towards steady states. The forcing terms on the right hand side of (5.37) and (5.44) are treated in a time consistent way in the

Runge-Kutta time stepping. The double derivative terms on the right hand side of (5.37) and (5.44) are approximated by standard central differences owing to their dissipative nature. The Poisson equation (5.43) is solved by direct Gauss elimination for one spatial dimension and by Successive Over-Relaxation (SOR) or the Conjugate Gradient (CG) method for two spatial dimensions. Initial conditions are chosen as $n = n_d$ for the concentration, $T = T_0$ for the temperature, and $u = v = 0$ for the velocities. A continuation method is used to reach the steady state: the voltage bias is taken initially as zero and is gradually increased to the required value, with the steady state solution of a lower biased case used as the initial condition for a higher one.

We simulate a two dimensional MESFET of the size $0.6 \times 0.2 \mu m^2$. The source and the drain each occupies $0.1 \mu m$ at the upper left and the upper right, respectively, with the gate occupying $0.2 \mu m$ at the upper middle (Fig.5.14, top left). The doping is defined by $n_d = 3 \times 10^{17} cm^{-3}$ in $[0, 0.1] \times [0.15, 0.2]$ and in $[0.5, 0.6] \times [0.15, 0.2]$, and $n_d = 1 \times 10^{17} cm^{-3}$ elsewhere, with abrupt junctions (Fig.5.14, top right). A uniform grid of 96×32 points is used. Notice that even if we may not have shocks in the solution, the initial condition $n = n_d$ is discontinuous, and the final steady state solution has a sharp transition around the junction. With the relatively coarse grid we use, the non-oscillatory shock capturing feature of the ENO algorithm is essential for the stability of the numerical procedure.

We apply, at the source and drain, a voltage bias $vbias = 2V$. The gate is a Schottky contact, with a negative voltage bias $vgate = -0.8V$ and a very low concentration value $n = 3.9 \times 10^5 cm^{-3}$. The lattice temperature is taken as $T_0 = 300°K$. The numerical boundary conditions are summarized as follows (where $\Phi_0 = \frac{k_b T}{e} \ln\left(\frac{n_d}{n_i}\right)$ with $k_b = 0.138 \times 10^{-4}$, $e = 0.1602$, and $n_i = 1.4 \times 10^{10} cm^{-3}$ in our units):

- At the source ($0 \leq x \leq 0.1, y = 0.2$): $\Phi = \Phi_0$ for the potential; $n = 3 \times 10^{17} cm^{-3}$ for the concentration; $T = 300°K$ for the temperature; $u = 0 \mu m/ps$ for the horizontal velocity; and Neumann boundary condition for the vertical velocity v (i.e. $\frac{\partial v}{\partial n} = 0$ where n is the normal direction of the boundary).
- At the drain ($0.5 \leq x \leq 0.6, y = 0.2$): $\Phi = \Phi_0 + vbias = \Phi_0 + 2$ for the potential; $n = 3 \times 10^{17} cm^{-3}$ for the concentration; $T = 300°K$ for the temperature; $u = 0 \mu m/ps$ for the horizontal velocity; and Neumann boundary condition for the vertical velocity v.
- At the gate ($0.2 \leq x \leq 0.4, y = 0.2$): $\Phi = \Phi_0 + vgate = \Phi_0 - 0.8$ for the potential; $n = 3.9 \times 10^5 cm^{-3}$ for the concentration; $T = 300°K$ for the temperature; $u = 0 \mu m/ps$ for the horizontal velocity; and Neumann boundary condition for the vertical velocity v.
- At all other parts of the boundary ($0.1 \leq x \leq 0.2, y = 0.2$; $0.4 \leq x \leq 0.5, y = 0.2$; $x = 0, 0 \leq y \leq 0.2$; $x = 0.6, 0 \leq y \leq 0.2$; and $0 \leq x \leq 0.6, y = 0$), all variables are equipped with Neumann boundary conditions.

The boundary conditions chosen are based upon physical and numerical considerations. They may not be adequate mathematically, as is evident from some serious boundary layers observable in the concentration (see pictures in [41]). ENO methods, owing to their upwind nature, are robust to different boundary conditions (including over-specified boundary conditions) and do not exhibit numerical difficulties in the presence of such boundary layers, even with the extremely low concentration prescribed at the gate (around 10^{-12} relative to the high doping). We point out, however, that boundary conditions affect the global solution significantly. We have also simulated the same problem with different boundary conditions, for example with Dirichlet boundary conditions everywhere for the temperature, or with Neumann boundary conditions for all variables except for the potential at the contacts. The numerical results (not shown here) are noticeably different. This indicates the importance of studying adequate boundary conditions, from both a physical and a mathematical point of view.

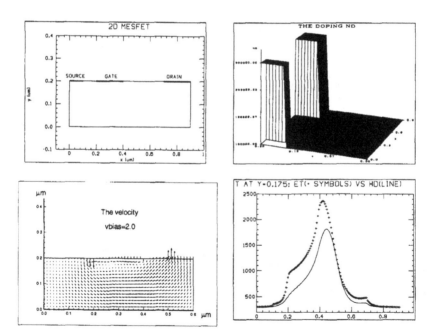

Fig. 5.14: Two dimensional MESFET. Top left: the geometry; Top right: the doping n_d; Bottom left: the velocity (u, v) obtained by the HD model; Bottom right: comparison of the hydrodynamic (HD) model (solid line) and the energy transport (ET) model (plus symbols), cut at the middle of the high doping blobs $y = 0.175$, the temperature T.

The velocity vectors resulting from the hydrodynamic model simulation are presented in Fig.5.14, bottom left. In Fig.5.14, bottom right, we compare the temperature at $y = 0.175$ from the simulations of the hydrodynamic model and of the ET model. Clearly there is a significant difference between these two models for this 2D case.

There are also new models in semiconductor device simulation (e.g. [14], [15]), which are worthy of investigations. ENO and WENO schemes provide robust and reliable tools for carrying out such investigations.

References

1. R. Abgrall, *On essentially non-oscillatory schemes on unstructured meshes: analysis and implementation*, Journal of Computational Physics, v114 (1994), pp.45–58.
2. N. Adams and K. Shariff, *A high-resolution hybrid compact-ENO scheme for shock-turbulence interaction problems*, Journal of Computational Physics, v127 (1996), pp.27–51.
3. H. Atkins and C.-W. Shu, *GKS and eigenvalue stability analysis of high order upwind scheme*, in preparation.
4. G. R. Baker and M. J. Shelley, *On the connection between thin vortex layers and vortex sheets*, Journal of Fluid Mechanics, v215 (1990), pp.161–194.
5. M. Bardi and S. Osher, *The nonconvex multi-dimensional Riemann problem for Hamilton-Jacobi equations*, SIAM Journal on Mathematical Analysis, v22 (1991), pp.344–351.
6. J. Bell, P. Colella and H. Glaz, *A Second Order Projection Method for the Incompressible Navier-Stokes Equations*, Journal of Computational Physics, v85, 1989, pp.257–283.
7. B. Bihari and A. Harten, *Application of generalized wavelets: an adaptive multiresolution scheme*, Journal of Computational and Applied Mathematics, v61 (1995), pp.275–321.
8. W. Cai and C.-W. Shu, *Uniform high-order spectral methods for one- and two-dimensional Euler equations*, Journal of Computational Physics, v104 (1993), pp.427–443.
9. C. Canuto, M.Y. Hussaini, A. Quarteroni and T. A. Zang, *Spectral Methods in Fluid Dynamics*, Springer-Verlag, 1988.
10. M. Carpenter and C. Kennedy, *Fourth-order 2N-storage Runge-Kutta schemes*, NASA TM 109112, NASA Langley Research Center, June 1994.
11. J. Casper, *Finite-volume implementation of high-order essentially nonoscillatory schemes in two dimensions*, AIAA Journal, v30 (1992), pp.2829–2835.
12. J. Casper and H. Atkins, *A finite-volume high-order ENO scheme for two dimensional hyperbolic systems*, Journal of Computational Physics, v106 (1993), pp.62–76.
13. J. Casper, C.-W. Shu and H. Atkins, *Comparison of two formulations for high-order accurate essentially nonoscillatory schemes*, AIAA Journal, v32 (1994), pp.1970–1977.
14. C. Cercignani, I. Gamba, J. Jerome and C.-W. Shu, *Applicability of the high field model: an analytical study via asymptotic parameters defining domain decomposition*, VLSI Design, to appear.

15. C. Cercignani, I. Gamba, J. Jerome and C.-W. Shu, *Applicability of the high field model: a preliminary numerical study*, VLSI Design, to appear.

16. S. Christofi, *The study of building blocks for ENO schemes*, Ph.D. thesis, Division of Applied Mathematics, Brown University, September 1995.

17. B. Cockburn, *An introduction to the discontinuous Galerkin method for convection-dominated problems*, this volume.

18. B. Cockburn and C.-W. Shu, *TVB Runge-Kutta local projection discontinuous Galerkin finite element method for conservation laws II: general framework*, Mathematics of Computation, v52 (1989), pp.411–435.

19. B. Cockburn, S.-Y. Lin and C.-W. Shu, *TVB Runge-Kutta local projection discontinuous Galerkin finite element method for conservation laws III: one dimensional systems*, Journal of Computational Physics, v84 (1989), pp.90–113.

20. B. Cockburn, S. Hou and C.-W. Shu, *The Runge-Kutta local projection discontinuous Galerkin finite element method for conservation laws IV: the multidimensional case*, Mathematics of Computation, v54 (1990), pp.545–581.

21. B. Cockburn and C.-W. Shu, *The Runge-Kutta discontinuous Galerkin method for conservation laws V: multidimensional systems*, to appear in Journal of Computational Physics.

22. M. Crandall and P. Lions, *Viscosity solutions of Hamilton-Jacobi equations*, Transactions of the American Mathematical Society, v277 (1983), pp.1–42.

23. M. Crandall and P. Lions, *Two approximations of solutions of Hamilton-Jacobi equations*, Mathematics of Computation, v43 (1984), pp.1–19.

24. A. Dolezal and S. Wong, *Relativistic hydrodynamics and essentially non-oscillatory shock capturing schemes*, Journal of Computational Physics, v120 (1995), pp.266–277.

25. R. Donat and A. Marquina, *Capturing shock reflections: an improved flux formula*, Journal of Computational Physics, v125 (1996), pp.42–58.

26. W. E and C.-W. Shu, *A numerical resolution study of high order essentially non-oscillatory schemes applied to incompressible flow*, Journal of Computational Physics, v110 (1994), pp.39–46.

27. G. Erlebacher, Y. Hussaini and C.-W. Shu, *Interaction of a shock with a longitudinal vortex*, Journal of Fluid Mechanics, v337 (1997), pp.129–153.

28. E. Fatemi, J. Jerome and S. Osher, *Solution of the hydrodynamic device model using high order non-oscillatory shock capturing algorithms*, IEEE Transactions on Computer-Aided Design of Integrated Circuits and Systems, v10 (1991), pp.232–244.

29. S. Gottlieb and C.-W. Shu, *Total variation diminishing Runge-Kutta schemes*, Mathematics of Computation, v67 (1998), pp.73-85.

30. B. Gustafsson, H.-O. Kreiss and A. Sundstrom, *Stability theory of difference approximations for mixed initial boundary value problems, II*, Mathematics of Computation, v26 (1972), pp.649–686.

31. E. Harabetian, S. Osher and C.-W. Shu, *An Eulerian approach for vortex motion using a level set regularization procedure*, Journal of Computational Physics, v127 (1996), pp.15–26.

32. A. Harten, *The artificial compression method for computation of shocks and contact discontinuities III: self-adjusting hybrid schemes*, Mathematics of Computation, v32 (1978), pp.363–389.

33. A. Harten, *High resolution schemes for hyperbolic conservation laws*, Journal of Computational Physics, v49 (1983), pp.357–393.

34. A. Harten, *Preliminary results on the extension of ENO schemes to two dimensional problems*, in Proceedings of the International Conference on Hyperbolic Problems, Saint-Etienne, 1986.

35. A. Harten, *ENO schemes with subcell resolution*, Journal of Computational Physics, v83 (1989), pp.148–184.

36. A. Harten, J. Hyman and P. Lax, *On finite difference approximations and entropy conditions for shocks*, Communications in Pure and Applied Mathematics, v29 (1976), pp.297–322.

37. A. Harten and S. Osher, *Uniformly high-order accurate non-oscillatory schemes, I*, SIAM Journal on Numerical Analysis, v24 (1987), pp.279–309.

38. A. Harten, B. Engquist, S. Osher and S. Chakravarthy, *Uniformly high order essentially non-oscillatory schemes, III*, Journal of Computational Physics, v71 (1987), pp.231–303.

39. A. Harten, S. Osher, B. Engquist and S. Chakravarthy, *Some results on uniformly high order accurate essentially non-oscillatory schemes*, Applied Numerical Mathematics, v2 (1986), pp.347–377.

40. A. Iske and T. Soner, *On the structure of function spaces in optimal recovery of point functionals for ENO-schemes by radial basis functions*, Numerische Mathematik, v74 (1996), pp.177–201.

41. J. Jerome and C.-W. Shu, *Energy models for one-carrier transport in semiconductor devices*, in IMA Volumes in Mathematics and Its Applications, v59, W. Coughran, J. Cole, P. Lloyd and J. White, editors, Springer-Verlag, 1994, pp.185–207.

42. J. Jerome and C.-W. Shu, *Transport effects and characteristic modes in the modeling and simulation of submicron devices*, IEEE Transactions on Computer-Aided Design of Integrated Circuits and Systems, v14 (1995), pp.917–923.

43. G. Jiang and C.-W. Shu, *Efficient implementation of weighted ENO schemes*, Journal of Computational Physics, v126 (1996), pp.202–228.

44. G. Jiang and S.-H. Yu, *Discrete shocks for finite difference approximations to scalar conservation laws*, SIAM Journal on Numerical Analysis, to appear.

45. G. Jiang and D. Peng, *Weighted ENO schemes for Hamilton-Jacobi equations*, to appear in SIAM Journal on Scientific Computing.

46. D. A. Kopriva, *A Practical Assessment of Spectral Accuracy for Hyperbolic Problems with Discontinuities*, Journal of Scientific Computing, v2, 1987, pp.249–262.

47. R. Krasny, *A study of singularity formation in a vortex sheet by the point-vortex approximation*, Journal of Fluid Mechanics, v167 (1986), pp.65–93.

48. R. Krasny, *Desingularization of periodic vortex sheet roll-up*, Journal of Computational Physics, v65 (1986), pp.292–313.

49. F. Ladeinde, E. O'Brien, X. Cai and W. Liu, *Advection by polytropic compressible turbulence*, Physics of Fluids, v7 (1995), pp.2848–2857.

50. F. Lafon and S. Osher, *High-order 2-dimensional nonoscillatory methods for solving Hamilton-Jacobi scalar equations*, Journal of Computational Physics, v123 (1996), pp.235–253.

51. P. D. Lax and B. Wendroff, *Systems of conservation laws*, Communications in Pure and Applied Mathematics, v13 (1960), pp.217–237.

52. R. J. LeVeque, *Numerical Methods for Conservation Laws*, Birkhauser Verlag, Basel, 1990.

53. X.-D. Liu, S. Osher and T. Chan, *Weighted essentially nonoscillatory schemes*, Journal of Computational Physics, v115 (1994), pp.200–212.

54. X.-D. Liu and S. Osher, *Convex ENO high order multi-dimensional schemes without field by field decomposition or staggered grids*, preprint.

55. A. Majda, J. McDonough and S. Osher, *The Fourier Method for Nonsmooth Initial Data*, Mathematics of Computation, v32, 1978, pp.1041–1081.

56. H. Nessyahu and E. Tadmor, *Non-oscillatory central differencing for hyperbolic conservation laws*, Journal of Computational Physics, v87 (1990), pp.408–463.

57. S. Osher, *Riemann solvers, the entropy condition, and difference approximations*, SIAM Journal on Numerical Analysis, v21 (1984), pp.217–235.

58. S. Osher and S. Chakravarthy, *Upwind schemes and boundary conditions with applications to Euler equations in general geometries*, Journal of Computational Physics, v50 (1983), pp.447–481.

59. S. Osher and J. Sethian, *Fronts propagating with curvature-dependent speed: algorithms based on Hamilton-Jacobi formulation*, Journal of Computational Physics, v79 (1988), pp.12–49.

60. S. Osher and C.-W. Shu, *High-order essentially nonoscillatory schemes for Hamilton-Jacobi equations*, SIAM Journal on Numerical Analysis, v28 (1991), pp.907–922.

61. C.S. Peskin, *Numerical analysis of blood flow in the heart*, Journal of Computational Physics, v25 (1977), pp.220–252.

62. P. L. Roe, *Approximate Riemann solvers, parameter vectors, and difference schemes*, Journal of Computational Physics, v43 (1981), pp.357–372.

63. A. Rogerson and E. Meiberg, *A numerical study of the convergence properties of ENO schemes*. Journal of Scientific Computing, v5 (1990), pp.151–167.

64. J. Sethian, *Level Set Methods: Evolving Interfaces in Geometry, Fluid Dynamics, Computer Vision, and Material Science*, Cambridge Monographs on Applied and Computational Mathematics, Cambridge University Press, New York, New York, 1996.

65. C.-W. Shu, *TVB uniformly high order schemes for conservation laws*, Mathematics of Computation, v49 (1987), pp.105–121.

66. C.-W. Shu, *Total-Variation-Diminishing time discretizations*, SIAM Journal on Scientific and Statistical Computing, v9 (1988), pp.1073–1084.

67. C.-W. Shu, *Numerical experiments on the accuracy of ENO and modified ENO schemes*, Journal of Scientific Computing, v5 (1990), pp.127–149.

68. C.-W. Shu, *Preface to the republication of "Uniform high order essentially non-oscillatory schemes, III," by Harten, Engquist, Osher, and Chakravarthy*, Journal of Computational Physics, v131 (1997), pp.1–2.

69. C.-W. Shu and S. Osher, *Efficient implementation of essentially non-oscillatory shock capturing schemes*, Journal of Computational Physics, v77 (1988), pp.439–471.

70. C.-W. Shu and S. Osher, *Efficient implementation of essentially non-oscillatory shock capturing schemes II*, Journal of Computational Physics, v83 (1989), pp.32–78.

71. C.-W. Shu, T.A. Zang, G. Erlebacher, D. Whitaker, and S. Osher, *High order ENO schemes applied to two- and three- dimensional compressible flow*, Applied Numerical Mathematics, v9 (1992), pp.45–71.

72. C.-W. Shu and Y. Zeng, *High order essentially non-oscillatory scheme for viscoelasticity with fading memory*, Quarterly of Applied Mathematics, v55 (1997), pp.459-484.

73. K. Siddiqi, B. Kimia and C.-W. Shu, *Geometric shock-capturing ENO schemes for subpixel interpolation, computation and curve evolution*, Computer Vision Graphics and Image Processing: Graphical Models and Image Processing (CVGIP:GMIP), to appear.

74. J. Smoller, *Shock Waves and Reaction-Diffusion Equations*, Springer-Verlag, New York, 1983.

75. G. A. Sod, *Numerical Methods in Fluid Dynamics*, Cambridge University Press, Cambridge, 1985.

76. J. Strikwerda, *Initial boundary value problems for the method of lines*, Journal of Computational Physics, v34 (1980), pp.94–107.

77. M. Sussman, P. Smereka, S. Osher, *A level set approach for computing solutions to incompressible two phase flow*, Journal of Computational Physics, v114 (1994), pp.146–159.

78. P. K. Sweby, *High resolution schemes using flux limiters for hyperbolic conservation laws*, SIAM Journal on Numerical Analysis, v21 (1984), pp.995–1011.

79. B. van Leer, *Towards the ultimate conservative difference scheme V. A second order sequel to Godunov's method*, Journal of Computational Physics, v32 (1979), pp.101–136.

80. F. Walsteijn, *Robust numerical methods for 2D turbulence*, Journal of Computational Physics, v114 (1994), pp.129–145.

81. J.H. Williamson, *Low-storage Runge-Kutta schemes*, Journal of Computational Physics, v35 (1980), pp.48–56.

82. P. Woodward and P. Colella, *The numerical simulation of two-dimensional fluid flow with strong shocks*, Journal of Computational Physics, v54, 1984, pp.115-173.

83. H. Yang, *An artificial compression method for ENO schemes, the slope modification method*, Journal of Computational Physics, v89 (1990), pp.125–160.

C.I.M.E. Session on "Advanced numerical approximation of nonlinear hyperbolic equations"

List of participants

T. BARTH, T27B-1, NASA Ames R.C., Moffett Field, CA 94040, USA
E. BASSANO, DETEC, Fac. Ing., P.le Tecchio 80, 80125 Napoli, Italy
E. BELARDINI, Dip.to di Energetica, Via S. Marta 3, 50139 Firenze, Italy
L. BERGAMASCHI, Dip.to di Matematica, Via Belzoni 7, 35131 Padova, Italy
S. BERRONE, Via S. Rocco 57, 15040 Frassineto Po (AL), Italy
E. BERTOLAZZI, Dip.to di Ingegneria Meccanica, Fac. Ing., Via Mesiano 77, 38050 Trento, Italy
F. BIANCO, Via San Rocco 254, San Vito Chietino, 66038 Chieti, Italy
F. BOSISIO, Dip.to di Matematica del Politecnico, Via Bonardi 9, 20133 Milano, Italy
M. BRYANI, Via Anna Magnani 25, 00197 Roma, Italy
A. BUFFA, Via Lombroso 3, 27100 Pavia, Italy
G. CORNETTI, CNR-IAN, Via Abbiategrasso 209, 27100 Pavia, Italy
G. CURRELLIL, Dip.to di Mecc. e Aero., Via Eudossiana 18, 00184 Roma, Italy
G. DE FELICE, Univ. di Napoli "Federico II", Piazzale Tecchio 80, 80125 Napoli, Italy
G. DELUSSU, CRS4, Via N. Sauro 10, 09100 Cagliari, Italy
W. EGARTNER, IWR, Univ. of Heidelberg, Im Neuenheimer Feld 368/3, D-69120 Heidelberg, Germany
M. FALCONE, Dip.to di Matematica, P.le A. Moro 5, 00185 Roma, Italy
L. FATONE, Dip.to di Matematica dell'Università, Via Saldini 50, 20133 Milano, Italy
C: FRIES, Inst. für Mathematik, Templergraben 55, 52062 Aachen, Germany
S. GAMBELLI, Dip.to di Matematica del Politecnico, Via Bonardi 9, 20133 Milano, Italy
I. GERACE, Dip.to di Matematica dell'Università, Via Saldini 50, 20133 Milano, Italy
R. GUGLIELMANN, Dip.to di Matematica dell'Univ., Via Saldini 50, 20133 Milano, Italy
A. IOLLO, INRIA, Projet SINUS, 2004 route des Lucioles, 06902 Sophia Antipolis Cedex, France
A. JANNELLI, Univ. degli Studi di Messina, Salita Sperone, Contrada Papardo, Messina, Italy
T. KATSAOUNIS Dept. of Math., Univ. of Crete, GR-714 09 Heraklion, Greece
A. MANCINI, Via P. Grassi 1, 52025 Montevarchi (AR), Italy
M. MANNA, Dip.to di Ing. Meccanica, Via Claudio, 21, 80125 Napoli, Italy
G. MANZINI, IAN-CNR, Via Abbiategrasso 209, 27100 Pavia, Italy
F.S. MARRA, Univ. di Napoli "Federico II", Piazzale Tecchio 80, 80125 Napoli, Italy
L. MAZZEI, Escondido Village 80-B, Stanford, CA 94305-7140, USA
C. MEOLA, Univ. di Napoli "Federico II", Piazzale Tecchio 80, 80125 Napoli, Italy
S: MICHELETTI, Dip.to di Matematica del Politecnico, Via Boanrdi 9, 20133 Milano, Italy
G. NALDI, Dip.to di Matematica, Via Abbiategrasso 215, 27100 Pavia, Italy
F. NOTARNICOLA, IRMA-CNR, Via Amendola 122/I, 70126 Bari, Italy
L. PARESCHI, Dip.to di Matematica, Via Machiavelli 35, 44100 Ferrara, Italy
A. PASCARELLI, Univ. di Napoli "Federico II", P.le Tecchio 80, 80125 Napoli
S. PEROTTO, Dip.to di Matematica del Politecnico, Via Bonardi 9, 20133 Milano, Italy
G. PUPPO, Dip.to di Matematica del Politecnico, Corso Duca degli Abruzzi 24, 10129 Torino, Italy
V. ROMANO, Dip.to di Matematica, Viale A. Doria 6, 95125 Catania, Italy
G. RUSSO, Dip.to di Matematica, Via Vetoio, Loc. Coppito, 67010 L'Aquila, Italy
R. SACCO, Dip.to di Matematica del Politecnico, Via Bonardi 9, 20133 Milano, Italy
F. SALERI, Dip.to di Matematica del Politecnico, Via Bonardi 9, 20133 Milano, Italy
G. SANGALLI, IAN-CNR, Corso Carlo Alberto 5, 27100 Pavia, Italy
M. SARDELLA, Dip.to di Matematica del Politecnico, Corso Duca degli Abruzzi 24, 10129 Torino, Italy
C. SIGNANI, Via Marcantonio Bragadin 50, 00136 Roma, Italy
A. VENEZIANI, Dip.to di Matematica del Politecnico, P.zza L. da Vinci 32, 20133 Milano, Italy
C. VON TORNE, Inst. für Ang. Math., Univ. Bonn, Wegelerstr. 10, D-53115 Bonn, Germany
G. ZOURARIS, Dept. of Math., Univ. of Crete, GR-714-09 Heraklion, Greece

434

1983 - 90. Complete Intersections	(LNM 1092) Springer-Verlag
91. Bifurcation Theory and Applications	(LNM 1057) "
92. Numerical Methods in Fluid Dynamics	(LNM 1127) "
1984 - 93. Harmonic Mappings and Minimal Immersions	(LNM 1161) "
94. Schrödinger Operators	(LNM 1159) "
95. Buildings and the Geometry of Diagrams	(LNM 1181) "
1985 - 96. Probability and Analysis	(LNM 1206) "
97. Some Problems in Nonlinear Diffusion	(LNM 1224) "
98. Theory of Moduli	(LNM 1337) "
1986 - 99. Inverse Problems	(LNM 1225) "
100. Mathematical Economics	(LNM 1330) "
101. Combinatorial Optimization	(LNM 1403) "
1987 - 102. Relativistic Fluid Dynamics	(LNM 1385) "
103. Topics in Calculus of Variations	(LNM 1365) "
1988 - 104. Logic and Computer Science	(LNM 1429) "
105. Global Geometry and Mathematical Physics	(LNM 1451) "
1989 - 106. Methods of nonconvex analysis	(LNM 1446) "
107. Microlocal Analysis and Applications	(LNM 1495) "
1990 - 108. Geometric Topology: Recent Developments	(LNM 1504) "
109. H$_\infty$ Control Theory	(LNM 1496) "
110. Mathematical Modelling of Industrial Processes	(LNM 1521) "
1991 - 111. Topological Methods for Ordinary Differential Equations	(LNM 1537) "
112. Arithmetic Algebraic Geometry	(LNM 1553) "
113. Transition to Chaos in Classical and Quantum Mechanics	(LNM 1589) "
1992 - 114. Dirichlet Forms	(LNM 1563) "
115. D-Modules, Representation Theory, and Quantum Groups	(LNM 1565) "
116. Nonequilibrium Problems in Many-Particle Systems	(LNM 1551) "

FONDAZIONE C.I.M.E.
CENTRO INTERNAZIONALE MATEMATICO ESTIVO
INTERNATIONAL MATHEMATICAL SUMMER CENTER

"Dynamical Systems and Small Divisors"

is the subject of the first 1998 C.I.M.E. Session.

The session, sponsored by the Consiglio Nazionale delle Ricerche (C.N.R.), the Ministero dell'Università e della Ricerca Scientifica e Tecnologica (M.U.R.S.T.) and the European Community, will take place, under the scientific direction of Prof. STEFANO MARMI (Università di Firenze) and Prof. JEAN-CHRISTOPHE YOCCOZ (Université de Paris Sud, Orsay) at Grand Hotel San Michele, Cetraro (Cosenza), **from 13 to 20 June, 1998.**

Courses

a) **KAM-Theory for Linear Quasi-Periodic Systems** (8 lectures in English)
Prof. L. Hakan ELIASSON (KTH, Stockholm)

Abstract

The systems we shall consider are linear differential/difference-equations which have a quasi-periodic dependence in the unknown variable - often thought of as "time". These systems have a multiplicative part consisting of multiplication by a quasi-periodic scalar or matrix and a differential/difference part.

Two kinds of perturbation theory can be considered. If the multiplicative part is very small we perturb from a constant coefficient system which can be solved explicitly. In this case the perturbation theory aims to constructing quasi-periodic solutions. If the multiplicative part is very large this can often be thought of as having a very small differential/difference part, in which case we perturb from a pure multiplicative system. In this case one looks for "localized" solutions.

There is a common feature of these problems: the unperturbed part is an operator with a dense point spectrum and the problem is to construct point eigenvalues or even a dense point spectrum for the perturbed operator. The basic difficulty in such a perturbation theory turns out to be to control the "almost multiplicities" of the eigenvalues.

The standard example that will be considered is the discrete one-dimensional quasi-periodic Schrödinger equations but also other examples will be discussed.

Important contributions for these problems have been made by Dinaburg-Sinai, Moser-Pöschel, Krikorian, Frölich-Spencer-Wittver, Chulcavsky-Sinai and others.

For references one can consult my paper "One-dimensional quasi-periodic Schrödinger operators - dynamical systems and spectral theory" in the proceedings of ECM2 in Budapest 1996.

b) **Invariant Tori** (8 lectures in English)
Prof. Michael HERMAN (Université de Paris 7, Ecole Polytechnique)

Contents

Invariant tori are a crucial feature not only of hamiltonian dynamics, but also of volume - preserving dynamics and even general dynamics. Hamilton's implicit function theorem in Fréchet spaces provides an efficient and flexible setting in which to study the existence of such tori. It allows to give simple proofs of classical results, such as the existence of diophantine invariant lagrangian tori in Hamiltonian dynamics, as well as less classical but very basic ones, such as the existence of codimension one translated diophantine tori in general dynamics.

Many important applications, in the hamiltonian as well as the volume preserving context, will be considered.

References

[1] Jean-Benoit Bost, *Tores invariants des systèmes dynamiques hamiltoniens* (d'après Kolmogorov, Arnold, Moser, Rüssmann, Zehnder, Herman, Pöschel, ...), Séminaire Bourbaki, exposé n. 639 (1985), Astérisque 133/134, 113-157.
[2] Jean-Christophe Yoccoz, *Travaux de Herman sur les tores invariants*, Séminaire Bourbaki, exposé n. 754 (1992), Astérique 206 (1992), 311-340.

c) **Geometrical Methods in Small Divisors Problems** (8 lectures in English)
Prof. Jean-Christopher YOCCOZ (Université de Paris-Sud, Orsay)

Contents

- Diophantine approximation, continued fraction, Brjuno's condition, the arithmetical condition for analytic linearization of analytic circle diffeomorphisms.
- Circle diffeomorphisms: a short review on the topological and smooth theory.
- Analytic circle diffeomorphisms: topological stability and analytic linearizability, periodic orbits in the neighbourhood, statement of the main conjugacy result.
- Geometric renormalization: what is the "continued fraction" of an analytic circle diffeomorphism.
- The inverse construction. Non linearizable diffeomorphisms.

- Renormalization with parameters: how Hörmander's L^2 estimates on $\overline{\partial}$ is a substitute to one complex variable theory.

References

[1] R. Perez-Marco, *Solution complète au problème de Siegel de linéarisation d'une application holomorphe aux voisinage d'un point fixe* (d'après J.-C. Yoccoz), Séminaire Bourbaki, exposé n. 753 (1992), Astérisque 206 (1992), 273-310.

[2] J.-C. Yoccoz, *An introduction to small divisors problems* in "From number theory to physics", edited by M. Waldschmidt, P. Moussa, J.-M. Luck and C. Itzykson, Berlin, Springer Verlag, 1992, pp. 659-679.

[3] J.-C. Yoccoz, *Théorème de Siegel, nombres de Brjuno et polynômes quadratiques*, Astérisque 231 (1995), 3-88.

FONDAZIONE C.I.M.E.
CENTRO INTERNAZIONALE MATEMATICO ESTIVO
INTERNATIONAL MATHEMATICAL SUMMER CENTER

"Mathematical Problems in Semiconductor Physics"

is the subject of the second 1998 C.I.M.E. Session.

The session, sponsored by the Consiglio Nazionale delle Ricerche (C.N.R.), the Ministero dell'Università e della Ricerca Scientifica e Tecnologica (M.U.R.S.T.) and the European Community, will take place, under the scientific direction of Professors MARCELLO ANILE (Università di Catania), PIERRE DEGOND (Université Paul Sabatier, Toulouse) and PETER A. MARKOWICH (TU, Berlin) at Grand Hotel San Michele, Cetraro (Cosenza), **from 15 to 22 July, 1998.**

Courses

a) **Drift Diffusion Equations and Applications** (6 lectures in English)
   Prof. Walter ALLEGRETTO (Univ. of Alberta, Canada)

- Drift Diffusion Equations (DDES): Introduction and mathematical results
- Numerical schemes for DDES: Introduction and analysis
- Recent mathematical results for DDES
- Practical microsensor applications, related and open mathematical problems

Background references:

[1] Mock, *Analysis of mathematical models for semiconductor devices*, Boole Press, 1983.
[2] Selberherr, *Analysis and simulation of semiconductor devices*, Springer, 1984.
[3] Markowich, Ringhofer and Schmeiser, *Semiconductor equations*, Springer, 1990.
[4] Ristic, *Sensor technology and devices*, Artech House, 1994.
[5] Jerome and Kerkhoven, *A finite element approximation theory for the drift diffusion semiconductor model*, SIAM J. Num. Anal. (1991), 403-422.
[6] Fang and Ito, *Global solutions of the time dependent drift-diffusion semiconductor equations*, J. Diff. Eqs (1995), 523-566.
[7] Gajewski and Gartner, *On the discretization of van Roosbroeck's equations with magnetic field*, ZAMM (1996), 247-264.
Other relevant references will be presented at the talks.

b) **An Introduction to Kinetic Theory** (6 lectures in English)
   Prof. David LEVERMORE (University of Arizona, Tucson)

1. Basic Structure
   - 1.1. Kinetic Regimes: Mean Free Paths; Boltzmann-Grad Limit; Boltzmann Equation
   - 1.2. Boltzmann Collision Operator: Galilean Symmetry; Boltzmann Identity; Conservation; Entropy, Dissipation and Equilibria
   - 1.3. Abstract Collision Operators: Galilean Symmetry; Conservation; Entropy, Dissipation and Equilibria
   - 1.4. Examples and Generalizations: Germions and Bosons; Generalized BKG Operators; Fokker-Planck Operators; Polyatomic Molecules; Multispecies Mixtures

2. Fluid Dynamical Limits
   - 2.1. Fluid Dynamical Regimes: Knudsen Number; Compressible Euler Limit; Compressible Euler Equations

Some backround reading

[1] C. Cercignani, R. Illner, and M. Pulvirenti, *The mathematical theory of dilute gases*, Appl. Math. Sciences 106, Springer Verlag, New York 1994.

[2] C. Bardos, F. Golse, D. Levermore, *Fluid dynamic limits of kinetic equations I: Formal derivations*, J. Stat. Phys. 63 (1991), 323-344.

[3] C. D. Levermore, *Moment closure hierarchies for kinetic theories*, J. Stat. Phys. 83 (1996), 1021-1065.

[4] P. Markowich, C. Ringhofer, and C. Schmeiser, *Semiconductor equations*, Springer Verlag, New York 1990.

[5] F. Poupaud, *Runaway phenomena and fluid approximation under high fields in semiconductor kinetic theory*, Z. Angew. Math. Mech. 72 (1992), 359-372.

c) **Transport Modelling in Semiconductors** (6 lectures in English)
Prof. Frederick POUPAUD (Université de Nice)

The goal of this course is to give a precise mathematical derivation of the various models used to describe transport phenomena in semiconductors.

The master equations of relativistic quantum dynamics are the Dirac equations. The main two asymptotics which can be performed starting from these equations are the (semi-) non relativistic limit (c, the velocity of light, tends to infinity) and the (semi-) classical limit (h, the Planck constant, tends to zero).

We first focus on the non relativistic limit which can be handled with classical tools of mathematical analysis. We show that in the limit the Dirac equations become the Schrödinger equation. We also prove that a better approximation is given by the Pauli equations.

Then we introduce new mathematical formalisms which allow to perform easily semiclassical limits of any hyperbolic system of equations. These formalisms are based on Wigner measures introduced independently by P. Gerard and P. L. Lions, T. Pauli in 91. We give the main properties of these objects and we state the general result obtained for hyperbolic systems.

Finally we apply this theory in the context of semiconductors. We show that the classical limit of Dirac or Schrödinger equations is the Vlasov Maxwell or Vlasov Poisson equation. The semiconductor medium is modelized at the quantum level by a periodic potential. In this case we show how the band structure appears and derive the semiclassical transport equation of semiconductors.

References

[1] Gerard, *Mesures semiclassiques et ondes de Blach*, Seminaire Ecole Polytechnique 90-91, exposé XVI.

[2] Lions, T. Paul, *Sur les mesures de Wigner*, Revista Mat. Iberoamericana, 9 (1993), 553-618.

[3] Markowich, N. J. Mauser, *The classical limit of a self-consistent quantum Vlasov equation in 3D*, Math. Meth. Mod. in Appl. Sci. 9 (1993), 109-124.

[4] Markowich, N. J. Mauser and F. Poupaud, *A Wigner function approach to (semi) classical limits: electrons in a periodic potential*, J. Math. Phys. 35 (1994), 1066-1094.

[5] Poupaud, C. Ringhofer, *Quantum hydrodynamic models in a crystal*, Appl. Math. Lett. 8 (6) (1995), 55-59.

[6] Poupaud and C. Ringhofer, *Semiclassical limits in a crystal with external potentials and effective mass theorems*, Comm. in P.D.E. 21 (1996), 1897-1918.
[7] Gérard, P. A. Markowich, N. Mauser and F. Poupaud, *Homogenization limits and Wigner transforms*, Comm. Pure Appl. Math. 50 (1997), 323-380.

d) Foundations of Mathematical Models for Semiconductor (5 lectures in English)
Prof. Christian RINGHOFER (Arizona State University, Tempe)

- Multi-particle systems, the Liouville and Vlasov equation, Schrödinger equation, Maxwell's equation (Refs: [2,9,11,16]).
- From multi-particle systems to single particle systems: The BBGKY hierarchy, the classical and semiclassical Boltzmann equation, semiconductors vs. gases, Bloch decomposition and energy bands, density matrices and Wigner functions (Refs: [1,11,14,17]).
- Properties of the Boltzmann equation, collision operators, macroscopic limits, the Hilbert expansion and the Chapman-Enskog expansion, the drift-diffusion system, the hydrodynamic equations, energy transport models (Refs: [2,3,7,8,9,12,13,14]).
- Properties of the Heisenberg and Wigner equations, classical limits, the effective mass approximation, quantum hydrodynamic models, relaxation time approximations and the Fokker-Planck collision term (Refs: [5,6,10,12,14,17,19]).
- Boundary conditions for the simulation of devices, rough surfaces, charge and current conservation considerations, wave absorption at contacts and insulators (Refs: [4,5,9,11,15,18]).

References

[1] A. Arnold, P. Degond, P. Markowich, H. Steinrueck, *The Wigner Poisson equation in a crystal*, Appl. Lect. Lett. 2 (1989), 187-191.
[2] N. Ashcroft, M. Mermin, *Solid state physics*, Holt-Saunders, New York, 1976.
[3] G. Baccarani, M. Wordeman, *An investigation of steady state velocity overshoot effects in Si and GaAs devices*, Solid State Electr. 28 (1985), 407-416.
[4] B. Engquist, A. Majda, *Absorbing boundary conditions for the numerical simulation of waves*, Math. Comp. 31 (1977), 629-651.
[5] D. Ferry. H. Grubin, *Modelling of quantum transport in semiconductor devices*, Solid State Phys. 49 (1995), 283-448.
[6] C. Gardner, *The quantum hydrodynamic model for semiconductor devices*, SIAM J. Appl. Math. 54 (1994), 409-427.
[7] H. Grad. *On the kinetic theory of rarefied gases*, Comm. Pure Appl. Math. 2 (1949), 331-407.
[8] H. Grad, *Principles of the kinetic theory of gases*, Handbooks Phys. 12 (1958), 205-294.
[9] A. Kersch, W. Morokoff, *Transport simulation in microelectronics*, Birkhäuser, Basel, 1995.
[10] P. Markowich, C. Ringhofer, *An analysis of the quantum Liouville equation*, ZAMM 69 (1989), 121-127.
[11] P. Markowich, C. Ringhofer, C. Schmeiser, *Semiconductor equations*, Springer, Berlin, 1990.
[12] P. Markowich, N. Mauser, F. Poupaud, *A Wigner function approach to semiclassical limits*, J. Math. Phys. 35 (1994), 1066-1094.
[13] F. Poupaud, *Diffusion approximation of the linear Boltzmann equation: Analysis of boundary layers*, Asympt. Anal. 4 (1991), 293-317.
[14] F. Poupaud, C. Ringhofer, *Quantum hydrodynamic models in semiconductor crystals*, Appl. Math. Lett. 8 (1995), 55-59.
[15] C. Ringhofer, D. Ferry, N. Kluksdahl, *Absorbing boundary condition for the simulation of quantum transport phenomena*, Transp. Theory and Stat. Phys. 18 (1989), 331-346.
[16] S. Selberherr, *Analysis of semiconductor devices*, 2nd ed., Wiley, NewYork, 1981.
[17] V. Tatarski, *The Wigner representation of quantum mechanics*, Soviet Phys. Uspekhi 26 (1983), 311-372.
[18] M. Taylor. *Pseudodifferential operators*, Princeton University Press, Princeton, 1981.
[19] E. Wigner, *On the quantum correction for thermodynamic equilibrium*, Phys. Rev. 40 (1932), 749-759.

FONDAZIONE C.I.M.E.
CENTRO INTERNAZIONALE MATEMATICO ESTIVO
INTERNATIONAL MATHEMATICAL SUMMER CENTER

"Stochastic PDE's and Kolmogorov Equations in Infinite Dimensions"

is the subject of the third 1998 C.I.M.E. Session.

The session, sponsored by the Consiglio Nazionale delle Ricerche (C.N.R.), the Ministero dell'Università e della Ricerca Scientifica e Tecnologica (M.U.R.S.T.) and the European Community, will take place, under the scientific direction of Professor GIUSEPPE DA PRATO (S. N. S., Pisa) at Grand Hotel San Michele, Cetraro (Cosenza), **from August 24 to September 1, 1998.**

Courses

a)   **Kolmogorov equations** (8 lectures in English)
      Prof. N. V. KRYLOV (Univ. of Minnesota, Minneapolis)

1.   Solvability of Itô's stochastic equations.
2   The Markov property of diffusion processes
3.   A conditional version of Kolmogorov's equation
4.   Differentiability of solutions of stochastic equations with respect to initial data
5.   Kolmogorov's equations in the whole space
6-7. Some integral approximations of differential operators
8.   Kolmogorov's equations in domains in the sense of distributions

References

[1]   N. K. Krilov, *Introduction to the theory of diffusion processes*, American Math Soc. Providence RI, 1995

b) L^p **-analysis of finite and infinite dimensional diffusion operators** (8 lectures in English)
      Prof. M. RÖCKNER (Universität Bielefeld)

1.   Kolmogorov equations in $L^p(R^d;\mu)$:-the symmetric case
2.   Some recent results on invariant measures
3.   Kolmogorov equations in $L^p(R^d;\mu)$:- the non-symmetric case
4.   Kolmogorov equations in $L^p(E\ ;\mu)$ for Banach spaces E
5.   Uniqueness results
6.   Relation to martingale problems and applications to stochastic quantization
7.   Kolmogorov equations in $L^p(E\ ;\mu)$ for infinite dimensional manifolds E : a case study from continuum statistical mechanics.
8.   Ergodicity

References

(Note: SFB--Preprints are available via internet SFB-Webpage:
http://www.mathematik.uni-bielefeld.de/sfb343/preprints)

[1]   S. Albeverio, V.I. Bogachev, M. Röckner, *On uniqueness of invariant measures for finite and infinite*

dimensional diffusions, SFB-343--Preprint 1997

[2] S.Albeverio, Y.G. Kondratiev, M. Röckner, *Dirichlet operators via stochastic analysis,* J. Funct. Anal. 128, 102--138 (1995).

[3] S.Albeverio, Y.G. Kondratiev, M. Röckner, *Ergodicity of L^2-semigroups and extremality of Gibbs states,* J. Funct. Anal. 144, 394--423 (1997).

[4] S.Albeverio, Y.G. Kondratiev, M. Röckner, *Geometry and Analysis on configuration spaces,* SFB--343--Preprint 1997. To appear in J. Funct. Anal.

[5] S. Albeverio, Y.G. Kondratiev, M. Röckner, *Geometry and analysis on configuration spaces. The Gibbsian case,* SFB--343--Preprint (1997). To appear in: J. Funct. Anal.

[6] V.I. Bogachev, N. Krylov, M. Röckner, *Elliptic regularity and essential self--adjointness of Dirichlet operators on R^d,* SFB--343--Preprint (1996). To appear in: Ann. Scuola Norm. di Pisa .

[7] V. I. Bogachev, M. Röckner, *Regularity of invariant measures on finite andinfinite dimensional spaces and applications,* J. Funct. Anal. 133, 168--223 (1995).

[8] V. I. Bogachev, M. Röckner, *A generalization of Hasminskii's theorem on existence of invariant measures,* SFB-343--Preprint 1997.

[9] V.I. Bogachev, M. Röckner, T.S. Zhang, *Existence and uniqueness of invariant measures: an approach via sectorial forms,* SFB-343--Preprint 1997.

[10] A. Eberle, *Uniqueness and non--uniqueness of singular diffusion operators,* Doktorate-Thesis. (1997), (available via http://www.mathematik.uni-bielefeld.de/~eberle ; e-mail: eberle@mathematik.uni-bielefeld.de)

[11] Z.M. Ma, M. Röckner, *An introduction to the theory of (non--symmetric) Dirichlet forms,* Berlin: Springer 1992.

[12] V. Liskevich, M. Röckner, *Strong uniqueness for a class of infinite dimensional Dirichlet operators and applications to stochastic quantization,* SFB-343--Preprint (1997).

[13] M. Röckner, T.S. Zhang, *Uniqueness of generalized Schrödinger operators and applications.* J. Funct. Anal. 105, 187--231 (1992).

[14] M. Röckner, T.S. Zhang, *Uniqueness of generalized Schrödinger operators -- Part II,* J. Funct. Anal. 119, 455--467 (1994).

[15] W. Stannat, *(Non--symmetric) Dirichlet operators on L^1. existence, uniqueness and associated Markov processes,* SFB-343--Preprint (1997).

[16] W. Stannat, *The theory of Generalized Dirichlet forms and its Applications in Analysis and Stochastics,* Doktorate--Thesis. SFB-343--Preprint (1997).

c) **Kolmogorov equations with infinite numbers of variables** (8 lectures in English)
Prof. J. ZABCZYK (Polskiej Akademii Nauk, Warszawa)

1. Infinite dimensional stochastic diffusions
2. Heat equations on Hilbert spaces
3. Kolmogorov equations corresponding to Ornstein-Uhlenbeck processes
4. Bismut-Elworthy formula and its consequences
5. Equations in open subsets of Hilbert spaces
6. Kolmogorov semigroups and invariant measures
7. Bellman equations of stochastic control
8. Variational parabolic inequalities

References

[1] L.Gross, *Potential Theory in Hilbert spaces,* J. Funct. Anal. 1(1965), 123-189.

[2] G. Da Prato and J. Zabczyk, *Stochastic Equations in Infinite Dimensions,* Cambridge University Press, 1992.

[3] G. Da Prato and J. Zabczyk, *Ergodicity for Infinite Dimensional Systems,* Cambridge University Press, 1996.

[4] G. Da Prato, B. Goldys and J. Zabczyk, *Ornstein-Uhlenbeck semigroups in open sets of Hilbert spaces,* C.R. Acad. Sci. Paris, t.325, serie I (1997), 433- 438.

[5] G.Da Prato and J. Zabczyk, *Differentiability of the Feynman-Kac semigroup and a control application,* Rend. Mat. Acc. Lincei. s.9, v. 8, 183--188.

FONDAZIONE C.I.M.E.
CENTRO INTERNAZIONALE MATEMATICO ESTIVO
INTERNATIONAL MATHEMATICAL SUMMER CENTER

"Filtration in Porous Media and Industrial Applications"

is the subject of the fourth 1998 C.I.M.E. Session.

The session, sponsored by the Consiglio Nazionale delle Ricerche (C.N.R.), the Ministero dell'Università e della Ricerca Scientifica e Tecnologica (M.U.R.S.T.) and the European Community, will take place, under the scientific direction of Prof. ANTONIO FASANO (Università di Firenze) and Prof. HANS VAN DUIJN (University of Amsterdam) at Grand Hotel San Michele, Cetraro (Cosenza), **from August 24 to September 1, 1998.**

Courses

a)   **Mathematical Models for Oil Reservoirs Engineering** (6 lectures in English)
       Prof. Magne S. ESPEDAL (University of Bergen)

b)   **Filtration Processes in Various Industrial Problems** (4 lectures in English)
       Prof. Antonio FASANO (Università di Firenze)

1. Flows in porous media with migrating solid particles
2. Non-isothermal flows with a wetting front
3. Some free boundary problems in the manufacturing of composite materials
4. Filtration in media with hydrophile granules

Prerequisites: Basic theory of p.d.e.'s

References

[1]   I. Barenblatt, V. M. Entov, V. M. Ryzhik, *Theory of fluid flows through natural rocks*, Kluwer 1990.
[2]   J. Bear, *Duynamics of fluids in porous media*, American Elsevier, New York 1972.
[3]   J. Bear, A. Verruijt, *Modelling ground water flow and pollution*. Reidel, Dordrecht 1987.
[4]   A. Fasano, *Some non-standard one-dimensional filtration problems*, Bull Fac. Ed. Chiba Univ. 44 (1996), 5-29.
[5]   L. Preziosi, *The theory of deformable porous media and its applications to composite material manufacturing*, Surveys Math. Ind. 6 (1996), 167-214.

c)   **Reactive Transport Processes in Porous Media** (3 lectures in English)
       Prof. Peter KNABER (Universität Erlangen-Nürnberg)

1. Modelling of Reactive Transport Processes in Porous Media.
       Abstract: Transport by diffusion, convection and dispersion, homogeneous and inhomogeneous reactions; equilibrium and non-equilibrium; complexation; adsorption; ion exchange; precipitation/dissolution.

2. Qualitative Properties of the Diffusion-Convection-Adsorption Equation.
       Abstract: Classification of isotherms, general theory: existence, uniqueness, regularity, comparison principle; special solutions: similarity solutions, travelling waves; finite speed of propagation; asymptotic states.

3. Identification of Constitutive Laws by Breakthrough Experiments.
       Abstract: Ill-posed problems; experiment design and identifiability; output least-squares approach; regularization; multilevel algorithms.

Prerequisites: Basic non-linear functional analysis and PDE's.

d) **Homogenization Theory and Applications to Filtration Processes** (6 lectures in English)
   Prof. A. MIKELIC (Université Lyon I)

1. An introduction to the multi-scale expansions for the equations of fluid mechanics in a porous medium.
   Abstract: We present the derivation of Darcy's law and the double porosity models for the single phase incompressible flow through periodic porous media by homogenization.

2. Non-Newtonian single phase flow through a porous media.
   Abstract: We consider the flow of the incompressible quasi-Newtonian fluid through a porous medium. The viscosity of the fluid is a monotone function of the norm of the deformation tensor. We derive the effective flow equation and rigorously establish the "duality" constitutive relations used for the directional flows in engineering literature.

3. Inertia effects for the single phase flow through a porous medium.
   Abstract: We consider the fast incompressible single phase flow through a porous medium. It is shown that the effective filtration law is a non-local two-pressures generalization of the Navier-Stokes system. After establishing the analiticity, we show how this complicated law can be approximated by polynomial relations between the filtration velocity and the pressure gradient. As a consequence we show that Forchheimer's filtration law is not acceptable.

4. The interface law of Beavers, Joseph and Saffman.
   Abstract: We consider finding of the boundary conditions for the interface between free fluid flow and flow in a porous medium. The corresponding boundary layers are presented in details. The modification of Beavers-Joseph law, due to Saffman, is justified and the parameters in the law are determined using the boundary layers.

5. Filtration through a porous medium with the permeable grains.
   Abstract: We consider filtration through a strongly heterogeneous porous material being made of components ("fluid" and "grains") with different permeabilities k_f and k_F, respectively. Fluid penetrates partly the grains and we have a large number of propagating wetting fronts. We find the effective filtration laws.

6. On the modelling of compressible flows in a porous medium.
   Abstract: We consider the compressible Navier-Stokes system in a porous medium and obtain variants of the porous medium equation as the effective filtration model.

Prerequisites: Basic applied non-linear functional analysis and PDE's. Basic knowledge of PDE's of hydrodynamics.

References

[1] H. Hornung, ed.: *Homogenization and porous media*, Interdisciplinary Applied Mathematics Series, Springer, New York, 1996. The chapter *"One-phase newtonian flow"* by G. Allaire is recommended as an introduction to the course.
[2] A. E. Scheidegger, *Hydrodynamics in porous media*, in ."Encyclopedia of Physics", vol. VIII/2, ed. S. Flügge, pp. 625-662, Springer, Berlin, 1963.

e) **Some Nonlinear Models Arising in Subsurface Transfport** (6 lectures in English):
   Prof. Hans VAN DUIJN (Delft University of Technology)

Abstract
   In three lectures, each of two hours, the following topics will be discussed:

1. Elliptic free boundary problems in salt water intrusion models
2. Uniqueness conditions in a hyperbolic model for oil recovery by steam drive
3. Stability methods for density induced porous media flow
   In the first two lectures, we introduce the interface between fresh and salt groundwater as the free boundary in the mathematical description. We discuss the problem of an upconed interface when wells are present in a two or three dimensional reservoir. We also present and discuss a salt water intrusion problem in a two dimensional vertical strip. We show existence, uniqueness various types of monotonicity and other qualitative properties.

   In the second pair of lectures we deal with a 2x2 hyperbolic system describing steam injection into a one dimensional reservoir (a porous column) which originally contains oil. Since steam condenses into water, we are led to consider a three-phase flow problem. We present a solution in terms of shock and rarefaction waves. Non-uniqueness occurs, which is resolved by introducing a particular regularization. It will be demonstrated that different regularisation lead to different solutions in the hyperbolic limit.

The third pair of lectures deals with an evaporation problem that describes the groundwater movement below salt lakes. We give two stability criteria. One is based on a linearization technique, the other on an energy method. The analytical results will be compared with both numerical and experimental resuls. Disregarding diffusion, we present a mathematical formulation which describes the occurrence of fingering as a continuous transition between the fluids.

References

[1] H. W. Alt & C. J. van Duijn, *A stationary flow of fresh and salt groundwater in a coastal aquifer*, Nonlinear Analysis TMA 14 (1990), 625-656.
[2] H. W. Alt & C. J. van Duijn, *A free boundary problem involving a cusp. Part 1: Global analysis*, European J. Appl. Math. 4 (1993), 39-63.
[3] J. Smoller, *Shock Waves and Reaction-Diffusion Equations*, New York, Springer, 1963.
[4] B. Straughan, *The energy method, stability and nonlinear convection*, New York, Springer, 1992.

FONDAZIONE C.I.M.E.
CENTRO INTERNAZIONALE MATEMATICO ESTIVO
INTERNATIONAL MATHEMATICAL SUMMER CENTER

"Optimal Shape Design"

is the subject of the first 1998 C.I.M.E.- C.I.M. (Centro International de Matematica, Coimbra, Portugal) Session.

The Session, sponsored by E. U. under TMR Programme, by C.I.M. and by C.I.M.E. will take place under the scientific direction of Professors Arrigo Cellina (University of Milan, Italy) and Antonio Ornelas (Universidade de Evora, Portugal) in Tróia, Portugal, **from June 1 to June 6, 1998**

Courses

a)   **Some nonconvex optimal shape problems** (4 lectures in English)
     Prof. Bernd KAWOHL, (Univ. Koeln, Germany)

Contents:
Newton's Problem of minimal resistance: how to find the shape of a body of minimal resistance moving in a rarefied fluid (1 hour);
The opaque square and the opaque circle: find a curve C of minimal length inside a square or a circle, having the property that every straight line transversing the square or the circle intersects C (1 hour);
 Variational approaches to proving symmetry (1 hour).

References
[1] G.Buttazzo, B.Kawohl: On Newton's problem of minimal resistance, Mathematical Intelligencer, 15 No.4 (1993) pp. 7-12.
[2] G.Buttazzo V.Ferone, B.Kawohl: Minimum problems over sets of concave functions and related questions, Mathematische Nachrichten 173   (1995), pp. 71--89.
[3] F.Brock, V.Ferone, B.Kawohl: A symmetry problem in the calculus of variations, Calculus of Variations 4 (1996) pp. 593-599.
[4] B.Kawohl, C.Schwab: Convergent finite elements for a class of nonconvex variational problems, IMA J. Numer. Anal., in print

b)   **Shape Control and Optimal Shape Design** (4 Lectures in English)
      Olivier PIRONNEAU, Analyse Numerique,(Paris 6, France)

Contents:
We consider the connections between shape design and equivalent boundary control design.
A new branch of optimal shape design is emerging along with what is known as "active control": the shapes are now time dependent and the control is preferably closed loop. In particular we shall deal with: sound control; drag control and lift control

References
[1]   Y. Achdou: Effect of a metallized coating on the reflection of an electromagnetic wave. Note CRAS, 1992.
[2]   Y. Achdou, O.Pironneau, F. Valentin, *Etude de lois de paroi d'ordre 1 et 2 pour des parois rugueuses par dècomposition de domaine*, INRIA report submitted also to JCP, 1997.
[3]   N.V. Banichuk , Introduction to optimization of structures. Springer series, 1990.
[4]   C. Bardos, O. Pironneau: Etudes prèliminaires pour le traitement des petites perturbations en aèro-èlasticitè. M2AN, 1993.
[5]   J. Cea, A.Gioan, J. Michel: Some results on domain identification. Calcolo 3/4 , 1973.
[6]   E.J. Haug, J. Cea: *Optimization of distributed parameter structures*, vol I and II , Sijthoff and Noordhoff, 1981
[7]   Ph. Morice: Optimisation de forme pour un problème de structure, INRIA report, 1976
[8]   F. Murat, J. Simon: Etude de problèmes d'optimum design. Proc. 7th IFIP conf. Lecture notes in Computer sciences, **41**, pp.54-62, 1976
[9]   O. Pironneau , Optimal shape design for elliptic systems', Springer-Verlag, 1984.

c)   **Homogenization methods in Optimal Design.** (4 Lectures in English)
        Prof. Luc TARTAR,(Carnegie Mellon University, USA)

References
[1]   L. Tartar: Control problems in the coefficients of PDE. In Lecture notes in Economics and Math systems. A. Bensoussan ed. Springer, 1974.

d)   **Explicit solutions in Elastic Optimization** (4 lectures in English)
        Prof. Piero VILLAGGIO,(Università di Pisa, Italy)

Contents:
The explicit solution of the boundary of an elastic body satisfying a given criterion of optimality is not available in general. However, some exact solutions (still useful for practical purposes) are obtained in plane elasticity by narrowing the class of boundaries in which the optimum is sought.

References
[1]   N. V. Banichuk, Introduction to Optimization of Structures, Springer 1990
[2]   H Neuber, Kerbspannungslehre, third Ed. Springer 1985
[3]   P. Villaggio, Mathematical Models for Elastic Structures, Cambridge University Press 1997

e)   **Optimal Shape Design: Theory, Modelling, Numerical Algorithms** (4 lectures in English)
        Prof. Jean Paul ZOLESIO, CNRS, Sophia Antipolis, France

Contents
Applications will be given to find the optimal shape of a ship; to the optimal shaping of the mirror of the European Very Large telescope; to the optimal design of plates.

References
[1]   J. Sokolowski, J.P. Zolesio: Introduction to Shape
[2]   Optimization. Springer Series in Computational Mathematics, 1991.

Springer
and the
environment

At Springer we firmly believe that an
international science publisher has a
special obligation to the environment,
and our corporate policies consistently
reflect this conviction.
We also expect our business partners –
paper mills, printers, packaging
manufacturers, etc. – to commit
themselves to using materials and
production processes that do not harm
the environment. The paper in this
book is made from low- or no-chlorine
pulp and is acid free, in conformance
with international standards for paper
permanency.

 Springer

Printing: Weihert-Druck GmbH, Darmstadt
Binding: Buchbinderei Schäffer, Grünstadt